COMPUTATIONAL NUCLEAR ENGINEERING AND RADIOLOGICAL SCIENCE USING PYTHON™

COMPUTATIONAL NUCLEAR ENGINEERING AND RADIOLOGICAL SCIENCE USING PYTHON™

RYAN G. MCCLARREN

Department of Aerospace and Mechanical Engineering
University of Notre Dame, Notre Dame, IN, USA

ACADEMIC PRESS

An imprint of Elsevier

Academic Press is an imprint of Elsevier
125 London Wall, London EC2Y 5AS, United Kingdom
525 B Street, Suite 1800, San Diego, CA 92101-4495, United States
50 Hampshire Street, 5th Floor, Cambridge, MA 02139, United States
The Boulevard, Langford Lane, Kidlington, Oxford OX5 1GB, United Kingdom

Notices

Knowledge and best practice in this field are constantly changing. As new research and experience broaden our
understanding, changes in research methods, professional practices, or medical treatment may become necessary.

Practitioners and researchers must always rely on their own experience and knowledge in evaluating and using any
information, methods, compounds, or experiments described herein. In using such information or methods they should be
mindful of their own safety and the safety of others, including parties for whom they have a professional responsibility.

To the fullest extent of the law, neither the Publisher nor the authors, contributors, or editors, assume any liability for any
injury and/or damage to persons or property as a matter of products liability, negligence or otherwise, or from any use or
operation of any methods, products, instructions, or ideas contained in the material herein.

Library of Congress Cataloging-in-Publication Data
A catalog record for this book is available from the Library of Congress

British Library Cataloguing-in-Publication Data
A catalogue record for this book is available from the British Library

ISBN: 978-0-12-812253-2

For information on all Academic Press publications
visit our website at https://www.elsevier.com/books-and-journals

Working together
to grow libraries in
developing countries

www.elsevier.com • www.bookaid.org

Publisher: Joe Hayton
Acquisition Editor: Joe Hayton
Editorial Project Manager: Kattie Washington
Production Project Manager: Sruthi Satheesh
Designer: Victoria Pearson

Typeset by VTeX

This book is for my wife, Katie. With tolerance and patience, she has supported and encouraged me.

Contents

II

NUMERICAL METHODS

7. Gaussian Elimination

8. LU Factorization and Banded Matrices

9. Iterative Methods for Linear Systems

10. Interpolation

11. Curve Fitting

About the Author

Ryan G. McClarren first tried to use computers to solve scientific problems in middle school when he thought his self-taught BASIC programming skills might make his algebra homework easier. Currently, he is Associate Professor of Aerospace and Mechanical Engineering at the University of Notre Dame. He obtained his Ph.D. (nuclear engineering and radiological sciences) from the University of Michigan. He is an active researcher in numerical methods for radiation transport problems and uncertainty quantification. Prior to joining Notre Dame, he was Assistant Professor of Nuclear Engineering in the Dwight Look College of Engineering at Texas A&M University, and was a scientist at Los Alamos National Laboratory in the Computational Physics and Methods Group (CCS-2). He is the author of over 40 publications appearing in peer-reviewed journals, including the Journal of Computational Physics, Nuclear Science and Engineering, Physics of Plasmas, and the Journal of Computational and Theoretical Transport. He also has extensive experience in applied data science and has consulted for a variety of large firms applying computational science to problems in the retail, banking, and entertainment spaces. He lives in Indiana with his wife, Katie and their four children: Beatrix, Flannery, Lowry, and Cormac.

Preface

This book is intended to serve two purposes: one to introduce students in nuclear and radiological engineering to Python and to use Python as a pedagogical tool for numerical methods relevant to their studies. The audience for this book is intended to be junior and senior undergraduate students. Most of the material is, however, suitable for sophomore students if appropriate background is provided for the nuclear reactor and radiation physics. The book arose out of a set of lecture notes for a course at Texas A&M University that was for juniors who had previously taken a course in nuclear reactor theory.

The first part of the book serves as the introduction to Python 3 and the relevant libraries for scientific computing (namely NumPy and Matplotlib). The use of the library SciPy is scrupulously minimized. This is not because the library is not useful (I find it particularly useful). Rather, learning the numerical methods needed for engineering problems would be minimized if students had such methods delivered on an Argentine platter. Additionally, this book does not cover object-oriented programming with Python. While this will be a useful skill for those students that will develop engineering software as a career, there is simply not room to do it justice in a single text.

The second part of the text introduces traditional engineering numerical methods and applies them to engineering problems relevant to the audience. On the whole, the applications do not assume a great deal of nuclear or radiological engineering background. The exception is chapters 18–20, where some knowledge of diffusion theory for neutral particles is assumed. Nevertheless, references to the relevant background are given.

The final part of the text covers the important topic of Monte Carlo methods for particle transport, in particular neutron transport. The discussion mentions neutrons specifically, but of the techniques directly apply to gamma or x-ray transport, the eigenvalue discussion in Chapter 23 notwithstanding. For the Monte Carlo chapters, the discussion walks a fine line between demonstrating the full power of Monte Carlo methods and minimizing the length of code listings. The idea is to show the reader how complications could be added to Monte Carlo codes, without having each code have the totality of the functionality discussed.

This book adopts the philosophy that all the elements of the text should, where possible, be included in the flow of the discussion and not to treat figures and code listings as floating objects that can appear far from where they are mentioned. Additionally, in the early chapters, the code to generate the figures is included to demonstrate how one makes such figures using Python.

The exercises in this book have been chosen to demonstrate the features of the numerical methods or Python code features discussed. The solutions are intended to include a large amount of discussion and critical analysis of the results. This is especially true for the programming projects. Ideally, the solutions provided by students for these problems are mini-lab reports, because in these problems, the students are performing numerical experiments.

Finally, this book seeks to serve the needs of students by making Python a tool for them to use to solve engineering problems. Many of problems are designed to teach a student how to set up a problem and then solve it with a known algorithm. The primary goal is to know how to apply the method. My view is that a deep understanding of numerical techniques is preferable, but not *de rigeur* for contemporary students.

Those students who do go on to be computational scientists will deepen their understanding in additional courses and reading. I fully realize that this point of view is not universally adopted. I only point out that using calculators without understanding the circuit boards inside did not make previous generations of students lesser scientists.

My ultimate goal is that this book generates excitement in students for computational science.

Ryan G. McClarren
April 27, 2017

Acknowledgment

This text would not be possible without the many students who asked questions, pointed out mistakes, or told me what I said that did not make sense during lectures or after reading my notes. In particular I want to thank Patrick Behne for catching many errors in a very early draft of my notes, and Logan Harbour for helping with the solutions to almost of the problems herein. I would also like to thank Dr. Jean Ragusa for allowing me to present a modified version of some of his problems from when he taught a course using Fortran.

I would also like to acknowledge the great technology of Jupyter notebooks for creating a means to express code in such a way as to interweave it with the reasoning behind it. I only hope that I could do the medium justice.

Ryan G. McClarren
April 27, 2017

INTRODUCTION TO PYTHON FOR SCIENTIFIC COMPUTING

You can be shaped, or you can be broken. There is not much in between. Try to learn. Be coachable. Try to learn from everybody, especially those who fail. This is hard. ... How promising you are as a Student of the Game is a function of what you can pay attention to without running away.

David Foster Wallace, Infinite Jest

CHAPTER POINTS

- Python is a computer programming language that we can use to solve engineering problems.

- One stores information in variables and can make computations and comparisons with those variables.

- Branching executes different parts of a code depending on conditions the programmer defines.

- Iteration execute the same block of code repeatedly under controlled conditions.

Computational Nuclear Engineering and Radiological Science Using Python
DOI: 10.1016/B978-0-12-812253-2.00002-9

3

1.1 WHY PYTHON?

In our study of computational nuclear engineering we are going to use Python, specifically version 3 of Python. Python is a powerful and widely accepted programming language that can do just about anything lower-level programming languages like Fortran, C, and C++ can. By learning to program in Python, you will learn the skills you need to program in any other computer language with relative ease.

While this book uses Python to explore computational nuclear engineering, it is not an exhaustive description of the Python language and how to use it. We will cover the topics needed for our computer simulation and numerical methods only. As a general computer programming language, Python can be used to analyze large data sets, write computer games, control devices, etc. The techniques we cover, and the approach we use to tackle problems using a computer will be applicable to these other fields as well.

The best way to start to learn a programming language is to actually use it to solve a problem. In almost any computer language the first thing you do is create a program called "Hello world!", where you make the computer say, in text, "Hello World!" (That is, after installing a way to write and run programs in the language. When installing Python on a machine, install Python 3 if you want to repeat the examples in this book. For those new to coding, a Python distribution such as Anaconda might be the easiest installation to begin with.) In Python you simply start a Python session and type:

```
In [1]: print("Hello World!")

Hello World!
```

This is the first command we will learn in Python, the `print` command:

BOX 1.1 PYTHON PRINCIPLE

The `print` command takes a comma-separated list of objects to print to the screen. Most commonly these are strings of charac- ters contained inside either single quotes or double quotes. When printing to the screen, each object is separated by a space by default.

The code to type in to your Python interpreter is the part that follows `In [1]:` and the output is directly below it separate by a blank line. We could have it print any string of characters. The string of characters could be something simple such as

```
In [2]: print("Saw 'em off")

Saw 'em off
```

to more exotic characters:

```
In [3]: print("Søren Kierkegaard and Jean-Fraçois Lyotard")

Søren Kierkegaard and Jean-Fraçois Lyotard
```

Note how Python supports the unicode character set so that we can get those fancy characters. Actually typing those characters in from a standard US keyboard is trickier, but if you do manage to input them, Python can handle it (I used copy and paste).

These results were obtained by running the code in interactive mode via a Jupyter notebook, which means the result of each line is displayed when I enter the line, and the result of the last line of input is printed to the screen. It is more common to put your code into a separate file and then execute it. These files can be executed either on the command line by typing `python codename.py` where "`codename.py`" is the name of your file, or by running it in an integrated development environment, such as IDLE or Spyder.

1.1.1 Comments

A comment is an annotation in your code to

- inform the reader, often yourself, of what a particular piece of code is trying to do,
- indicate the designed output, result, etc. of a part of the code, and
- make the code more readable.

A comment can be anything to tell you or another reader of the source code what is going on in the code. Comments can also be useful to remind you to come back and clean up an ugly part of the code, or explain to your future self why the code is written in such a way.

Comments are your friend. They can be time consuming to add, but never have I looked back at old code and regretted adding them. Any code you write that other people *might* read should be well commented. This includes code you may write for a course on Python.

You use the pound (aka hashtag) # to comment out the rest of a line: everything that follows the # is ignored by Python. Therefore, you can put little notes to yourself or others about what is going on in the code.

```
In [4]:  #This next line should compute 9*9 + 19 = 100
         print(9*9 + 19)

         #You can also make comments inside a line
         print(9*9 #+ 19)

100
81
```

There are also ways to make multiline comments by using a triple quote "'

```
In [5]:  "'The beginning of a multiline comment.
         This comment will be followed by meaningless
         code.  Enjoy "'
         print("I am the very model of a modern major general.")

I am the very model of a modern major general.
```

BOX 1.2 PYTHON PRINCIPLE

A single line comment is everything on a line after a pound (or hashtag) character, #. Multiple lines can be commented by using triple quote at the start and at the end of the comment. Whatever is inside a commented block of code is ignored by the Python interpreter.

Later we will discuss some standard formats for comments at the beginning of a function. For now we will use comments as needed to illustrate what particular snippets of code are doing.

1.1.2 Errors

In any code you write longer than a few lines, you will make a mistake. In the parlance of our times these errors are called bugs. Now there are good bugs and bad bugs. (The term bug for an error or defect goes back to at least Thomas Edison in 1878 describing an error in an invention. The most celebrated use of the word was from Grace Hopper regarding an instance in 1947 when a moth lodged itself inside one of the components of the room-sized computers of the day, and caused a malfunction.) The good bugs get caught by Python and it will complain when it finds them. The bad bugs are insidious little beings that make your code do the wrong thing, without you knowing it. Good bugs are easier to find because Python will alert you to the error. Bad bugs can exist in a code for a long time (decades even) before being unearthed. Yes, decades: Microsoft Windows reportedly had a 17-year-old bug (http://www.computerworld.com/article/2523045/malware-vulnerabilities/microsoft-confirms-17-year-old-windows-bug.html).

Even experienced programmers write code with bugs. There are many different procedures to try to rid a code of bugs, but even the most sophisticated software quality assurance techniques will not catch every one.

We will now look at a good bug and a bad bug in the following code:

```
In [6]: #This is a good bug because the Python interpreter complains
        9*9 +


        File "<iPython-input-8-64b47963658c>", line 2
        9*9 +
             ^
    SyntaxError: invalid syntax
```

Notice that Python printed a whole host of mumbo jumbo to the screen, but if you look at it closely it tells you what exactly went wrong: in line 2 of the code, there was a plus sign without anything on the right of it. This bug is good because the code didn't run and you know to go back in and fix it.

A bad bug does the wrong thing, at least according to what you want it to do, and the user and the person writing the code may be none the wiser, as in this example:

```
In [7]:  #This is a bad error because
         #it doesn't do what you might think
         #Say you want to compute (3 + 5)^2 = 8^2 = 64,
         #but you actually input
         print(3 + 5**2)
         #You don't get the correct answer,
         # and no one tells you that you're wrong.

28
```

This is an example of the power and feebleness of computers. A computer can do anything you tell it to, but it does not necessarily do what you want it to. Always keep this in mind: just because the computer did something, that does not mean it did what you wanted.

Later, we will talk in more detail about bugs, finding bugs (called debugging), and testing of code.

BOX 1.3 LESSON LEARNED

All codes that are longer than a few lines have bugs. This does not mean that those bugs meaningfully affect the program output, or that those bugs are ever encountered in the typical usage of the code. The bugs are there, however.

1.1.3 Indentation

Python is, by design, very picky about how you lay out your code. It requires that code blocks be properly indented. We will discuss what code blocks are later, but your code needs to be properly aligned to work.

```
In [8]:  #If I improperly indent, my code won't work
         print("Not indented")
             print("indented, but it shouldn't be")


             File "<iPython-input-15-0b1b509e390c>", line 3
         print("indented, but it shouldn't be")
         ^
     IndentationError: unexpected indent
```

Notice that none of the code executed because of the indentation error. In Python when you indent something it tells the interpreter that the indented code is part of a code block that is executed differently than other levels of indentation. Only at certain times can one indent and it make sense. This sounds pretty abstract and nebulous right now, but it should become clear as we go through further examples.

1.2 NUMERIC VARIABLES

Almost every code needs to store information at some point in its execution. When this information is stored by a program in the computers memory, we call the identifier or name of the information a **variable**. Information, or data, is stored in variables using the equals sign. There are different types of variables for different types of data and we will discuss several of them here. Variable type means what type of information the variable stores. A simple example is storing a number versus text.

We will discuss numeric variables, i.e., variables that store a number, first. Later we will discuss how to store text and more exotic variables.

BOX 1.4 PYTHON PRINCIPLE

A Python expression of the form

```
variable_name = expression
```

will store in a variable named "variable_name" the value that expres-sion evaluates to. The type of a variable indicates what kind of data the variable holds. The type function will identify a variable's type:

```
type(variable_name)
```

1.2.1 Integers

Integers are whole numbers, including the negatives. They never have a fractional, or decimal, part and should only be used for data that is a count.

```
In [9]: #assign the value 2 to x
        x = 2
        print(x*2)

        #check that x is an integer
        print(type(x))

4
int
```

The function $type(x)$ returns the name of the type of the variable x. Notice that Python abbreviates the term integer to "int".

Integers are useful for things that can be counted: perhaps the number of times we execute a loop, the number of elements of a vector, or the number of students in a class.

1.2.2 Floating Point Numbers

Floating point numbers are numbers that do have a fractional part. Most numbers in engineering calculations are floating point types.

```
In [10]:#Now make some other floating point variables
         y = 4.2
```

```
print("x =",x)
print(type(y))
#note that exponentiation is **
z = (x / y)**3
print("(2 / 4.2)**3 =",z)
```

```
x = 2
<class 'float'>
(2 / 4.2)**3 = 0.10797969981643449
```

The way that floating point numbers are represented on a computer has only a finite precision: there are only a finite number of bytes in the computer memory to hold the digits in the number. That means we cannot represent a number exactly in many cases. In fact floating point numbers are actually rational numbers (fractions) in the computer's internal workings. We will see later an example of how floating point accuracy can make a difference in a calculation.

1.2.2.1 Built-in Mathematical Functions

Having the ability to store floating point numbers and manipulate them with simple algebra would be of limited use to us without some common mathematical functions that we use repeatedly in engineering calculations. For instance, every time we wanted to evaluate the cosine function, we would have to program some approximation to the function, perhaps a Taylor series about a known value. Thankfully, almost every common mathematical function you might need is already built-in with Python. To use these functions you have to import the math functions using the command `import`.

BOX 1.5 PYTHON PRINCIPLE

In a code where you will be doing numerical calculations it is useful to start the code with `import math` to make a the wide range of common mathematical functions available by typing

```
math.[function]
```
where `[function]` is the name of the function.

In the code below, I set it up so that to use a math function you use the syntax `math.[function]` where `[function]` is the name of the function you want to call.

See https://docs.Python.org/3.4/library/math.html for the complete list of built-in mathematical functions.

The following code snipped uses the built-in Python function for computing the cosine of the number.

```
In [11]:import math
        #take cosine of a number close to pi
        theta = 3.14159
        trig_variable = math.cos(theta)
        print("cos(",theta,") =",trig_variable)
```

```
#use the exponential to give e
e = math.exp(1)
print("The base of the natural logarithm is",e)

#Python has a built-in pi as well
print("The value of pi is",math.pi)
```

```
cos( 3.14159 ) = -0.9999999999964793
The base of the natural logarithm is 2.718281828459045
The value of pi is 3.141592653589793
```

Notice how in the print statements, if I give it multiple arguments, it prints each with a space in between. This is useful for combining static text with calculations, as we did above.

To evaluate logarithms we note an idiosyncrasy in the way that Python names the relevant functions. The natural logarithm is just math.log and the base 10 logarithm is math.log10.

```
In [12]:print("The natural log of 10 is",math.log(10))
         print("The log base-10 of 10 is",math.log10(10))
```

```
The natural log of 10 is 2.302585092994046
The log base-10 of 10 is 1.0
```

There are two non-obvious mathematical operators, integer division: //, and the modulus (or remainder): %

```
In [13]: # 7 / 3 is 2 remainder 1
         print("7 divided by 3 is",7//3,"remainder",7%3)
         print("851 divided by 13 is",851//13,"remainder",851%13)
```

```
7 divided by 3 is 2 remainder 1
851 divided by 13 is 65 remainder 6
```

1.2.3 Complex Numbers

Python can handle complex numbers, that is, numbers that have a real and imaginary part. We denote complex numbers using two floats: one for the real part and one for the complex part, multiplied by 1j. The one is necessary so that Python knows you are not referring to a variable named j. Also, when Python prints complex variables, it typically surrounds them in parentheses. One can also do arithmetic with complex numbers using standard operators:

```
In [14]: z1 = 1.0 + 3.14 * 1j
         z2 = -6.28 + 2*1j
         print(z1,"+", z2,"=", z1+z2)
         print(z1,"-", z2,"=", z1-z2)
         print(z1,"*", z2,"=", z1*z2)
         print(z1,"/", z2,"=", z1/z2)
```

```
(1+3.14j) + (-6.28+2j) = (-5.28+5.140000000000001j)
(1+3.14j) - (-6.28+2j) = (7.28+1.1400000000000001j)
(1+3.14j) * (-6.28+2j) = (-12.56-17.7192j)
(1+3.14j) / (-6.28+2j) = (-0-0.5j)
```

To use common mathematical functions on complex numbers, we need to import the module cmath. With cmath, the common special functions and trigonometric functions can be applied to complex numbers. To illustrate this, we will compute the quadratic formula to find the roots of the polynomial

$$x^2 + \left(2 - \sqrt{2}\right)x - 2\sqrt{2} = \left(x - \sqrt{2}\right)(x + 2).$$

```
In [15]: import cmath
         a = 1.0
         b = (2 - math.sqrt(2))
         c = -2*math.sqrt(2)
         root1 = (-b + cmath.sqrt(b*b - 4*a*c))/(2*a)
         root2 = (-b - cmath.sqrt(b*b - 4*a*c))/(2*a)
         print("Roots are",root1,root2)

Roots are (1.4142135623730954+0j) (-2+0j)
```

Notice that this example used cmath.sqrt when taking the square root of a number that could be negative.

In cmath the constants cmath.e and cmath.pi are defined. We can use this to demonstrate Euler's famous relation:

```
In [16]: print(cmath.exp(cmath.pi*1j))

(-1+1.2246467991473532e-16j)
```

Here we see the effects of finite precision arithmetic in that this does not evaluate to exactly -1.

1.3 STRINGS AND OVERLOADING

A string is a data type that is a collection of characters, and needs to be inside quotes (you can use single or double quotes to enclose strings as the examples here will indicate):

```
In [17]: #This is a string
         aString = "Coffee is for closers."
         print(aString)
         print("aString")

Coffee is for closers.
aString
```

Anything inside the quotes is taken literally by Python. That is why the second print statement above just printed the literal text aString.

You can also subset, that is get some of the characters in a string, using brackets. Putting a single number in a bracket gives you the character in that position. Note, Python starts numbering at 0 so that 0 is the first character in a string.

BOX 1.6 PYTHON PRINCIPLE

A string is a collection of characters. The characters can be accessed individually using the string name followed by the character you want to access in square brackets. To access multiple characters use : indexing.

```
In [18]: aString[0]

Out[18]: 'C'

In [19]: aString[5]

Out[19]: 'e'
```

You can also get a range of characters in a string using the colon. The colon operator is non-intuitive in that $[a:b]$ says give me the elements in the string from position a to position $b - 1$.

- Python defines its ranges this way so that if the string is of length N, $[0:N]$ returns the whole string.
- Negative subsets start at the end of the string (i.e., -1 is the last character of the string).

BOX 1.7 PYTHON PRINCIPLE

For strings the first character has index 0. To access multiple characters use the syntax

```
str_variable[a:b]
```

where `str_variable` is the name of the string and `a` is the index of first character re-turned and b-1 is the index of last character returned. The first character in the string has an index of 0 and the last character has an index of -1.

Here are some examples of more advanced string indexing.

```
In [20]: aString[1:6]

Out[20]: 'offee'

In [21]:aString[-1]

Out[21]: '.'

In [22]: aString[-5:-2]

Out[22]: 'ser'
```

With characters (and strings) the + operator is overloaded. What we mean by overloaded is that the operator is defined so that the idea of what addition means is conferred to strings.

```
In [23]: 'Negative Ghostrider: ' + 'the pattern is full'

Out[23]: 'Negative Ghostrider: the pattern is full'

In [24]: 'a' + 'b' + 'c'

Out[24]: 'abc'
```

The + operator concatenates (or smushes together) the strings/characters it operates on. The multiplication operator is similarly overloaded, though it is not as general. It is simple to think about 'The roof,' * 3

```
In [25]: 'The roof, ' * 3

Out[25]: 'The roof, The roof, The roof, '
```

However, 'The roof,' * 3.15 is not defined and will give an error:

```
In [26]: 'The roof, ' * 3.15


---------------------------------------------------------------
    TypeError               Traceback (most recent call last)

      <iPython-input-28-b68ac6dcc130> in <module>()
----> 1 'The roof, ' * 3.15


        TypeError: can't multiply sequence by non-int of type 'float'
```

The upshot is that only an integer times a character/string makes sense. The order of operations is respected by the operators

```
In [27]: 'The roof, ' * 3 + 'is on fire...'

Out[27]: 'The roof, The roof, The roof, is on fire...'
```

Minus makes sense, but only sometimes, so it is not allowed. In this instance, even though subtraction makes sense to us, Python does not allow it:

```
In [28]: 'The roof, ' - 'oof'


---------------------------------------------------------------
    TypeError               Traceback (most recent call last)

      <iPython-input-30-c999e0505465> in <module>()
----> 1 'The roof, ' - 'oof'


        TypeError: unsupported operand type(s) for -: 'str' and 'str'
```

The principle of operator overloading will be useful later when we talk about matrices and vectors. These objects have, for example, addition defined so that the user can add two vectors by using the plus sign +.

1.4 INPUT

There are times when you want the user of the program to interact with the program while it is running. For the purposes of engineering calculations our interactions will be fairly simple and through text input. The means that we can ask the user for input from the keyboard and record it. In Python we can prompt the user for input using the `input` command. To illustrate how this command works, we ask the user for a number and then double that number.

```
In [29]: user_value = input("Enter a number: ")
         user_value = float(user_value)
         print("2 *", user_value,"=", 2*user_value)

Enter a number: 9.95
2 * 9.95 = 19.9
```

When the program encounters an `input` command, it waits for the user to type something in and press the enter or return key. In this example, the user entered `9.95` as the input.

BOX 1.8 PYTHON PRINCIPLE

The function

```
input_variable = input(str_variable)
```

prints `str_variable` to the screen and waits for the user to enter a value. It will store a string of the characters the user entered in the variable `input_variable`. In Python 3 the input function always returns a string, even when the user enters a number.

Finally, this example also introduces the function `float`. This function takes a variable and changes it into a `float`. This is necessary because the `input` function always returns a character string variable.

1.5 BRANCHING (IF STATEMENTS)

Sometimes you want to execute code differently based on the value of a certain variable. This is most commonly done in if-else constructs. Here is an example that takes input from the keyboard and then executes different lines of code based on the response.

```
In [30]: instructors_op=input("What is your opinion of student? ")
         grade = ' '
         if (instructors_op == 'annoying'):
             grade = 'F+'
         elif (instructors_op == 'Not annoying'):
             grade = 'B+'
         else:
             grade = 'A'
         print(grade)
```

```
What is your opinion of student? Not Annoying
A
```

What this codes says is that if the value of `instructors_opinion` is "annoying", the grade will be "*F+*", otherwise or else if (`elif` in Python-speak) `instructors_opinion` is "Not annoying" the grade will be "*B+*", and anything else will be a grade of "*A*". In the example I typed in "Not Annoying" and the `if` statement and the `elif` statement require that the string exactly match, so it executed the `else` part of the code.

BOX 1.9 PYTHON PRINCIPLE

The if-else construct allows the code to execute different branches based on the value of expressions. The code

```
if expression1:
        [some code]
elif expression2:
        [some other code]
else:
        [something else]
```

will execute the block of code [some code] if expression1 evaluates to true, execute [some other code] if expression1 evaluates to false *and* expression2 evaluates to true, or will execute [something else] if both expression1 *and* expression2 evaluate to false. There could be more than one `elif` condition, or the `else` and `elif` statements could not be there at all. That is, it is possible to have an `if` without an `elif` or an `else`.

It is important to remember that when you want to check equality between two things you need to use == and not a single equals sign. A single equals sign is what you use when you want to assign something to a variable. You can compare numbers using the standard greater than, less than, and other operators. See Box 1.10 for a list of commonly used operators.

BOX 1.10 PYTHON PRINCIPLE

To compare numbers, and other variables, we can use the following operators to make comparisons:

- a `>` b — a greater than b
- a `>=` b — a greater than or equal to b
- a `==` b — a equal to b
- a `<` b — a less than b
- a `<=` b — a less than or equal to b
- `not(a)` — a not true

Each statement will evaluate to true or false.

In Python, when an expression evaluates to true, it evaluates to the integer 0; a false expression evaluates to 0. Therefore, we can treat a false expression as a zero and a true expression as non-zero, as we will do in later examples.

Python also has a `not` operator. This operator will return true if its argument is false (or zero); it will return false is the argument is true or nonzero. For example, `not(0)` will evaluate to true, and `not(1)` and `not(2.005)` will both evaluate to false. The not operator can be combined with other expressions to make complex conditional statements. As an example, the mathematical statement $a \neq b$ can be written in Python as `not(a == b)`.

It is often common to have a condition where one checks if a number is close to another within some tolerance.

```
In [31]: import math
         pi_approx = 22/7
         if math.fabs(pi_approx - math.pi) < 1.0e-6:
             print("Yes, that is a good approximation")
         else:
             print("No,",pi_approx,
                   "is not a good approximation of",
                   math.pi,".")
```

```
No, 3.142857142857143 is not a good approximation of
3.141592653589793.
```

The function `math.fabs` is the float version of the absolute value function. In this case we were checking to see if an approximation is within 10^{-6} of π. Here the number 10^{-6} is written as `1.0e-6` which is shorthand for 1.0×10^{-6}.

Branching statements are most powerful when combined with iteration, as we will now explore.

1.6 ITERATION

Iteration executes a piece of code repeatedly, based on some criteria. In this example we will try to find a good approximation to π by trying many different values in succession.

```
In [32]: #this code finds a decent approximation to pi
         converged = 0
         guess = 3.14
         iteration = 0
         #Define tolerance for approximating pi
         eps = 1.0e-6
         #converged will be 0 if false, 1 if true
         converged = math.fabs(guess - math.pi) < eps
         while (converged == 0):
             guess = guess + eps/2
             converged = math.fabs(guess - math.pi) < eps
             iteration += 1 #same as iteration = iteration + 1
         print("Our approximation of pi is", guess)
         print("It took us", iteration,"guesses to approximate pi")
```

```
Our approximation of pi is 3.1415920000002227
It took us 3184 guesses to approximate pi
```

In this code, as long as `converged == 0` the code in the while block—the indented code below `while (converged == 0):`—will execute over and over. When the value of our guess is within 10^{-6} to π in absolute value, `converged` will become 1 and the `while` loop will not start executing the code block again.

I did something tricky, but useful, in this example. In Python when a conditional expression like `a > b` is true it evaluates to an integer of 1, and evaluates to an integer of 0 when false. We will make use of this trick later on and it is good to see it now to help you get accustomed to it.

The idea of a while loop is not unique to Python, and can even be found in children's movies. The seven dwarfs in *Snow White* used the logic of a `while` loop to provide a sound-track to their labors in a small mining operation, though they did not use Python:

```
while (working):
    [whistle]
```

BOX 1.11 PYTHON PRINCIPLE

The while loop is written in Python as

```
while expression:
    [some code]
```

This will execute the code in the block `[some code]` as long as `expression` evaluates to true when the loop returns to the top of the code block.

We can modify our code by tightening the tolerance to 10^{-8}, and we will change the condition for the `while` loop to show that there are multiple ways of accomplishing the same task.

```
In [33]: guess = 3.14
         iteration = 0
         eps = 1.0e-8
         converged = abs(guess - math.pi) >= eps
         while (converged==1):
             guess = guess + eps/2
             converged = abs(guess - math.pi) >= eps
             iteration += 1
         print("Our approximation of pi is", guess)
         print("It took us", iteration,"guesses to approximate pi")

Our approximation of pi is 3.141592644990321
It took us 318529 guesses to approximate pi
```

The `while` loop is an important form of iteration. Another type is the `for` loop which executes a set number of times based on a set condition. We will talk about that loop in the next chapter.

THE GREAT BEYOND

We have only scratched the surface of what Python can do. For example, we can generate graphs with relative ease, and we will cover this in detail in a few chapters. This will allow us to visualize our calculations easily as both a check of our computation and a means to report our results. Here is an example of how simply we can generate the graph of a function:

```
In [34]: import matplotlib.pyplot as plt
         import numpy as np
         x = np.linspace(0, 3*np.pi, 5000)
         fig = plt.figure(figsize=(8,6), dpi=600)
         plt.plot(x, np.sin(x**2))
         plt.title('A simple chirp');
```

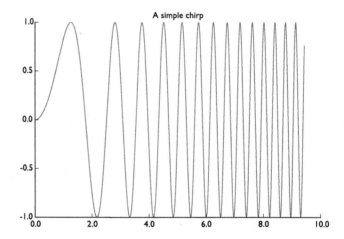

This example uses several features of Python that we have not discussed yet, namely numpy and matplotlib. These extensions of Python, called libraries, allow us to make a plot in just a few lines of code.

FURTHER READING

There are a number of reference books on programming in Python. An advanced text on using Python to write software to solve science problems is the work of Scopatz and Huff [1]. A more gentle introduction to Python and thinking like a coder is the work of Guttag [2].

PROBLEMS

Short Exercises

1.1. Ask the user for two numbers. Print to the user the value of the first number divided by the second. Make your code such that it never divides by zero.

1.2. Ask the user for an integer. Tell the user if the number is prime. You will need to use a while loop with a counting variable that goes up by one each time through the loop.

1.3. Ask the user for the length of two sides of a right triangle that form the right angle. Tell the user what the length of the hypotenuse is and the number of degrees in each of the other two angles.

1.4. When a object is not at rest, its mass is increased based on the formula $m = \gamma m_0$ where m is the mass, m_0 is the rest mass, and γ is a relativistic factor given by

$$\gamma = \frac{1}{\sqrt{1 - \frac{v^2}{c^2}}}.$$

The rest mass of a baseball is 145 g and the speed of light, c, is 2.99792458×10^8 m/s. What is the mass of a baseball when thrown at $v = 169.1$ km/h (the fastest recorded pitch)? How fast does it have to move to have a mass of 1.45 kg?

Programming Projects

1. Harriot's Method for Solving Cubics

The cubic equation

$$x^3 + 3b^2 x = 2c^3$$

has a root given by

$$x = d - \frac{b^2}{d},$$

where

$$d^3 = c^3 + \sqrt{b^6 + c^6}.$$

This method was developed by Thomas Harriot (1560–1621), who also introduced the less than and greater than symbols.

Write a program that prompts the user for coefficients of a general cubic,

$$Ax^3 + Bx^2 + Cx + D,$$

and determines if the cubic can be solved via Harriot's method (i.e., $B = 0$). If it can be solved via Harriot's method, then print the solution. Also, print the residual from the root, that is the value you get when you plug each into the original equation. Make sure that you do not divide by zero, and that your method can handle imaginary roots.

2

Digging Deeper Into Python

The secret, I don't know... I guess you've just gotta find something you love to do and then... do it for the rest of your life. For me, it's going to Rushmore.

–"Max Fischer" in the movie Rushmore

CHAPTER POINTS

- For loops make performing a fixed number of iterations simple.

- Python has data structures that can contain various groups of items and manipulate them efficiently.

- Floating point numbers are not exact, and this can lead to unexpected behavior.

2.1 A FIRST NUMERICAL PROGRAM

Thus far we have talked about the basics of Python and some of the rudimentary building blocks of a program. We have used these to make toy codes that did not do much useful. Now we will make things a bit more concrete and show how we can use Python to perform numerical calculations. We will start out very basic and expand our repertoire as we go.

To begin, we will consider a common calculation in many applications (movie and computer graphics for computing the reflection angles for lighting and shading, for example):

$$y = \frac{1}{\sqrt{x}}.$$

Assuming that we cannot just use the built-in square root function, we could find the value of y such that

$$y\sqrt{x} - 1 = 0.$$

One way we could solve this equation is to start with a guess that we know is either high (or low), and then decrement (or increment) the guess until it is accurate enough. This is not a particularly good way of solving this problem, but it is easy to understand. To tell when we are close enough to an answer, we will evaluate the residual. In simple terms the residual for a guess y_i is given by

$$\text{residual} = y_i\sqrt{x} - 1.$$

When the guess is equal to the solution y, the residual will be zero. Similarly, when the residual is small, we are close to the answer.

For our problem we know that

$$x > \frac{1}{\sqrt{x}}$$

as long as x is greater than 1. So we could start at x and decrease the guess until we get the answer we desire. This is an example of exhaustive enumeration. It is called enumeration because we list (i.e., enumerate) possible solutions to the problem until we get one close enough. It is exhaustive because we list "all" the possible solutions up to some precision. As we will see, it is also exhaustive because it is a lot of work to solve a simple problem.

Here is Python code to solve this problem. The user inputs an x and the code computes $1/\sqrt{x}$.

```
In [1]: #this code computes 1/sqrt(x), for x > 1
        import math
        x = float(input("Enter a number greater than 1: "))
        if (x<=1):
            print("I said a number greater than 1")
        else:
            converged = 0
            answer = x #initial guess is x
            eps = 1.0e-6 #the residual tolerance
            converged = math.fabs(answer * math.sqrt(x) - 1.0) < eps
            iteration = 0
            while not(converged):
                answer = answer - 0.5*eps
                converged = math.fabs(answer * math.sqrt(x) - 1.0) < eps
                iteration += 1
            print("1/sqrt(",x,") =",answer)
            print("It took",iteration,"guesses to get that answer.")
```

```
Enter a number greater than 1: 3
1/sqrt( 3.0 ) = 0.5773504997552141
It took 4845299 guesses to get that answer.
```

Before discussing how the algorithm performs, it is worthwhile to discuss how it works. Notice that the code is using an if-else statement to check that the input provided by the user is greater than one. Users can enter any number of wacky things, so it is always a good idea to check user-specified input.

After assuring that the value of x is greater than or equal to one, the code starts with a guess at the answer of x. It then checks if the absolute value of the residual is smaller than a specified tolerance. If the absolute value of the residual is larger than the tolerance, then it enters the while loop and decreases the value of the answer by one-half times the tolerance until the residual is small enough in magnitude.

This is a really slow algorithm: it can easily take millions of guesses. Though it is easy to deride this simplistic algorithm, starting with a very basic algorithm that is slow, but that we know will work, is a good idea. In other words, having a slow, working algorithm is better than nothing. To paraphrase, an algorithm in code is worth two in your head.

If we can bracket the answer, that is say it is between two numbers, we can improve the algorithm by bisecting (dividing in two) the interval and zooming in on the answer. To start, we can bracket the solution to our equation because we know that the answer is between $1/x$ and x for $x > 1$. The following code starts out by defining this interval and checking on which half of the interval the solution is in.

```
In [2]: #this code computes 1/sqrt(x), for x > 1
        import math
        x = float(input("Enter a number greater than 1: "))
        if (x<=1):
            print("I said a number greater than 1")
        else:
            converged = 0
            upper_bound = x
            lower_bound = 1.0/x
            answer = (upper_bound + lower_bound)*0.5
            eps = 1.0e-6
            converged = math.fabs(answer * math.sqrt(x) - 1.0) < eps
            iteration = 0
            while not(converged):
                mid = answer
                if (mid < 1.0/math.sqrt(x)):
                    lower_bound = mid
                else:
                    upper_bound = mid
                answer = (upper_bound + lower_bound)*0.5
                #print(upper_bound,lower_bound)
                converged = (math.fabs(answer * math.sqrt(x) - 1.0)
                            < eps)
                iteration += 1
            print("1/sqrt(",x,") =",answer)
            print("It took",iteration,"guesses to get that answer.")
```

```
Enter a number greater than 1: 3
1/sqrt( 3.0 ) = 0.5773506164550781
It took 20 guesses to get that answer.
```

This method is called the bisection method, and we will take a closer look at it in the future. The way it works is it takes a range that brackets the root of an equation, and then looks at the midpoint of that range. Based on the value of the function at the midpoint, you then know if the root is in the lower half or upper half of the range. We will explain this in more detail in a future chapter when we talk about solving nonlinear equations.

Notice that it took a lot fewer guesses using bisection compared to exhaustive enumeration. Exhaustive enumeration is not very good for problems where we are trying to find a continuous variable.

2.2 FOR LOOPS

While loops are great, and they can do everything we need for iteration. Nevertheless, there are instances when we need to iterate a fixed number of times, it is useful to have a shorthand for this type of iteration structure. For instance, if we wanted to execute a block of code a set number of times, we have to define a counting variable and increment it by hand:

```
In [3]: #Some code that counts to ten
        count = 1
        while (count <= 10):
            print(count)
            count += 1

1
2
3
4
5
6
7
8
9
10
```

The for loop is built for such a situation. The way we typically use it is with the range function. This function takes 3 input parameters: range(start, stop, [step]). The stop parameter to the range function tells range to go to the number *before* stop. The parameter step is in brackets because it is optional. If you do not define it, Python assumes you want to count by 1 (i.e., step by 1). What range returns is a sequence that starts at start and counts up to stop - 1 by step. The next example demonstrates this:

```
In [4]: print(list(range(1,10)))
        #the list command tells Python to write out the range

[1, 2, 3, 4, 5, 6, 7, 8, 9]
```

Also, if you just give range one parameter, it treats that as stop and assumes you want to start at 0:

```
In [5]: #These should be the same
        print(list(range(0,10)))
        print(list(range(10)))

[0, 1, 2, 3, 4, 5, 6, 7, 8, 9]
[0, 1, 2, 3, 4, 5, 6, 7, 8, 9]
In [6]: #Here's something using the step parameter
        print(list(range(0,10,2)))

[0, 2, 4, 6, 8]
```

With the range command we can have a for loop assign a variable a value in the range, in order, each time the code block of the for loop executes:

```
In [7]: for i in range(10):
            print(i+1)

1
2
3
4
5
6
7
8
9
10
```

BOX 2.1 PYTHON PRINCIPLE

The range function

```
range(start, stop, [step])
```

creates a list of integers that begins at start, increments by step, and stops before stop. The step parameter is assumed to be one if it not included. The range function can be called with one parameter:

```
range(stop)
```

This is equivalent to the three-parameter version with start equal to 0, and step equal to 1, that is, range(0,stop,1).

Suppose we want to add a number to itself seven times. To do this we could use a for loop:

```
In [8]:  number = 10
         sum = 0
         for i in range(7):
             sum += number
         print(sum)
```

70

We could also do this using a `while` loop, but it takes two extra lines: one to initialize a variable, and another to increment it.

```
In [9]:   #while loop version
          number = 10
          sum = 0
          i = 0
          while (i<7):
              sum += number
              i += 1
          print(sum)
```

70

BOX 2.2 PYTHON PRINCIPLE

The `for` loop is written in Python as

```
for i in X:
    [some code]
```

The code in the block [`some code`] will execute once for each item in the object X and each time through the code block i will take on the value of an item in X, in order.

Here is, perhaps, a more practical use of a for loop: to compute π. We do this using random numbers picked between -1 and 1 using the `random` library that comes with Python. In that library there is a function called uniform that gives a uniformly distributed random number between two endpoints.

```
In [10]:  "'Compute pi by picking random points between x = -1 and 1,
          y = -1 and 1. The fraction of points
          such that x^2 + y^2 < 1, compared with the total number
          of points is an approximation to pi/4"'
          import random
          number_of_points = 10**5
          number_inside_circle = 0
          random.seed() #this seeds the random number generator
          for point in range(number_of_points):
              x = random.uniform(-1,1) #pick random number between -1 and 1
              y = random.uniform(-1,1) #pick random number between -1 and 1
              if x**2 + y**2 < 1: #is the point in the circle
                  number_inside_circle += 1
          pi_approx = 4.0*number_inside_circle/number_of_points
          print("With",number_of_points,
              "points our approximation to pi is",pi_approx)
```

```
With 100000 points our approximation to pi is 3.1406
```

This works because the ratio of the number of points inside the circle to the total number of points will converge to the ratio of the area of the circle (π) to the total area of the square (4). In particular, `number_inside_circle/number_of_points` will converge to $\pi/4$ as the number of points chosen goes to infinity. The code above is our first example of a Monte Carlo method where we use random numbers to compute fixed quantities, and we will return to these methods in the last part of this text.

The random numbers generated by Python are actually items in a really long list of numbers that seem random (such random numbers are called pseudorandom numbers). In the code above we set where we start in the list using `random.seed()`, which then uses the system time to pick a starting point, so that each time the code is run, it starts somewhere different.

Using NumPy, which we have not covered yet, we can do this in an even fancier way. If we use Matplotlib, another Python library, we can get nice graphs as well. The code below picks 1000 random points to estimate π and plots the points on a graph in a particular color, and draws the circle in another color (the print version of the book will have the points in gray and the circle in black).

```
In [11]: import numpy as np
         import matplotlib.pyplot as plt
         #pick our points
         number_of_points = 10**3
         x = np.random.uniform(-1,1,number_of_points)
         y = np.random.uniform(-1,1,number_of_points)
         #compute pi
         pi_approx = 4.0*np.sum(x**2 + y**2 <= 1)/number_of_points
         #now make a scatter plot
         maize = "#ffcb05"
         blue = "#00274c"
         fig = plt.figure(figsize=(8,6), dpi=600)
         #scatter plot with hex color
         plt.scatter(x, y, alpha=0.5, color=maize)
         #draw a circle of radius 1 with center (0,0)
         circle = plt.Circle((0,0),1,color=blue, alpha=0.7,
                             fill=False, linewidth=4)
         #add the circle to the plot
         plt.gca().add_patch(circle)
         #make sure that the axes are square so that our circle is circular
         plt.axis('equal')
         #set axes bounds: axis([min x, max x, min y, max y])
         plt.axis([-1,1,-1,1])
         #make the title have the approximation to pi
         plt.title("$\\pi \\approx $" + str(pi_approx))
         #show the plot
         plt.show()
```

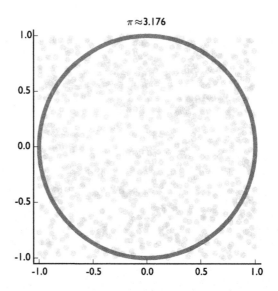

You can also simply accomplish other tasks `for` loops. For instance, you can have the control variable take on non-numeric things. If we have a list of strings, numbers, or whatever we can loop over all the elements of that sequence. In the case below we have the `for` loop make the control variable contain each string in a sequence.

```
In [13]: #silly hat code
         hats = ["fedora","trilby","porkpie","tam o'shanter",
             "Phrygian cap","Beefeaters' hat","sombrero"]
         days = ["Monday","Tuesday","Wednesday","Thursday",
             "Friday","Saturday","Sunday"]
         count = 0
         for today in hats:
             print("It is",days[count],"and I will wear a",today)
             count += 1

It is Monday and I will wear a fedora
It is Tuesday and I will wear a trilby
It is Wednesday and I will wear a porkpie
It is Thursday and I will wear a tam o'shanter
It is Friday and I will wear a Phrygian cap
It is Saturday and I will wear a Beefeaters' hat
It is Sunday and I will wear a sombrero
```

Notice what this code did: we defined a list called `days` that contained strings for the names of the days of the week. Inside our for loop we had a numeric variable that kept track of what number the day of the week was (0 for Monday in this case). Then when we access `days[count]` it returns the string in position count.

We can go one step beyond and plan our haberdashery decisions a month in advance using random numbers.

```
In [14]: #silly hat code
         import random
```

```
        hats = ["fedora","trilby","porkpie","tam o'shanter",
                "Phrygian cap","Beefeaters' hat","sombrero"]
        days = ["Monday","Tuesday","Wednesday","Thursday",
                "Friday","Saturday","Sunday"]
        for count in range(30):
            hat_choice = round(random.uniform(0,6))
            print("It is",days[count % 7],"and I will wear a",
                  hats[hat_choice])
```

```
It is Monday and I will wear a fedora
It is Tuesday and I will wear a Phrygian cap
It is Wednesday and I will wear a tam o'shanter
It is Thursday and I will wear a Beefeaters' hat
It is Friday and I will wear a Phrygian cap
It is Saturday and I will wear a Beefeaters' hat
It is Sunday and I will wear a Beefeaters' hat
It is Monday and I will wear a Phrygian cap
It is Tuesday and I will wear a Beefeaters' hat
It is Wednesday and I will wear a Phrygian cap
It is Thursday and I will wear a sombrero
It is Friday and I will wear a sombrero
It is Saturday and I will wear a trilby
It is Sunday and I will wear a Beefeaters' hat
It is Monday and I will wear a porkpie
It is Tuesday and I will wear a trilby
It is Wednesday and I will wear a Phrygian cap
It is Thursday and I will wear a sombrero
It is Friday and I will wear a tam o'shanter
It is Saturday and I will wear a porkpie
It is Sunday and I will wear a Phrygian cap
It is Monday and I will wear a fedora
It is Tuesday and I will wear a Beefeaters' hat
It is Wednesday and I will wear a Phrygian cap
It is Thursday and I will wear a Beefeaters' hat
It is Friday and I will wear a Beefeaters' hat
It is Saturday and I will wear a porkpie
It is Sunday and I will wear a Beefeaters' hat
It is Monday and I will wear a porkpie
It is Tuesday and I will wear a tam o'shanter
```

We will use the ability to iterate over a list in future codes to do things like investigate radioactive decay.

2.3 LISTS AND TUPLES

2.3.1 Lists

In the previous section we defined a list of strings as

```
In [15]: hats = ["fedora","trilby","porkpie","tam o'shanter","Phrygian cap",
                 "Beefeaters' hat","sombrero"]
         print(hats)
```

```
['fedora', 'trilby', 'porkpie', "tam o'shanter", 'Phrygian cap',
 "Beefeaters' hat", 'sombrero']
```

A list is a Python object that contains several other objects. A list is defined by enclosing the list members in square brackets. In the case above the objects were strings, but they could have been numbers or other objects. We can access items in a list using square brackets:

```
In [16]: hats[2]
```

```
Out[16]: 'porkpie'
```

Slicing using the colon is allowed on lists just as it is in strings, like we did in the previous chapter:

```
In [17]: print(hats[3:7])
```

```
["tam o'shanter", 'Phrygian cap', "Beefeaters' hat", 'sombrero']
```

You can also add items to a list using the append function. To do this you use the name of the list followed by .append:

```
In [18]: hats.append("toque")
         print(hats)
```

```
['fedora', 'trilby', 'porkpie', "tam o'shanter", 'Phrygian cap',
 "Beefeaters' hat", 'sombrero', 'toque']
```

You can delete an item from a list using the remove function in the same way we used the append function:

```
In [19]: hats.remove('trilby')
         print(hats)
```

```
['fedora', 'porkpie', "tam o'shanter", 'Phrygian cap',
 "Beefeaters' hat", 'sombrero', 'toque']
```

A feature of lists is that they can contain different types of items. For example you can mix strings and numbers:

```
In [20]: my_list = ["Item 0", "Item 1", 2]
         print(my_list)
```

```
['Item 0', 'Item 1', 2]
```

In many cases we would like to know how many elements are in a list. To find the length of a list use the len function:

```
In [21]: len(my_list)
```

```
Out[21]: 3
```

To see if an value is contained in a list use the `in` operator. This operator will return true if its argument is in the list, and false otherwise:

```
In [22]: "Item 2" in my_list

Out[22]: False

In [23]: 2 in my_list

Out[23]: True
```

The plus operator is overloaded so that we can use it to concatenate two lists, much like we did with strings:

```
In [24]: print(my_list + hats)

['Item 0', 'Item 1', 2, 'fedora', 'porkpie', "tam o'shanter",
 'Phrygian cap', "Beefeaters' hat", 'sombrero', 'toque']
```

BOX 2.3 PYTHON PRINCIPLE

A list in Python is a collection of data that can be changed after creation. The syntax to define a list is

```
list_var = [item1, item2, ..., itemN]
```

where the `...` indicates a number of other items separated by commas. Items in a list can be indexed using square brackets, and slice indexing using `:` is also supported. To add or remove items from the list use the `append` and `remove` commands with syntax of the form

```
list_var.append(item_add)
list_var.remove(item_remove)
```

where `item_add` and `item_remove` are items to add or remove from a list, respectively. Appending to a list adds the element to the end.

The `len` function will return the size of the list,

```
len(list_var)
#return the length of list_var
```

The `in` operator can be used to test if an item is in the list:

```
test_item in list_var
```

will evaluate to true if `test_item` is in the list `list_var`.

The plus operator, +, is overloaded so that

```
list_var1 + list_var2
```

will concatenate `list_var2` to the end of `list_var1`.

2.3.2 Tuples

A tuple is very similar to a list with two exceptions. The first exception is minor in that you define a tuple using regular parentheses:

```
In [25]: hats_tuple = ("fedora","trilby","porkpie",
                       "tam o'shanter","Phrygian cap",
                       "Beefeaters' hat","sombrero")
         print(hats_tuple)

('fedora', 'trilby', 'porkpie', "tam o'shanter", 'Phrygian cap',
 "Beefeaters' hat", 'sombrero')
```

You can access an element in a tuple with square brackets the same way you do a list

```
In [26]: hats_tuple[1]

Out[26]: 'trilby'
```

The major difference between lists and tuples is that tuples cannot be modified once created. The technical term for this is that tuples are immutable, whereas lists are mutable. If we try to change a tuple, we get an error:

```
In [27]: hats_tuple[1] = 'deer stalker'


-------------------------------------------------------------------
TypeError                 Traceback (most recent call last)

    <ipython-input-27-60a4d89f7f83> in <module>()
----> 1 hats_tuple[1] = 'deer stalker'


    TypeError: 'tuple' object does not support item assignment
```

The main point of tuples is to give the programmer the ability to group items together in a lightweight manner without all the behind-the-scenes infrastructure that a mutable type (like a list) has. It also can be useful if you want to define an object that you want to assure will not change. We will not have many uses for tuples in our work, but it is useful to know that they exist.

BOX 2.4 PYTHON PRINCIPLE

A tuple in Python is a collection of data that cannot be changed after creation. Parentheses are used to define a tuple:

```
tuple_var = (item1, item2, ..., itemN)
```

where the ... indicates a number of other items separated by commas. Items in a tu-ple can be indexed using square brackets, and slice indexing using : is also supported.

The len function will return the size of the tuple,

```
len(tuple_var)
#return the length of tuple_var
```

2.4 FLOATS AND NUMERICAL PRECISION

Previously, it was mentioned that floating point numbers are not exact because a computer has a finite number of bits to represent the numbers (i.e., we cannot have an infinite number of decimal places in the number). Typically, this is not a major issue, but when it does matter, it can cause problems. In our work, one particular time that this matters is when we want to have a stopping criteria on a floating point number, for instance this is a bad idea:

```
In [28]: import random
         import math
         iteration = 0
         guess = 0
         closest = 0
         target = 0.3
         while (guess != target) and (iteration < 10**6):
             guess = random.random()#same as random.uniform(0,1)
             if (math.fabs(guess - target) <
                 math.fabs(closest - target)):
                 closest = guess
             iteration += 1
         print("In", iteration,
             "the closest random number we got to",
             target,"is", closest)

In 1000000 the closest random number we got to 0.3
is 0.3000001943632269
```

To most people, and for most computations, 0.3000001943632269 is close enough to 0.3, but the computer just knows that these two numbers are not equal.

What we probably wanted is something like

```
In [29]: import random
         import math
         iteration = 0
         guess = 0
         tolerance = 1.0e-6
         while (math.fabs(guess - 0.3) > tolerance and
                 (iteration < 10**6)):
             guess = random.random()
             iteration += 1
         print("It took", iteration,
             "guesses to get within",tolerance,"of 0.3")
         print("The number we ended with is",guess)

It took 128626 guesses to get within 1e-06 of 0.3
The number we ended with is 0.29999952310331457
```

The bottom line, is that one should not use equality tests with floating point numbers. That is why in our exhaustive enumeration example in Section 2.1 we only tried to get within a tolerance rather than match the number exactly.

Even very simple equality tests with floating point numbers can fail.

```
In [30]: (0.1+0.1+0.1) == 0.3
```

```
Out[30]: False
```

Wait, what? The number 0.1 is not exactly represented in the computer because a computer stores numbers using a base 2 representation and not a base 10 representation like we are used to. To demonstrate this we can print out 0.1 to 20 digits.

```
In [31]: print("%.20f" % 0.1)
```

```
0.10000000000000000555
```

There is a tiny error that is amplified when we compute $0.1 + 0.1 + 0.1$.

```
In [32]: 0.1 + 0.1 + 0.1
```

```
Out[19]: 0.30000000000000004
```

Later, especially in solving linear systems, we will see that the numerical precision can have a large effect on our answers, if we formulate algorithms that are sensitive to these errors. For now, let us agree not to use equality with floating point numbers.

FURTHER READING

A excellent example of the power and limitations of finite precision arithmetic can be found in the SIAM 100-Digit Challenge [3]. For further reading on Monte Carlo methods, Kalos and Whitlock is the standard reference [4].

PROBLEMS

Short Exercises

2.1. Write a Python code that asks the user to input a string. Then print back to the user their string backwards.

2.2. If we list all the natural numbers below 10 that are multiples of 3 or 7, we get 3, 6, 7, and 9. The sum of these multiples is 25. Find the sum of all the multiples of 3 or 7 below 10000.

2.3. Prompt the user for a number. Report back the cube root of the number. Test the code with some numbers that are perfect cubes, e.g., 125. How accurate is the answer?

2.4. The public domain folk song "Dem Bones" is a song that describes in a pseudo-scientific manner the layout of the bones of the human body. An example couplet of this song reads as

The foot bone connected to the leg bone
The leg bone connected to the knee bone

which can be generalized to

> The b_i bone connected to the b_{i+1} bone
> The b_{i+1} bone connected to the b_{i+2} bone

Write a Python code that uses a `for` loop and a `list` that prints out an entire verse where the bones, in order, are

```
b = ["foot", "leg", "knee", "thigh", "back", "neck", "head"]
```

Programming Projects

1. Nuclear Reaction Q Values

Write a Python code that asks the user to input the masses (in amu) of two reactants in a nuclear reaction and the two products of the reaction. The code will output the Q value of the reaction in MeV. Here is an example reaction:

$$^{A_X}_{Z_X} X + ^{A_x}_{Z_x} x \rightarrow ^{A_Y}_{Z_Y} Y + ^{A_y}_{Z_y} y.$$

The user should be able to enter zero for the mass in case the reaction is a decay (i.e., has only one reactant) or is a reaction that has only one product. Use 1 amu $= 931.494061$ MeV/c^2.

2. Calculating e, the Base of the Natural Logarithm

The series

$$\frac{1}{0!} + \frac{1}{1!} + \frac{1}{2!} + \frac{1}{3!} + \frac{1}{4!} + \frac{1}{5!} \cdots = e,$$

is a means of approximating e from the Taylor expansion of e^x. We can write a partial sum as

$$e_{\text{approximate}} = \sum_{i=0}^{N} \frac{1}{i!}.$$

2.1. Using a `for` loop, compute an approximation using

$$N = 1, 2, 3, 4, 5, 6, 7, 8, 9, 10, 100, 1000, 10000.$$

2.2. Describe how the solution converges to the exact answer as a function of N. That is, how does the error in the estimate change as a function of N?

2.3. How many digits of

$$e = 2.71828182845904523536028747135266249775724709369995$$

can you compute correctly?

3

Functions, Scoping, Recursion, and Other Miscellany

There is no way that this winter is *ever* going to end as long as this groundhog keeps seeing his shadow. I don't see any other way out. He's got to be stopped. And I have to stop him.

–"Phil Connors" in the movie Groundhog Day

CHAPTER POINTS

- Defining functions makes code reuseable.

- Docstrings are special comments that indicate how to use a function, and can be accessed using the `help` function.

- Functions have their own variables, as defined by scoping rules.

- Functions can be bundled into a module and called from many programs.

3.1 FUNCTIONS

In this chapter we are going to define our own functions to make life easier on ourselves. Defining functions will make our code more robust, less prone to errors, and more usable. When we define a function what we want to do is to create an abstract version of a concrete set of steps that we want to execute in our code. By creating abstract versions we will be able to run the same lines of code repeatedly without typing them over and over. Motivating why this is necessary for writing good code is a bit of an uphill battle because cutting and pasting code repeatedly can seem pretty easy.

To demonstrate why we might want to define a function, we will solve in code a particular example of a simple system of linear equations. Our example system is

$$4.5x + 3y = 10.5$$

$$1.5x + 3y = 7.5.$$

We could find the values of x and y that solve these equations with the following Python code that eliminates a variable by combining the equations to solve for y and then solving for x. Read the comments to see what is happening.

```
In [1]: """Python code to solve
        4.5 x + 3 y = 10.5
        1.5 x + 3 Y = 7.5
        by solving the second equation for y first,
        and then solving for x"""
        #step 1 solve for y, multiply equation 2 by
        #-3, and add to first equation
        LHS_coefficient = -3*3 + 3 #the coefficient for y
        RHS = -3*7.5 + 10.5 #the right-hand side
        #now divide right-hand side by left-hand side coefficient
        y = RHS / LHS_coefficient
        #plug y into first equation
        x = (10.5 - 3*y)/4.5
        #print the solution, note \n produces a linebreak
        print("The solution to:\n4.5 x + 3 y = 10.5\n
            1.5 x + 3 y = 7.5\n is x =",x,"y=",y)

The solution to:
4.5 x + 3 y = 10.5
1.5 x + 3 y = 7.5
 is x = 1.0 y= 2.0
```

Our code appears to work (you can check by plugging in the values into the system). Given that we have put the effort into solving that system, it is likely that we want to solve another 2 by 2 system with different coefficients. We could just take our old code and change out the coefficients and the right-hand sides, but there are many places that we need to change and it is likely that we will make a mistake.

What we would like to do is define a function that will solve the system for any coefficients and right-hand side (provided there is a solution). The definition of such a function to solve

$$a_1 x + b_1 y = c_1$$

$$a_2 x + b_2 y = c_2,$$

is given here

```
In [2]: def two_by_two_solver(a1,b1,c1,a2,b2,c2, LOUD=False):
            """Calculate the solution of the system
            a1 x + b1 y = c1,
            a2 x + b2 y = c2

            Args:
                a1: x coefficient in first equation (cannot be zero)
                b1: y coefficient in first equation
                c1: right-hand side in first equation
                a2: x coefficient in second equation
                b2: y coefficient in second equation
                c2: right-hand side in second equation
                LOUD: boolean that decides whether to print out the answer

            Returns:
                list containing the solution in the format [x,y]
            """
            #step one, eliminate x from the second equation
            #by multiplying first equation by -a2/a1
            #and then adding it to second equation
            new_b2 = b2 - a2/a1*b1
            new_c2 = c2 - a2/a1*c1
            #solve the new equation 2
            y = new_c2/new_b2
            #plug y into original equation 1
            x = (c1-b1*y)/a1

            if (LOUD):
                print("The solution to:\n",a1,"x +",b1,"y =",c1,"\n",a2,"x +",
                      b2,"y =",c2,"\n is x =",x,"y=",y)
            return [x,y]
```

After we define a function, we can call it to solve for the problem above by typing

```
In [3]:  two_by_two_solver(4.5,3,10.5,1.5,3,7.5,True)
```

This will give the output.

```
The solution to:
 4.5 x + 3 y = 10.5
 1.5 x + 3 y = 7.5
 is x = 1.0 y= 2.0

Out[3]:  [1.0, 2.0]
```

Given our function definition, when we type its name followed by the required input parameters, separated by commas, Python executes the code in the body of the function on those input parameters. Then at the end the function will output the values specified by the return statement. In this case the return statement creates a list that contains the values of x and y.

Functions are very flexible in both the inputs they can take, and the outputs they can return. Our use of Python to solve engineering problems will be rife with the use of functions, and we will see many examples of this flexibility. Before moving on, we will use this function to demonstrate more features of functions.

BOX 3.1 PYTHON PRINCIPLE

You can define a function using the syntax

```
def function_name([input_variables]):
    [code]
    return [output_variables]
```

In this definition the name of the function is function_name and the input variables to the function are entered, separated by commas, in the spot [input_variables]. This function returns the variable, or variables, listed in the spot [output_variables].

Because we have defined a function, we can also solve other systems by changing the parameters when the function is called. We can also solve simple systems.

```
In [4]:  two_by_two_solver(1,0,3,0,1,2,True)

The solution to:
 1 x + 0 y = 3
 0 x + 1 y = 2
 is x = 3.0 y= 2.0

Out[4]: [3.0, 2.0]
```

This function cannot solve systems where a_1 is zero because our function divides by a_1. If we wanted to handle this case, we would have to make some changes to the way our functions works. As it stands, if we give the function a system with $a_1 = 0$, we will get an error:

```
In [5]: two_by_two_solver(0,1,2,1,0,3,True)


-------------------------------------------------------------------
ZeroDivisionError                      Traceback (most recent call last)

    <iPython-input-23-e8717fed1588> in <module>()
----> 1 two_by_two_solver(0,1,2,1,0,3,True)


    <iPython-input-19-25039de1b80f> in
    two_by_two_solver(a1, b1, c1, a2, b2, c2, LOUD)
---> 18     new_b2 = b2 - a2/a1*b1
     19     new_c2 = c2 - a2/a1*c1
     20     #solve the new equation 2


    ZeroDivisionError: division by zero
```

We will develop a fix for this problem later when we talk about pivoting. Giving an error when dividing by zero is a nice feature of Python for engineering calculations: if we accidentally divide by zero, Python tells the user where it happened, rather than giving a nonsensical answer.

3.1.1 Calling Functions and Default Arguments

In the above examples, we called our function two_by_two_solver by listing out the arguments in the order that it expects them a1, b1, c1, a2, b2, c2, LOUD. Nevertheless, Python allows you to call them in any order, as long as you are explicit in what goes where. In the next snippet of code we will specify the left-hand side coefficients first, and then the right-hand sides:

```
In [6]: two_by_two_solver(a1 = 4.5, b1 = 3, a2 = 1.5, b2 = 3,
                          c1 = 10.5, c2 = 7.5, LOUD = True)

The solution to:
  4.5 x + 3 y = 10.5
  1.5 x + 3 y = 7.5
  is x = 1.0 y= 2.0

Out[6]: [1.0, 2.0]
```

In this example we gave the values of the parameters explicitly: we told the function what each parameter was, rather than relying on the order that the parameters was listed.

BOX 3.2 LESSON LEARNED

It is often a good idea to call a function explicitly: that way if you mess up the order of the arguments, it does not matter.

In this example, there is also an example of a default parameter. Notice that in the function definition, the argument LOUD has =False after it. This indicates that if the function is called without a value for LOUD, it assumes the caller does not what the function to "be loud" and print out extra detail. Here we call the function without the LOUD parameter

```
In [7]:  two_by_two_solver(a1 = 4.5, b1 = 3, a2 = 1.5,
                           b2 = 3, c1 = 10.5, c2 = 7.5)

Out[7]:  [1.0, 2.0]
```

Notice that it did not print out its spiel about the system. The default behavior of not printing out extra information is common, because if we were going to call this function as part of a larger code many times, we do not want the screen filled with text to the point where it is indecipherable.

3.1.2 Return Values

At the end of the function we have a `return` statement. This tells Python what the function is returning to the caller. In this case we return a list that has the solution for x and y. We can store this in a new variable, or do whatever we like with it.

```
In [8]:   answer = two_by_two_solver(a1 = 4.5, b1 = 3,
                                      a2 = 1.5, b2 = 3,
                                      c1 = 10.5, c2 = 7.5)
          #store in the variable x the first value in the list
          x = answer[0]
          #store in the variable y the first value in the list
          y = answer[1]
          print("The list",answer,"contains",x,"and",y)
```

```
The list [1.0, 2.0] contains 1.0 and 2.0
```

We can do even fancier things, if we are so bold. For example, we can grab the first entry in the list returned by the function using square brackets on the end of the function call. We can also assign the two entries in the list to two variables in a single line.

```
In [9]:  #just get x
         x = two_by_two_solver(a1 = 4.5, b1 = 3, a2 = 1.5,
                               b2 = 3, c1 = 10.5, c2 = 7.5)[0]
         print("x =",x)

         #assign variables to the output on the fly
         x,y = two_by_two_solver(a1 = 4.5, b1 = 3, a2 = 1.5,
                                 b2 = 3, c1 = 10.5, c2 = 7.5)
         print("x =",x,"y =",x)
```

```
x = 1.0
x = 1.0 y = 1.0
```

These examples are more advanced, and they are designed to show you some of the neat tricks you can do in Python.

3.2 DOCSTRINGS AND HELP

Our 2×2 solver code had a long, and a detailed comment at the beginning of it. This comment is called a docstring and it is meant to tell the user of the function how to call the function and what it does. The user will need to know, for example, what the function will return to prepare to use that information. The user can get the information from the docstring by using the help function:

```
In [10]: help(two_by_two_solver)
```

```
Help on function two_by_two_solver in module __main__:
```

```
two_by_two_solver(a1, b1, c1, a2, b2, c2, LOUD=False)
    Calculate the solution of the system
    a1 x + b1 y = c1,
    a2 x + b2 y = c2
```

```
Args:
    a1: x coefficient in first equation (cannot be zero)
    b1: y coefficient in first equation
    c1: right-hand side in first equation
    a2: x coefficient in second equation
    b2: y coefficient in second equation
    c2: right-hand side in second equation
    LOUD: boolean that decides whether to print out the answer

Returns:
    list containing the solution in the format [x,y]
```

The point of this long comment is to tell the client (or caller) of the function what the function expects, in terms of arguments, and what the client should expect in terms of what is going to be returned. In this example we can see that we need to provide at least 6 numbers, and possibly an optional boolean.

You may wonder why a docstring is important, when you have the code defining the function right in front of you. The answer is that the user of a function will not always have the definition of the function readily available (perhaps somebody else wrote it). If you want to call that function properly, you can refer to the docstring.

BOX 3.3 PYTHON PRINCIPLE

Docstrings are long comments at the beginning of the body of a function that tells the user what the function needs as input parameters and what the function returns. These useful comments can be obtained by a user of the function by calling

```
help(function_name)
```

where `function_name` is the name of a function.

Let us look at the docstring for some members of the `math` module and the `random` module.

```
In [11]:import math
         help(math.fabs)

Help on built-in function fabs in module math:

fabs(...)
    fabs(x)

    Return the absolute value of the float x.

In [12]:import random
         help(random.uniform)

Help on method uniform in module random:

uniform(a, b) method of random.Random instance
    Get a random number in the range [a, b) or [a, b] depending on rounding.
```

We do not have the source code for these functions in front of us, but if we want to know how to call them, the docstring tells us what to do.

The docstrings for `random.uniform` and `math.fabs` are a bit different that the one we used in our function for solving a linear system. The format that we used is derived from the Google coding standards for Python docstrings (https://google-styleguide.googlecode. com/svn/trunk/pyguide.html#Comments).

3.3 SCOPE

The variables that we define in our code store information in the computer's memory. The computer divides the memory that you access into different sections based on scoping rules. Scoping rules are, in essence, a way for the program to separate information in memory and control access to that information. Understanding scoping rules is important when we define functions. Functions have their own scope in memory that is different than the memory used by other parts of a code. That is the memory used by a function is separate from the rest of the program and only knows about the outside world through the parameters it gets passed. In practice, what this means is that any variables used by the function (including those passed to the function) are *completely different* than the variables outside the function. When a function is called, it creates its own copy of the variables that get passed to it.

Here is a simple, but illustrative, example of how a function makes its own copy of the data it gets passed.

```
In [13]: def scope_demonstration(input_variable):
             x = input_variable*3
             return x

         #now call the function after defining some variables
         x = "oui "
         y = "no "

         new_x = scope_demonstration(x)
         new_y = scope_demonstration(y)
         print("x =",x,"\nnew_x =",new_x)
         print("y =",y,"\nnew_y =",new_y)

x = oui
new_x = oui oui oui
y = no
new_y = no no no
```

Now let us analyze what happened in the code. Before we called the function, we defined a variable x to be the string "oui". Then we called `scope_demonstration`, passing it x. Notice that even though `scope_demonstration` defines a variable x as `input_variable*3`, the value of x that exists outside the function is not changed. This is because when I call `scope_demonstration` it creates its own memory space and any variable I create in there is different than in the rest of the program, even if the variables have the same name.

In this particular example, the function first copies the value passed to the function into the variable input_variable, and then manipulates that copy of the data.

There are many subtleties in scoping rules, but this example outlines the main pitfall for a neophyte programmer. There are extra rules we will need, but these will be covered as we need them.

3.4 RECURSION

The idea behind recursion is that a function can call itself. Recursion enables some neat tricks for the programmer and can lead to very short code for some complicated tasks. In many cases there is often a faster way of doing things than using recursion, but it can be a useful tool for a programmer. Here is an example of computing the factorial of n,

$$n! = 1 \times 2 \times \cdots \times (n-1) \times n,$$

with both a recursive and non-recursive implementation.

```
In [14]: def factorial(n, prev=1):
             if not((n==1) or (n==0)):
                 prev = factorial(n-1,prev)*n
             elif n==0:
                 return 1
             else
                 return prev

         def factorial_no_recursion(n):
             output = 1;
             #can skip 1 because x*1 = 1
             for i in range(2,n+1):
                 output *= i
             return output
         x = 12
         print(x,"! =",factorial(x))
         print(x,"! =",factorial_no_recursion(x))

12 ! = 479001600
12 ! = 479001600
```

We can time the functions to see which is faster. To make the amount of time it takes run large enough to measure well, we will compute the factorials of 0 through 20, 100,000 times. (These are timings on my computer, if you run this example for yourself you may see differences based on your computer hardware and other demands on the system.)

```
In [15]: for times in range(10**5):
             for n in range(21):
                 factorial(n)
```

The time it took to run this was 16 μs. Compare this to

```
In [16]: for times in range(10**5):
             for n in range(21):
                 factorial_no_recursion(n)
```

which takes 6 μs.

The no recursion version, while not as neat, is nearly 50% faster. Even though we're talking microseconds, if my code was going to do this millions of times, the difference would matter. Part of the difference is that every time a function is called, there is an overhead in terms of creating the new memory space, etc. With recursion this extra work has to be done when the function recursively calls itself.

Another drawback to recursion is that it is possible to have too many levels of recursion. The number of levels of recursion, that is, the number of times the function can call itself, is limited to make sure the computer has enough memory to keep track of all the functions that have been called. In this example we demonstrate such an error by having a function call itself about 1000 times:

```
In [17]: x = 1000
         #this won't work and prints ~1000 errors
         #the errors are not repeated here
         print(x,"! =",factorial(x))
```

```
In [18]: x = 1000
         #this works
         print(x,"! =",factorial_no_recursion(x))
```

```
1000 ! = 40238726007709377354370243392300398571937486421071
4632543799910429938512398629020592044208486969404800047998
8610197196058631666872994808558901323829669944459099742450
4087073759918823627727188732519779505950995276120087497546
2497043601418278094646496291056393887437886487337311918104
5825783647849977012476632889835955735432513185323958463307
5557409114262417474349347553428646576611667797396666882029
1207379143853719588249808126867838374559731746136085379953
4524221586593201928090878297308431392844403281231558611103
6976801357304216168747609675871348312025478589320767169113
2448426236131412508780208000261683151027341827977704784633
5868170164365024153691398281264810213092761244896359928706
5114964975419909342221566832572080821333186116811553615836
5469840467089756029009505376164758477284218896796462449496
5160765353408198901385442487984959953319101723355556602139
9450399736280750137837615307127761926849034352625200015883
8535147331611702103968175921510907788019393178114194545257
7223865554146106289218796022383897147608850627686296714667
4697562911234082439208160153780889893964518263243671616767
2179168909779911903754031274622289988005195444414282012186
7361745992642956658174662830295557029902432415318161721046
5832036786906117260158783520751516284225540265170483304227
6143974286933061690897968482590125458327168226458066526766
9958652682272807075781391858178889652208164348344825993267
6043367660176999612831860788386150279465955131156552036096
3988180612138558600301435694527224206344631797460594682573
1037900840244324384656572450144028218852524709351906209290
```

90231364932734975655139587205596542287497740114133 4696271
54228458623773875382304838656889764619273838149001 4076731
04466402598994902222217659043399018860185665264850 6179970
23561938970178600408118897299183110211712298459016 4192106
88843871218556461249607987229085192968193723886426 1483965
73822911231250241866493531439701374285319266498753 3721894
06942814341185201580141233448280150513996942901534 8307764
45690990731524332782882698646027898643211390835062 1709500
25973898635542771967428222487575867657523442202075 7363056
94988250879689281627538488633969099598262809561214 5099487
17012445164612603790293091208890869420285106401821 5439945
71568059418727489980942547421735824010636774045957 4178516
08292301353580818400969963725242305608559037006242 7124341
69090041536901059339838357779394109700277534720000 0000000
00000000000000000000000000000000000000000000000000 0000000
00000000000000000000000000000000000000000000000000 0000000
00000000000000000000000000000000000000000000000000 0000000
00000000000000000000000000000000000000000000000000 0000000
0000000000

3.5 MODULES

Often we have to define many functions and do not want to have one giant source file. Also, once we define a function, we do not want to have to copy and paste functions that we have already defined into each file we create. To make functions defined in one Python file available in others, we can import the functions from a file (called a module).

To demonstrate this we define several functions inside a Python file. In particular consider a file called sphere.py. This file defines two functions volume and surface_area that compute the volume and surface area of a sphere. Because the file is called sphere.py we can import those functions using import sphere leaving off the .py part of the name. The code inside of sphere.py is

```
def volume(radius):
    '''''compute volume of a sphere
    Args:
    radius: float giving the radius of the sphere

    Returns:
    volume of the sphere as a float
    """
    return 4.0/3.0*math.pi*radius**3

def surface_area(radius):
    '''''compute surface area of a sphere
    Args:
    radius: float giving the radius of the sphere

    Returns:
    surface area of the sphere as a float
```

```
    """
    return 4.0*math.pi*radius**2
```

After we import the module, we can pass it to the help function to find out what is in the module:

```
In [19]: import sphere
         help(sphere)

Help on module sphere:

NAME
    sphere

FUNCTIONS
    surface_area(radius)
        compute surface area of a sphere

        Args:
        radius: float giving the radius of the sphere

        Returns:
        surface area of the sphere as a float

    volume(radius)
        compute volume of a sphere

        Args:
        radius: float giving the radius of the sphere

        Returns:
        volume of the sphere as a float

FILE
    /Users/mcclarren/sphere.py
```

With the module imported, I can call the functions inside the module by preceding the function name by `sphere`.

```
In [20]: r = 1.0
         print("The volume of a sphere of radius",r,"cm is",
             sphere.volume(r),"cm**3")
         print("The surface area of a sphere of radius",r,"cm is",
             sphere.surface_area(r),"cm**2")

The volume of a sphere of radius 1.0 cm is 4.1887902047863905 cm**3
The surface area of a sphere of radius 1.0 cm is 12.566370614359172 cm**2
```

Modules will be useful for us because our numerical algorithms will build on each other. Using modules will make our source code manageable and eliminate copy/paste errors.

3.6 FILES

It will be useful for us occasionally to use files for input and output from our codes. In this section we will cover some rudimentary uses of files for reading and writing. This discussion is not exhaustive, but will give you the necessary ingredients for file input and output.

It is simple to read in text files in Python. Just about any file that is plain text can be read by Python. One simple way to read in a file is to read it in a line at a time. This can be done by having a `for` loop, loop over the file: each pass through the loop body will read in a new line of the file.

BOX 3.4 PYTHON PRINCIPLE

Using a `for` loop to iterate over a file, the looping variable will contain an entire line of the file each pass through the loop.

In the folder that the following code is saved in, I have created a file called `fifth_republic.txt` that has the names of the presidents of France's fifth republic, one per line. Using a `for` loop, we can read that file and print it to the screen line by line:

```
In [21]: #open fifth_republic.txt for reading ('r')
         file = open('fifth_republic.txt', 'r')
         for line in file:
             print(line)
         file.close()
```

```
Charles de Gaulle

Georges Pompidou

Valéry Giscard d'Estaing

François Mitterrand

Jacques Chirac

Nicolas Sarkozy

François Hollande

Emmanuel Macron
```

Notice how the for loop can iterate through each line of the file. You can also read a line at a time:

```
In [22]: #open fifth_republic.txt for reading ('r')
         file = open('fifth_republic.txt', 'r')
         first_line = file.readline()
         second_line = file.readline()
```

```
print(first_line)
print(second_line)
file.close()
```

Charles de Gaulle

Georges Pompidou

It is also possible to write to a file. A straightforward way of doing this is to open a file for writing using the open function. With the file open, we then can write to the file much like we use the print statement. To end a line we use \n. Also, note that if we open a file for writing and it exists, the file will be wiped clean (sometimes called clobbered) upon opening it.

```
In [23]: #open hats.txt to write (clobber if it exists)
         writeFile = open("hats.txt","w")
         hats = ["fedora","trilby","porkpie","tam o'shanter",
                 "Phrygian cap","Beefeaters' hat","sombrero"]
         for hat in hats:
             writeFile.write(hat + "\n") #add the endline
         writeFile.close()

         #now open file and print
         readFile = open("hats.txt","r")
         for line in readFile:
             print(line)
```

fedora

trilby

porkpie

tam o'shanter

Phrygian cap

Beefeaters' hat

sombrero

There are more sophisticated ways to use files for input and output, but these examples will enable the basic functionality we need for file manipulation.

PROBLEMS

Short Exercises

3.1. Write a function for Python that simulates the roll of two standard, six-sided dice. Have the program roll the dice a large number of times and report the fraction of rolls that are each possible combination, 2–12. Compare your numbers with the probability of each possible roll that you calculate by hand.

3.2. What is the value of x at the end of the following code snippet:

```
def sillyFunction(input_var):
    x = 1.0
    return input_var

x = 10.0
sillyFunction(x)
```

3.3. Write a function that takes as a parameter the name of the file, and a parameter that gives the number of lines to read from the file. The function should open the file, read the lines and print them to the screen, and then close the file.

Programming Projects

1. Monte Carlo Integration

The exponential integral, $E_n(x)$, is an important function in nuclear engineering and health physics applications. This function is defined as

$$E_n(x) = \int_1^\infty dt \, \frac{e^{-xt}}{t^n}.$$

One way to compute this integral is via a Monte Carlo procedure for a general integral,

$$\int_a^b dy \, f(y) \approx \frac{b-a}{N} \sum_{i=1}^N f(y_i),$$

where

$$y_i \sim U[a, b],$$

or in words, y_i is a uniform random number between a and b. For this problem you may use the random or numpy modules.

3.1. Make the substitution $\mu = 1/t$ in the integral to get an integral with finite bounds.

3.2. Write a Python function to compute the exponential integral. The inputs should be n, x, and N in the notation above. Give estimates for $E_1(1)$ using $N = 1, 10, 100, 1000$, and 10^4.

3.3. Write a Python function that estimates the standard deviation of several estimates of the exponential integral function from the function you wrote in the previous part. The formula for the standard deviation of a quantity g given L samples is

$$\sigma_g = \sqrt{\frac{1}{L-1} \sum_{l=1}^L (g_l - \bar{g})^2},$$

where $\bar{g}$ is the mean of the L samples. Using your function and L of at least 10, estimate the standard deviation of the Monte Carlo estimate of $E_1(1)$ using $N = 1, 10, 100, 1000, 10^4$, and 10^5.

Harry, I have no idea where this will lead us, but I have a definite feeling it will be a place both wonderful and strange.

–"Dale Cooper" in the television series Twin Peaks

CHAPTER POINTS

- NumPy is a library that provides a flexible means to define and manipulate vectors, matrices, and higher-dimensional arrays.

- Matplotlib enables the visualization of numerical results with only a few lines of code.

In this chapter we will cover two important libraries that are available for Python: NumPy and Matplotlib. In this lecture we will make particular reference to nuclear and radiological engineering applications. It is not essential to understand these applications, but these will help motivate our discussion.

53

4.1 NUMPY ARRAYS

NumPy is a collection of modules, called a library, that gives the programmer powerful array objects and linear algebra tools, among other things. In this section we will explore that arrays that NumPy supplies.

The basic unit in NumPy is a multi-dimensional array, sometimes called an N-dimensional or $N-D$ array. You can think of an array as a collection of pieces of data, most typically in our work the data we store in an array is a float. You have already seen arrays in other areas of mathematics: a one-dimensional (1-D) array you can think of as a vector, and a 2-D array is a matrix. We can generalize from there by thinking of a 3-D array is a vector of matrices, and so on. It is not too common in our work to go beyond a 3-D array, but one can define these more exotic data structures.

In the following code, we make a vector and a matrix. The first line tells Python that we want to use NumPy, but we do not want to type numpy every time we need to use a function from the library; we abbreviate to np.

```
In [1]: import numpy as np
        a_vector = np.array([1,2,3,4])
        a_matrix = np.array([(1,2,3),(4,5,6),(7,8,9)])
        print("The vector",a_vector)
        print("The matrix\n",a_matrix)

The vector [1 2 3 4]
The matrix
 [[1 2 3]
 [4 5 6]
 [7 8 9]]
```

Now that we have defined arrays, we want to work with them. Arrays have several "attributes" that you can use to find out information regarding a particular array. The following code blocks explore these attributes, as noted in the comments:

```
In [2]: #shape tells you the shape
        print("The shape of a_vector is ", a_vector.shape)
        print("The shape of a_matrix is ", a_matrix.shape)

The shape of a_vector is  (4,)
The shape of a_matrix is  (3, 3)

In [3]: #ndim tells you the dimensionality of an array
        print("The dimension of a_vector is ", a_vector.ndim)
        print("The dimension of a_matrix is ", a_matrix.ndim)

The dimension of a_vector is  1
The dimension of a_matrix is  2

In [4]: #size is the total number of elements = the product of
        # the number of elements in each dimension
        print("The size of a_vector is ", a_vector.size,"= ",
            a_vector.shape[0])
        print("The size of a_matrix is ", a_matrix.size,"=",
            a_matrix.shape[0],"*",a_matrix.shape[1])
```

```
The size of a_vector is  4 =  4
The size of a_matrix is  9 = 3 * 3
```

For an existing array, you can change the shape after creating it. You can "reshape" an array to have different dimensions as long as the size of the array does not change. Here is an example:

```
In [5]: A = np.array([2,4,6,8])
        print("A is now a vector",A)
        A = A.reshape(2,2)
        print("A is now a matrix\n",A,"\nSorcery!")

A is now a vector [2 4 6 8]
A is now a matrix
 [[2 4]
 [6 8]]
Sorcery!
```

Notice how I needed to assign A with the reshaped array in the third line of code.

4.1.1 Creating Arrays in Neat Ways

In the examples above, we created arrays by specifying the value of each element in the array explicitly. We would like to have a way to define an array without having to type in each element. This is especially useful when we want to fill an array with thousands or millions of elements.

The function arange is a NumPy variant of range, which we saw earlier. The difference is that the function will create a NumPy array based on the parameters passed to arange.

```
In [6]:  #let's make a vector from 0 to 2*pi in intervals of 0.1
         dx = 0.1
         X = np.arange(0,2*np.pi,dx)
         print(X)

[ 0.   0.1  0.2  0.3  0.4  0.5  0.6  0.7  0.8  0.9  1.   1.1  1.2  1.3  1.4
  1.5  1.6  1.7  1.8  1.9  2.   2.1  2.2  2.3  2.4  2.5  2.6  2.7  2.8  2.9
  3.   3.1  3.2  3.3  3.4  3.5  3.6  3.7  3.8  3.9  4.   4.1  4.2  4.3  4.4
  4.5  4.6  4.7  4.8  4.9  5.   5.1  5.2  5.3  5.4  5.5  5.6  5.7  5.8  5.9
  6.   6.1  6.2]
```

We can also generate a fixed number of equally spaced points between fixed endpoints (a linearly increasing set of points) with the linspace function.

```
In [7]:  X = np.linspace(start = 0, stop = 2*np.pi, num = 62)
         print(X)

[ 0.          0.10300304  0.20600608  0.30900911  0.41201215  0.51501519
  0.61801823  0.72102126  0.8240243   0.92702734  1.03003038  1.13303342
  1.23603645  1.33903949  1.44204253  1.54504557  1.64804861  1.75105164
  1.85405468  1.95705772  2.06006076  2.16306379  2.26606683  2.36906987
  2.47207291  2.57507595  2.67807898  2.78108202  2.88408506  2.9870881
  3.09009113  3.19309417  3.29609721  3.39910025  3.50210329  3.60510632
```

```
3.70810936  3.8111124   3.91411544  4.01711848  4.12012151  4.22312455
4.32612759  4.42913063  4.53213366  4.6351367   4.73813974  4.84114278
4.94414582  5.04714885  5.15015189  5.25315493  5.35615797  5.459161
5.56216404  5.66516708  5.76817012  5.87117316  5.97417619  6.07717923
6.18018227  6.28318531]
```

Notice how it starts and ends exactly where I told it to. The linspace function is very useful, and we will use it extensively.

BOX 4.1 NUMPY PRINCIPLE

The function

np.linspace(start, stop, num)

creates a NumPy array of length num starting at start and ending at stop.

There are other special arrays that you might want to define. Defining arrays to be all zeros or ones can be very useful for initializing arrays to a fixed value:

```
In [8]:  zero_vector = np.zeros(10) #vector of length 10
         zero_matrix = np.zeros((4,4)) #4 by 4 matrix
         print("The zero vector:",zero_vector)
         print("The zero matrix\n",zero_matrix)

The zero vector: [ 0.  0.  0.  0.  0.  0.  0.  0.  0.  0.]
The zero matrix
 [[ 0.  0.  0.  0.]
 [ 0.  0.  0.  0.]
 [ 0.  0.  0.  0.]
 [ 0.  0.  0.  0.]]

In [9]:  ones_vector = np.ones(10) #vector of length 10
         ones_matrix = np.ones((4,4)) #4 by 4 matrix
         print("The ones vector:",ones_vector)
         print("The ones matrix\n",ones_matrix)

The ones vector: [ 1.  1.  1.  1.  1.  1.  1.  1.  1.  1.]
The ones matrix
 [[ 1.  1.  1.  1.]
 [ 1.  1.  1.  1.]
 [ 1.  1.  1.  1.]
 [ 1.  1.  1.  1.]]
```

There are times when we want to define an array filled with random values. For these purposes NumPy has an additional module called random that generalizes the library we used earlier. The NumPy random module can create matrices with random entries between 0 and 1 with the function np.random.rand, random entries between endpoints using np.random.uniform, and random integers with np.random.randint.

```
In [10]: random_matrix = np.random.rand(2,3) #random 2 x 3 matrix
         print("Here's a random 2 x 3 matrix\n",random_matrix)
```

```
print("Another example")

#make a random array between two numbers
print(np.random.uniform(low=-5,high=6,size=(3,3)))

#make random integers
print(np.random.randint(low=1,high=6,size=10))
```

```
Here's a random 2 x 3 matrix
 [[ 0.13097005  0.52015702  0.94753032]
 [ 0.99428635  0.10108597  0.62091224]]
Another example
[[ 1.86706869 -1.76316942 -1.88067072]
 [-2.13386359  2.84442703 -2.04880365]
 [-2.06079464 -2.17357025 -3.70912541]]
[5 3 3 2 2 3 3 4 4 4]
```

It is also possible to automatically generate an identity matrix:

```
In [11]: #3 x 3 identity matrix
         identity33 = np.identity(3)
         print(identity33)
```

```
[[ 1.  0.  0.]
 [ 0.  1.  0.]
 [ 0.  0.  1.]]
```

There are many other ways to define matrices using NumPy, but the ones we have covered will be the most useful for us.

BOX 4.2 NUMPY PRINCIPLE

To create NumPy vectors in simple ways you can use the following functions

```
#create a vector of length 1
with entries 1.0
np.ones(1)
```

```
#create a vector of length 1
with entries 0.0
np.zeros(1)
#create a vector of length 1
w/random values in [0,1]
np.random.rand(1)
```

BOX 4.3 NUMPY PRINCIPLE

To create NumPy matrices in simple ways you can use the following functions

```
#create an 1 by m matrix
with entries 1.0
np.ones((1,m))
#create an 1 by m matrix
```

```
with entries 0.0
np.zeros((1,m))
#create an 1 by m matrix
w/random values in [0,1]
np.random.rand(1,m)
#create an 1 by 1 identity matrix
np.identity(1)
```

4.1.2 Operations on Arrays

Previously, we talked about operator overloading where we can apply common operators, such as arithmetic operations, to objects other than numbers. We would like to be able to take several arrays and do things like add them, multiply by a scalar, and perform other linear algebra operations on them. NumPy has defined most of these operations for us by overloading the common operators. We will explore these operations here.

NumPy defines arithmetic operations mostly in the way that you would expect. For example, addition is easily accomplished provided that the arrays are of the same size and shape. We will begin by demonstrating operations on vectors.

```
In [12]: #vector addition
         x = np.ones(3) #3-vector of ones
         y = 3*np.ones(3)-1 #3-vector of 2's
         print(x,"+",y,"=",x+y)
         print(x,"-",y,"=",x-y)

[ 1.  1.  1.] + [ 2.  2.  2.] = [ 3.  3.  3.]
[ 1.  1.  1.] - [ 2.  2.  2.] = [-1. -1. -1.]
```

Multiplication and division are "element-wise"; this means that the elements in the same position are multiplied together:

```
In [13]: y = np.array([1.0,2.0,3.0])
         print(x,"*",y,"=",x*y)
         print(x,"/",y,"=",x/y)

[ 1.  1.  1.] * [ 1.  2.  3.] = [ 1.  2.  3.]
[ 1.  1.  1.] / [ 1.  2.  3.] = [ 1.         0.5         0.33333333]
```

If you want the dot product, you have to use the dot function:

```
In [14]: print(x,".",y,"=",np.dot(x,y))

[ 1.  1.  1.] . [ 1.  2.  3.] = 6.0
```

Matrices work about the same way as vectors, when it comes to arithmetic operations.

```
In [15]: silly_matrix = np.array([(1,2,3),(1,2,3),(1,2,3)])
         print("The sum of\n",identity33,"\nand\n",
               silly_matrix,"\nis\n",identity33+silly_matrix)

The sum of
 [[ 1.  0.  0.]
 [ 0.  1.  0.]
 [ 0.  0.  1.]]
and
 [[1 2 3]
 [1 2 3]
 [1 2 3]]
is
 [[ 2.  2.  3.]
 [ 1.  3.  3.]
 [ 1.  2.  4.]]
```

Multiplication and division are also element-wise:

```
In [16]: identity33 * silly_matrix

Out[16]: array([[ 1.,  0.,  0.],
                [ 0.,  2.,  0.],
                [ 0.,  0.,  3.]])

In [17]: identity33 / silly_matrix

Out[17]: array([[ 1.       ,  0.       ,  0.        ],
                [ 0.       ,  0.5      ,  0.        ],
                [ 0.       ,  0.       ,  0.33333333]])
```

The dot function will give you the matrix product when the it is passed two matrices:

```
In [18]: print("The matrix product of\n",identity33,"\nand\n",
               silly_matrix,"\nis\n",
               np.dot(identity33,silly_matrix))

The matrix product of
 [[ 1.  0.  0.]
 [ 0.  1.  0.]
 [ 0.  0.  1.]]
and
 [[1 2 3]
 [1 2 3]
 [1 2 3]]
is
 [[ 1.  2.  3.]
 [ 1.  2.  3.]
 [ 1.  2.  3.]]
```

To compute the product of a matrix and vector, we use the dot function and pass it the matrix followed by the vector.

```
In [19]: #matrix times a vector
         print(silly_matrix,"times", y, "is")
         print(np.dot(silly_matrix,y))

[[1 2 3]
 [1 2 3]
 [1 2 3]] times [ 1.  2.  3.] is
[ 14.  14.  14.]
```

When we use the dot function, the matrices and vectors must have the appropriate sizes. If the sizes are incompatible, Python will give an error.

BOX 4.4 NUMPY PRINCIPLE

Standard arithmetic operations are overloaded so that they work on an element by element basis on NumPy arrays. This means that the arrays must have the same size. To use linear algebra operations such as matrix multiplication or the dot product of two vectors you will want to use the np.dot(a,b) function.

4.1.3 Universal Functions

It is common that we might want to interpret a vector as a series of points to feed to a function. For example, the vector could be a list of angles we want to compute the sine of. For these, and more general situations, NumPy provides universal functions that operate on each element of an array. Common mathematical functions are defined in this way, and are used in a similar way to the functions in the math module we used previously:

```
In [20]: #recall we defined X as a linspace from 0 to 2pi
         print(X)
         #taking the sin(X) should be one whole sine wave
         print(np.sin(X))
```

```
[ 0.          0.10300304   0.20600608   0.30900911   0.41201215   0.51501519
  0.61801823  0.72102126   0.8240243    0.92702734   1.03003038   1.13303342
  1.23603645  1.33903949   1.44204253   1.54504557   1.64804861   1.75105164
  1.85405468  1.95705772   2.06006076   2.16306379   2.26606683   2.36906987
  2.47207291  2.57507595   2.67807898   2.78108202   2.88408506   2.9870881
  3.09009113  3.19309417   3.29609721   3.39910025   3.50210329   3.60510632
  3.70810936  3.8111124    3.91411544   4.01711848   4.12012151   4.22312455
  4.32612759  4.42913063   4.53213366   4.6351367    4.73813974   4.84114278
  4.94414582  5.04714885   5.15015189   5.25315493   5.35615797   5.459161
  5.56216404  5.66516708   5.76817012   5.87117316   5.97417619   6.07717923
  6.18018227  6.28318531]
[  0.00000000e+00   1.02820997e-01   2.04552066e-01   3.04114832e-01
   4.00453906e-01   4.92548068e-01   5.79421098e-01   6.60152121e-01
   7.33885366e-01   7.99839245e-01   8.57314628e-01   9.05702263e-01
   9.44489229e-01   9.73264374e-01   9.91722674e-01   9.99668468e-01
   9.97017526e-01   9.83797952e-01   9.60149874e-01   9.26323968e-01
   8.82678798e-01   8.29677014e-01   7.67880446e-01   6.97944155e-01
   6.20609482e-01   5.36696194e-01   4.47093793e-01   3.52752087e-01
   2.54671120e-01   1.53890577e-01   5.14787548e-02  -5.14787548e-02
  -1.53890577e-01  -2.54671120e-01  -3.52752087e-01  -4.47093793e-01
  -5.36696194e-01  -6.20609482e-01  -6.97944155e-01  -7.67880446e-01
  -8.29677014e-01  -8.82678798e-01  -9.26323968e-01  -9.60149874e-01
  -9.83797952e-01  -9.97017526e-01  -9.99668468e-01  -9.91722674e-01
  -9.73264374e-01  -9.44489229e-01  -9.05702263e-01  -8.57314628e-01
  -7.99839245e-01  -7.33885366e-01  -6.60152121e-01  -5.79421098e-01
  -4.92548068e-01  -4.00453906e-01  -3.04114832e-01  -2.04552066e-01
  -1.02820997e-01  -2.44929360e-16]
```

Universal functions are useful for plotting when we define a vector of points for the x axis and apply the function to get the y axis. Using `matplotlib`, which we cover extensively in a later section in this chapter, we can use a universal function to plot the sine function:

```
In [21]: import matplotlib.pyplot as plt
         plt.plot(X,np.sin(X));
```

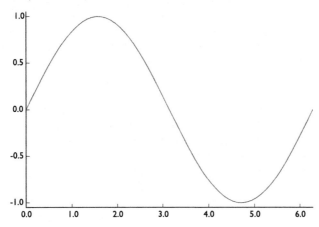

Therefore, if we wanted to plot the fundamental mode of a slab reactor of width 10, we could combine the arithmetic operators and universal functions defined by NumPy in the following way:

```
In [22]: X = np.linspace(0,10,100)
         Y = np.sin(np.pi*X/10)
         plt.plot(X,Y);
```

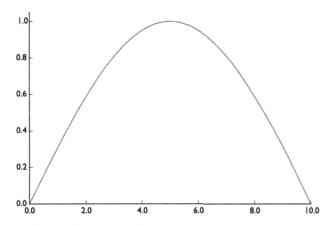

4.1.4 Copying Arrays and Scope

The assignment operator = behaves differently for NumPy arrays than for other data types we have used. When you assign a new variable name to an existing array, it is the same as giving two names for the object. It does not copy the array into a new array.

```
In [23]: a = np.array([1.0,2,3,4,5,6])
         print(a)
         #this will make a and b different names for the same array
         b = a
         #changing b at position 2, also changes a
         b[2] = 2.56
         print("The value of array a is",a)
         print("The value of array b is",b)

[ 1.  2.  3.  4.  5.  6.]
The value of array a is [ 1.    2.    2.56  4.    5.    6.  ]
The value of array b is [ 1.    2.    2.56  4.    5.    6.  ]
```

The reason for this behavior is that the array could have thousands or millions of elements, and creating new arrays in a pell mell fashion could quickly fill up the computer memory and slow down the program. To make a real copy of the entire elements of an array you need to explicitly tell Python that you want to make a copy by using the copy function on the array. This assures that you only create copies of an array when you truly want that to happen. We can modify the previous code snippet to copy the array:

```
In [24]: a = np.array([1.0,2,3,4,5,6])
         print(a)
         #this will make a and b different copies for the same array
         b = a.copy()
         #changing b at position 2, will not change a
         b[2] = 2.56
         print("The value of array a is",a)
         print("The value of array b is",b)

[ 1.  2.  3.  4.  5.  6.]
The value of array a is [ 1.  2.  3.  4.  5.  6.]
The value of array b is [ 1.    2.    2.56  4.    5.    6.  ]
```

BOX 4.5 NUMPY PRINCIPLE

To copy a NumPy array named origAr-ray to the NumPy array copyArray use the syntax

copyArray = origArray.copy()

The syntax

sameArray = origArray

will give an additional name, sameArray, for the array origArray.

Typically, when you pass a variable to a function it copies that variable into the function's memory scope. This does not happen with NumPy arrays. When you pass an array to a function, the function does not copy the array, it just assigns that array another name as in the previous example. This means that if the NumPy array is changed inside the function, it is also changed outside the function.

```
In [25]: def devious_function(func_array):
             #changes the value of array passed in
             func_array[0] = -1.0e6

         a = np.array([1.0,2,3,4,5,6])
         print("Before the function a =",a)
         devious_function(a)
         print("After the function a =",a)

Before the function a = [ 1.  2.  3.  4.  5.  6.]
After the function a =
[ -1.00000000e+06   2.00000000e+00   3.00000000e+00
   4.00000000e+00   5.00000000e+00   6.00000000e+00]
```

This is different than what we saw previously for passing floats, and integers to functions. The difference is that a NumPy array is a mutable object that could possibly be large. The technicalities are not of much interest here, but, because a NumPy array could be millions of elements, it is best for efficient memory usage if functions do not make multiple copies of millions of elements. This is same rationale that leads to the behavior of the assignment operator not automatically copying the entire array into a new array. For example, if I had an array with one billion elements, passing that array into a function could take a long time to make all of those copies into a new array, before the function even begins to do its work.

4.1.5 Indexing, Slicing, and Iterating

Oftentimes we will want to access or modify more than one element of an array at a time. We can do this by slicing. Slicing, in many ways, mirrors what we did with strings before. One feature we have not discussed is the plain : operator without a number on either side. This will give all the values in a particular dimension, as seen below.

```
In [26]: #bring these guys back
         a_vector = np.array([1,2,3,4])
         a_matrix = np.array([(1,2,3),(4,5,6),(7,8,9)])
         print("The vector",a_vector)
         print("The matrix\n",a_matrix)
         #single colon gives everything
         print("a_vector[:] =",a_vector[:])
         #print out position 1 to position 2 (same as for lists)
         print("a_vector[1:3] =",a_vector[1:3])
         print("For a matrix, we can slice in each dimension")
         #every column in row 0
         print("a_matrix[0,:] =",a_matrix[0,:])
         #columns 1 and 2 in row 0
         print("a_matrix[0,1:3] =",a_matrix[0,1:3])
         #every row in column 2
         print("a_matrix[:,2] =",a_matrix[:,2])

The vector [1 2 3 4]
The matrix
 [[1 2 3]
 [4 5 6]
```

```
   [7 8 9]]
a_vector[:] = [1 2 3 4]
a_vector[1:3] = [2 3]
For a matrix, we can slice in each dimension
a_matrix[0,:] = [1 2 3]
a_matrix[0,1:3] = [2 3]
a_matrix[:,2] = [3 6 9]
```

We can also use a for loop to iterate over an array. If the array is a vector, the iteration will be over each element. Iteration over a matrix will give you everything in a row. If you want to iterate over the columns of the matrix, take the transpose of the matrix when iterating.

```
In [27]: a_matrix = np.array([(1,2,3),(4,5,6),(7,8,9)])
         count = 0
         for row in a_matrix:
             print("Row",count,"of a_matrix is",row)
             count += 1

         count = 0
         for column in a_matrix.transpose():
             print("Column",count,"of a_matrix is",column)
             count += 1

Row 0 of a_matrix is [1 2 3]
Row 1 of a_matrix is [4 5 6]
Row 2 of a_matrix is [7 8 9]
Column 0 of a_matrix is [1 4 7]
Column 1 of a_matrix is [2 5 8]
Column 2 of a_matrix is [3 6 9]
```

To iterate over every element in the matrix you'll need two for loops: one to get the rows, and another to iterate over each element in the row. This is an example of nested for loops: a for loop with another for loop inside.

```
In [28]: a_matrix = np.array([(1,2,3),(4,5,6),(7,8,9)])
         row_count = 0
         col_count = 0
         for row in a_matrix:
             col_count = 0
             for col in row:
                 print("Row",row_count,"Column",col_count,
                       "of a_matrix is",col)
                 col_count += 1
             row_count += 1

Row 0 Column 0 of a_matrix is 1
Row 0 Column 1 of a_matrix is 2
Row 0 Column 2 of a_matrix is 3
Row 1 Column 0 of a_matrix is 4
Row 1 Column 1 of a_matrix is 5
Row 1 Column 2 of a_matrix is 6
Row 2 Column 0 of a_matrix is 7
Row 2 Column 1 of a_matrix is 8
Row 2 Column 2 of a_matrix is 9
```

4.1.6 NumPy and Complex Numbers

NumPy can handle complex numbers without much difficulty. A NumPy array can be specified to contain complex numbers by adding the parameter `dtype = "complex"` to the creation. For example, an array full of $0+0j$ can be created by

```
In [29]: cArray = np.zeros(5, dtype = "complex")
         print(cArray)

[ 0.+0.j  0.+0.j  0.+0.j  0.+0.j  0.+0.j]
```

The `dtype` argument tells NumPy what datatype will be in the array. This will work with other methods we discussed for creating arrays `ones`, `identity`, and `array`.

You do not have to specify `dtype` if it is obvious that you are using complex numbers. For example, this will create an array of complex numbers without the `dtype` command

```
In [30]: cArray2 = np.array([1+1j,-1])
         print(cArray2)

[ 1.+1.j -1.+0.j]
```

The time that you have to be careful with complex numbers and NumPy is when you have a non-complex array that evaluates to a complex number inside a function. For instance, the square root of a negative number is imaginary. Calling `np.sqrt` on a negative float will give an error, unless the `dtype = "complex"` argument is provided. If this argument is not provided, the result may be "not a number" or `nan`, and Python may throw an error. Here is an example of the wrong way and the right way to take a square root of an array that contains negative numbers:

```
In [31]: fArray = np.array([-1,1])
         print("Wrong way gives", np.sqrt(fArray))
         print("Right way gives", np.sqrt(fArray, dtype = "complex"))
         print(cArray2)

Wrong way gives [ nan   1.]
Right way gives [ 0.+1.j  1.+0.j]
```

This is an important consideration when using built-in mathematical functions with NumPy.

We have now covered the details of NumPy that we will need for our numerical investigations. Having numbers in an array is an important step in engineering analysis, but understanding what comes out of a calculation can be much easier if we can visualize the results. In the next section we discuss a method for this visualization.

4.2 MATPLOTLIB BASICS

Matplotlib is a library for Python that allows you to plot the arrays from NumPy, as well as many other features. It is designed to be intuitive and easy to use, and it mimic the plotting interface of MATLAB, a widely used toolkit and language for applied mathematics and computation. Therefore, if you have experience with MATLAB, using Matplotlib will be very familiar.

In the following example, a plot is created, and properly annotated using Matplotlib.

```
In [32]: import matplotlib.pyplot as plt
         import numpy as np
         #make a simple plot
         x = np.linspace(-100,100,1000)
         y = np.sin(x)/x
         #plot x versus y
         plt.plot(x,y)
         #label the y axis
         plt.ylabel("sinc(x) (arb units)");
         #label the x axis
         plt.xlabel("x (cm)")
         #give the plot a title
         plt.title("Line plot of the sinc function")
         #show the plot
         plt.show()
```

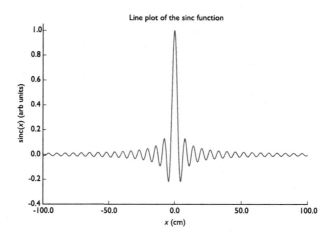

Reviewing what the code above did we see that the plot function took in two arguments that served as the data for the x and y axes. The ylabel and xlabel functions take in a string to print on the respective axes; the title function prints a string as the plot title. Finally, the show function tells Python to show us the plot. Notice how I labeled each axis and even gave the plot a title. I also included the units for each axis.

4.2.1 Customizing Plots

You can also change the plot type. Here we will change from line plotting to using red dots:

```
In [33]: x = np.linspace(-3,3,100)
         y = np.exp(-x**2)
         plt.plot(x,y,"ro"); #red dots on the plot
         plt.ylabel("$e^{-x^2}$ (arb units)");
         plt.xlabel("x (cm)")
         plt.title("Plot of $e^{-x^2}$")
         plt.show()
```

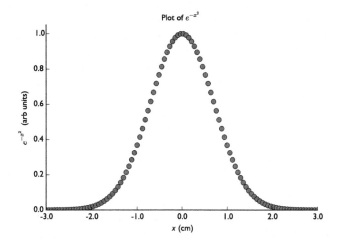

We could also use dots and a line in our plot:

```
In [34]: x = np.linspace(-3,3,100)
         y = np.exp(-x**2)
         plt.plot(x,y,"ko-"); #black dots and a line on the plot
         plt.ylabel("$e^{-x^2}$ (arb units)");
         plt.xlabel("x (cm)")
         plt.title("Plot of $e^{-x^2}$")
         plt.show()
```

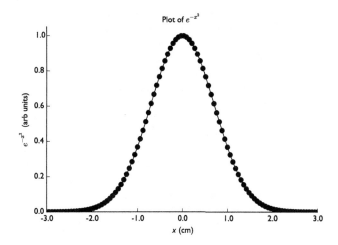

Notice these examples have included LATEX mathematics in the labels by enclosing it in dollar signs. LATEX mathematics is a markup language for mathematical characters. For example, you can make characters superscript by enclosing the characters in curly braces preceded by ^. In LATEX to use Greek letters you use a backslash before the name of the letter, and other usually obvious characters. We will use LATEX, extensively to annotate figures throughout the work.

It is also possible to plot several lines on a plot and include a legend. The legend text is passed into the `plot` function through the parameter `label` which takes in a string, potentially with LATEX. Calling the `legend` function will add the legend to the figure. Here is an example of multiple lines:

```
In [35]: x = np.linspace(-3,3,100)
         y = np.exp(-x**2)
         y1 = np.exp(-x**2/2)
         y2 = np.exp(-x**2/4)
         plt.plot(x,y,marker="", color="r",
                    linestyle="--", label="$\sigma^2 = 1$")
     plt.plot(x,y1,color="blue",
                    label="$\sigma^2 = 2$")
     plt.plot(x,y2,color="black",
                marker = "+", label="$\sigma^2 = 4$")
         plt.ylabel("$e^{-x^2/\sigma^2}$ (arb units)");
         plt.xlabel("x (cm)")
         plt.title("Plot of $e^{-x^2/\sigma^2}$")
         plt.legend()
         plt.show()
```

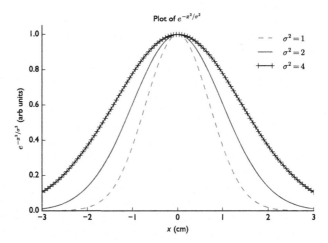

With MatPlotLib it is also possible to make plots that are more than just lines. This example here we make a contour plot of a function of two variables. We will not go into detail about these plots here, rather we will demonstrate additional plotting features as we need them.

```
In [36]: phi_m = np.linspace(0, 1, 100)
         phi_p = np.linspace(0, 1, 100)
```

```
X,Y = np.meshgrid(phi_p, phi_m)
Z = np.sin(X*2*np.pi)*np.cos(Y*2*np.pi)
CS = plt.contour(X,Y,Z, colors='k')
plt.clabel(CS, fontsize=9, inline=1)
plt.xlabel("x (cm)");
plt.ylabel("y (cm)");
plt.title("Second harmonic of $\phi$ (arb units)");
```

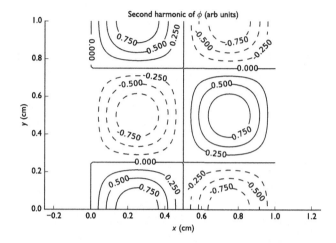

FURTHER READING

For a less nuclear engineering-specific tutorial on NumPy see http://wiki.scipy.org/. For a list, see http://matplotlib.org/users/pyplot_tutorial.html. LaTeX has several books available that are useful references such as [5,6]. Additionally, there are many tutorials available on the Internet, including one from the LaTeX project https://www.latex-project.org/.

PROBLEMS

Short Exercises

4.1. Write a simple dot product calculator. Ask the user for the size of the vector, and then to input the value for each element in the vector. Print to the user the value of the dot product of the two vectors.

4.2. Ask the user for a matrix size, N. Print to the user a random matrix of size N by N. Then print to the user the vector that contains the sum of each column.

4.3. Ask the user for a vector size, N. Print to the user a random vector of size N. Then print the sorted matrix. NumPy has a function for sorting a vector: x.sort(), where x is the name of a vector.

4.4. Ask the user for a vector size, N. Print to the user a random vector of size N. Then print the maximum value of the vector. NumPy has a function that returns the index of the maximum value: x.argmax(), where x is the name of a vector.

4.5. Ask the user for a vector size, N. Print to the user a random vector of size N where the elements of N are from a normal distribution with mean 0 and standard deviation 1. To generate this vector you should use np.random.normal. Then print the mean value of the vector and the standard deviation. NumPy has a function that returns the mean value and standard deviation: x.mean() and x.std(), where x is the name of a vector.

4.6. Ask the user for a vector size, N. Print to the user two random vectors of size N; call these vectors x and y. If we consider these vectors being the coordinates of points (x_i, y_i), compute the matrix **r** that contains the distances between each point and each of other points. This matrix will have elements

$$r_{ij} = \sqrt{(x_i - x_j)^2 + (y_i - y_j)^2}.$$

The diagonal of this matrix should be all zeros.

Programming Projects

1. Inhour Equation

The inhour equation (short for inverse hour) describes the growth or decay of the neutron population in a reactor as described by the point kinetics equations. The equation relates the reactivity, ρ, the mean generation time, Λ, the fraction of fission neutrons born from each of six delayed neutron precursor groups, β_i, and the decay constant for those groups, λ_i. The inhour equation is

$$\rho = s \left(\Lambda + \sum_{i=1}^{6} \frac{\beta_i}{s + \lambda_i} \right).$$

The seven values of s that satisfy this equation are used to express the neutron population as function of time, $n(t)$, as

$$n(t) = \sum_{\ell=1}^{7} A_\ell e^{s_\ell t}.$$

The decay constant of a delayed neutron precursor group is related to the half-life for neutron emission by the group is

$$\lambda_i = \frac{\ln 2}{t_{1/2}}.$$

Common values of the necessary constants in this equation are [7]: $\Lambda = 5 \times 10^{-5}$ s

$$\beta = \{0.00021, 0.00142, 0.00128, 0.00257, 0.00075, 0.00027\},$$

where the total delayed fraction $\bar{\beta} = 0.0065$, and

$$t_{1/2} = \{56, 23, 6.2, 2.3, 0.61, 0.23\} \text{ s}.$$

Using this data, plot the right-hand side of the inhour equation as a function of s, and plot a horizontal line corresponding to ρ to graphically illustrate the roots of the inhour equation. Do this for $\rho = -1, 0, 0.1\bar{\beta}, \bar{\beta}$, and discuss the results. Make sure that the scale of your plot makes sense given that there will be singularities in the plot.

2. Fractal Growth

In this problem you will code a program that grows a cluster of particles that stick together using a diffusion process known as a random walk. The resulting structure will be a fractal, that is a structure that is self-similar, where the structure looks the same at every scale. To build these structures we start with a single particle at the position $(0, 0)$, then introduce a particle at "infinity". The new particle undergoes a random walk where it jumps a random distance in a random direction. The particle undergoes random jumps until it strikes the center particle. Then, we repeatedly introduce particles at "infinity" and follow each until it sticks to the existing particles.

To make the algorithm work quickly, we will define "infinity" by placing a particle randomly on a circle that circumscribes the existing structure. We also, allow the distance traveled by the particle in each step to be the minimum distance between the particle and the structure. The following code will perform this algorithm, but there are expressions missing in key positions. The comments will tell you how to fill them in; these are denoted by ??.. Get the code working, and make plots of the resulting structures with matplotlib. Make figures of various sizes by changing the value of N, i.e., N = 100, 500, 1000, 500, 10000, and even higher values if the code is fast enough. Comment on the similarity of the structures of different size.

```
import matplotlib.pyplot as plt
import numpy as np
#initialize list of particles
#column 0 is x position
#column 1 is y position
particles = np.zeros([1,2])
#how many particles to simulate
N = 500
#how close do we need to be to "stick"
tol = 1.0e-2
#how many steps can a particle take before we stop tracking
maxsteps = 1e3
#radius of structure, initially small
r = 2.0*tol
#pick the angle on the circle in [0,2pi]
theta = np.random.uniform(low=0,
                          high=2*np.pi,
                          size = 1)
for i in range(N):
    #make the initial position of the new particles
    newposx = r*np.sin(theta)
```

```
newposy = r*np.cos(theta)
steps = 0
#how close is the new particle to the cluster
dist = r
while (dist > tol) and (steps<maxsteps):
    #how close is the new particle to the cluster
    dist_vect = np.sqrt(
              (particles[:,0]-newposx)**2 + (particles[:,1]-newposy)**2)
    dist = np.min(dist_vect)
    #compute the jump distance randomly in [0,dist]
    rho = ??
    #compute the direction of the jump randomly in [0,2 pi]
    theta = ??
    #move the particle
    newposx += ??
    newposy += ??
    steps += 1
#if the while loop is exited with fewer steps than
#the maximum, then add the particle to the list
#otherwise forget it
if (steps < maxsteps):
    tmp = np.ndarray(shape = [1,2])
    tmp[0,0] = newposx
    tmp[0,1] = newposy
    particles = np.append(particles, tmp,axis=0)
#make the starting point for the next particle
#by calculating the radius and angle
r = (1+tol)*np.max(np.sqrt(particles[:,0]**2 + particles[:,1]**2))
theta = ??
```

The figures below show the similarity of scales and were generated with this code, and $N = 5 \times 10^5$.

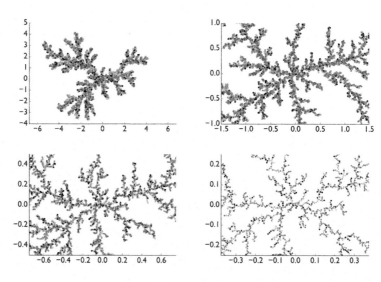

3. *Charges in a Plane*

Consider a collection of N charged point particles distributed in the xy plane. The particles have known masses, m, and charges, q. The Coulomb force on particle i is given by

$$F_i = \sum_{j=1, j \neq i}^{N} \frac{q_i q_j}{r^2},$$

where r is the distance between particle i and j. The x and y components of the force are

$$F_i^x = \sum_{j=1, j \neq i}^{N} \frac{q_i q_j}{r^2} \frac{x_i - x_j}{r}, \qquad F_i^y = \sum_{j=1, j \neq i}^{N} \frac{q_i q_j}{r^2} \frac{y_i - y_j}{r}.$$

From the forces, we can compute the acceleration in each direction using $\vec{F}_i = m_i \vec{a}$. Given the equations of motion of a particle

$$v_{x,i}^{l+1} = v_{x,i}^{l} + \Delta t a_{x,i}^{l}, \qquad v_{y,i}^{l+1} = v_{y,i}^{l} + \Delta t a_{y,i}^{l},$$

$$x_i^{l+1} = x_i^l + \Delta t v_{x,i}^{l+1}, \qquad y_i^{l+1} = y_i^l + \Delta t v_{y,i}^{l+1},$$

for $l = 1 \ldots L$ and $v_{x,i}^0 = v_{y,i}^0 = 0$. Where the superscripts indicate the time level, i.e., $x_i^l = x(l\Delta t)$.

Write a Python code that solves the above problem using $L = 1000$ and $\Delta t = 0.001$, and the following table of masses and initial positions as defined in the following code snippet:

```
#Code to advect point charges using Coulomb's law
import numpy as np
import matplotlib.pyplot as plt

#set up initial values of x,y,vx,vy,Fx,Fy
N = 20
x = np.zeros(N)
x[0:(N//2)] = 0*np.linspace(-N,-5,N//2)
x[(N//2):N] = np.linspace(1,10,N//2)
y = np.zeros(N)
y[0:(N//2)] = np.linspace(-10,10,N//2)
y[(N//2):N] = np.random.uniform(low = -1e-3, high = 1e-3, size = N//2)
mass = np.zeros(N)
mass[0:(N//2)] = 1e8
mass[(N//2):N] = 1
q = np.zeros(N)
q[0:(N//2)] = 10
q[(N//2):N] = 1
vx = np.zeros(N);
vx[0:(N//2)] = 0.0
vx[(N//2):N] = -10.0
vy = np.zeros(N)
Fx = np.zeros(N)
Fy = np.zeros(N)
```

To solve this problem you will need three loops: one to compute the forces, one to compute the velocities, and one to compute the new positions. These loops will be nested inside a loop that keeps track of the time.

Plot the solution at the final time for varying values of N: 20, 50, 100. Explain the differences in the results.

CHAPTER

5

Dictionaries and Functions as Arguments

DICTIONARY, n. A malevolent literary device for cramping the growth of a language and making it hard and inelastic. This dictionary, however, is a most useful work.

–Ambrose Bierce, **The Devil's Dictionary**

CHAPTER POINTS

- Dictionaries are lists where the elements are accessed via special names, called keys.

- In Python it is possible to pass the name of a function as an argument to another function. This allows functions, such as numerical integration, to operate on an arbitrary function.

- Lambda functions allow the programmer to define a function in one line and use all the variables in the current scope.

5.1 DICTIONARIES

Previously, we learned about lists as a sequence of items that we can access via position using square brackets. There may be cases where we do not want to access items based on a numerical index, rather we want to access them based on a name. A typical example of this might be the children in a family. You could have a list of children that you access via the order that they were born; however, this would be fairly impersonal and not a useful ordering for anyone other than the parents in that family. The solution to this problem in Python can be found in the dictionary. A dictionary is like a list in many ways, but you access it with the name of the item.

In technical terms, a dictionary is a set of key:value pairs. The key is the analog to the index of a list, and is, in effect, the name of the item. You define a dictionary using curly braces.

```
In [1]: #simple dictionary
        days_of_week = {"M":"Monday", "T":"Tuesday",
                        "W":"Wednesday", "R":"Thursday",
                        "F":"Friday", "S":"Saturday",
                        "U":"Sunday"}
        print("Key M gives", days_of_week["M"])
        print("Key R gives", days_of_week["R"])
        #is G a key in days_of_week?
        print("G" in days_of_week.keys())

Key M gives Monday
Key R gives Thursday
False
```

Instead of accessing the dictionary using a position, like we have done with strings, lists, and NumPy arrays, we use the key. This is useful because we then do not have to remember the order we have listed the values in. For example, in the days of the week above we don't have to remember how we ordered the days (e.g., Monday first or Sunday first).

Also, the above example used the in operator to indicate if a particular key is in a dictionary. In particular, it tells us that "G" is not a key in the dictionary.

BOX 5.1 PYTHON PRINCIPLE

A dictionary is a sequence of items that is accessed via a name called a key. The elements of the dictionary that the key refers to is called the value. To define a dictionary we use curly brackets as in the following example

```
my_dictionary ={key_1:value_1,
                key_2:value_2, ...}
```

A list of the keys in a dictionary can be obtained via the function d.keys(), where d is the name of a dictionary.

For a further example we will read in a comma-separated-values text file (often called a csv), using the module csv. The text file will be used to give the key:value pairs in a dictionary. The format of this file is chemical symbol, element name. The first few lines of the file are

```
Ac,Actinium
Ag,Silver
Al,Aluminum
Am,Americium
```

The following code reads in the file and uses the chemical symbol as the key and the chemical name as the value. It also asks the user to input a chemical symbol, and will return the name.

```
In [2]: import csv
        #create a blank dictionary
        element_dict = {}
        #this block will only execute if the file opens
        with open('ChemicalSymbols.csv') as csvfile:
            chemreader = csv.reader(csvfile)
            for row in chemreader: #have for loop that loops over each line
                element_dict[row[0]] = row[1] #add a key:value pair
        key = input("Enter a valid chemical symbol: ")
        if key in element_dict:
            print(key,"is",element_dict[key])
        else:
            print("Not a valid element")
```

```
Enter a valid chemical symbol: Pu
Pu is Plutonium
```

Dictionaries can be made even more powerful, if we make a dictionary of dictionaries. Yes, you read that correctly: the value in the key:value pair can be another dictionary. For many applications, this is where dictionaries become very useful. In the following example we use idea of a dictionary of dictionaries to store extra information about the days of the week.

```
In [3]: #simple dictionary of dictionaries
        days_of_week = {"M":{"name":"Monday","weekday":True,"weekend":False},
                        "T":{"name":"Tuesday","weekday":True,"weekend":False},
                        "W":{"name":"Wednesday","weekday":True,"weekend":False},
                        "R":{"name":"Thursday","weekday":True,"weekend":False},
                        "F":{"name":"Friday","weekday":True,"weekend":False},
                        "S":{"name":"Saturday","weekday":False,"weekend":True},
                        "U":{"name":"Sunday","weekday":False,"weekend":True}}
        print("The days that are weekdays:")
        for day in days_of_week: #for loop over dictionary, loops over keys
            if days_of_week[day]["weekday"] == True:
                print(days_of_week[day]["name"],"is a weekday.")

        for day in days_of_week: #for loop over dictionary, loops over keys
            if days_of_week[day]["weekend"] == True:
                print(days_of_week[day]["name"],"is a weekend, whoop.")
```

```
The days that are weekdays:
Thursday is a weekday.
Wednesday is a weekday.
Tuesday is a weekday.
Monday is a weekday.
Friday is a weekday.
Saturday is a weekend, whoop.
Sunday is a weekend, whoop.
```

Notice that when a dictionary is iterated over in a `for` loop, the loop variable will get each of the keys of the dictionary. Also, the order of the keys is not guaranteed to match the order in which they were input. In the above loop, the keys were not printed out in the order Monday through Sunday.

We can use the idea of a dictionary of dictionaries idea to make a code that can compute radioactive decay for us, automatically. To do this I will create a dictionary where the key is the atomic number (Z), and the value will be a dictionary with the element name and symbol. The file that is read below is of the format Z, `Symbol`, `Name`.

```
In [4]: import csv
        element_dict = {} #create a blank dictionary
        #this block will only execute if the file opens
        with open('ChemicalSymbolsZ.csv') as csvfile:
            chemreader = csv.reader(csvfile)
            #have for loop that loops over each row
            for row in chemreader:
                #add a key:value pair
                element_dict[row[0]]={"symbol":row[1],"name":row[2]}
        key = input("Enter a valid atomic number: ")
        if key in element_dict:
            print(key,"is",element_dict[key]["symbol"],
                  ":",element_dict[key]["name"])
        else:
            print("Not a valid element")

Enter a valid atomic number: 34
34 is Se : Selenium

In [5]: key = input("Enter a valid atomic number: ")
        if key in element_dict:
            print(key,"is",element_dict[key]["symbol"],
                  ":",element_dict[key]["name"])
        else:
            print("Not a valid element")

Enter a valid atomic number: 104
104 is Rf : Rutherfordium
```

Given that we have a dictionary where we can look up an element by its atomic number, we can write a function that computes the product of alpha decay of a particular nuclide. We will pass the function the atomic number, the mass number, and the dictionary of elements, and it will return the atomic number and the mass number of the product, along with printing some information to the screen.

```
In [6]: def alpha_decay(Z,A,elements):
            """Alpha decay a nuclide

            Args:
                Z: atomic number of nuclide
                A: mass number of nuclide
                elements: dictionary of elements
```

```
    Returns:
        Z and A of daughter nuclide (both ints)
    Side effects:
        Prints a descriptive string of the decay
    """
    newZ = int(Z) - 2 #lose two protons in alpha decay
    newA = int(A) - 4 #lose four nucleons in alpha decay
    print(elements[str(Z)]["name"],"-",A,"(",
        elements[str(Z)]["symbol"],"-",A,"), alpha decays to",
        elements[str(newZ)]["name"],"-",newA,"(",
        elements[str(newZ)]["symbol"],"-",newA,")")
    return newZ,newA
z_value = input("Enter the Z of the nuclide: ")
a_value = input("Enter the mass number of the nuclide: ")
Z,A = alpha_decay(z_value, a_value, element_dict)
```

```
Enter the Z of the nuclide: 94
Enter the mass number of the nuclide: 239
Plutonium - 239 ( Pu - 239 ), alpha decays to Uranium - 235 ( U - 235 )
```

Given that the function returns the atomic and mass numbers of the products, we can run `alpha_decay` in a loop.

```
In [7]: #alpha decay something 10 times
        Z = 94
        A = 239
        for decays in range(10):
            Z,A = alpha_decay(Z, A, element_dict)
```

```
Plutonium - 239 ( Pu - 239 ), alpha decays to Uranium - 235 ( U - 235 )
Uranium - 235 ( U - 235 ), alpha decays to Thorium - 231 ( Th - 231 )
Thorium - 231 ( Th - 231 ), alpha decays to Radium - 227 ( Ra - 227 )
Radium - 227 ( Ra - 227 ), alpha decays to Radon - 223 ( Rn - 223 )
Radon - 223 ( Rn - 223 ), alpha decays to Polonium - 219 ( Po - 219 )
Polonium - 219 ( Po - 219 ), alpha decays to Lead - 215 ( Pb - 215 )
Lead - 215 ( Pb - 215 ), alpha decays to Mercury - 211 ( Hg - 211 )
Mercury - 211 ( Hg - 211 ), alpha decays to Platinum - 207 ( Pt - 207 )
Platinum - 207 ( Pt - 207 ), alpha decays to Osmium - 203 ( Os - 203 )
Osmium - 203 ( Os - 203 ), alpha decays to Tungsten - 199 ( W - 199 )
```

This example does not check if such an alpha decay is possible or likely, however, with an appropriate modification to the dictionary we could add information about the decay mode for a particular nuclide. This dictionary would be much more complicated, because it might require a three-level hierarchy consisting of a top level dictionary where the keys are the atomic number and the values are a dictionary of dictionaries where the key is the mass number and the values are the decay modes. Setting this up and filling it with data would be messy, but in principle doable.

Another use of dictionaries would be to store information about the different fuel elements in a reactor. We will consider a reactor that has two types of fuel, high-enriched uranium (HEU) and low-enriched uranium (LEU). We will use a dictionary to describe the properties of each type of fuel. Then using this information, we plot the HEU fuel geometric cross-section.

```
In [8]:  fuel_types = {}
         fuel_types["heu"] = {"fuel":{"nu sigma_f":12.0,
                                      "D":3.0, "thickness":5.0},
                              "clad":{"nu sigma_f":0.0, "
                                      D":300.0, "thickness":0.5}}
         fuel_types["leu"] = {"fuel":{"nu sigma_f":8.5,
                                      "D":1.25, "thickness":4.25},
                              "clad":{"nu sigma_f":0.0,
                                      "D":300.0, "thickness":1.25}}
         #plot heu
         #heu fuel
         fuel_radius = fuel_types["heu"]["fuel"]["thickness"]
         clad_radius = fuel_radius + fuel_types["heu"]["clad"]["thickness"]
         fuel = plt.Circle((0,0),fuel_radius,
                           facecolor="white",label="Fuel", hatch="//")
         clad = plt.Circle((0,0),clad_radius,color='gray',label="Clad")
         fig = plt.figure(figsize=(8,6), dpi=600)
         plt.gca().add_patch(clad)
         plt.gca().add_patch(fuel)
         plt.title("HEU Fuel")
         plt.axis('equal')
         plt.legend()
         plt.axis([-clad_radius,clad_radius,-clad_radius,clad_radius])
         plt.show();
```

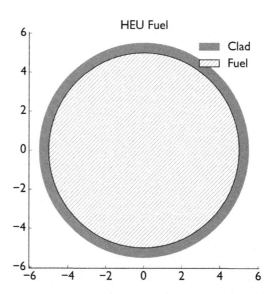

In a similar manner, the LEU fuel can be visualized by using the dictionary.

```
In [9]:  fuel_radius = fuel_types["leu"]["fuel"]["thickness"]
         clad_radius = fuel_radius + fuel_types["leu"]["clad"]["thickness"]
         fig = plt.figure(figsize=(8,6), dpi=600)
         fuel = plt.Circle((0,0),fuel_radius,
                           facecolor="white",label="Fuel", hatch="+")
```

```
clad = plt.Circle((0,0),clad_radius,color='gray',label="Clad")
plt.gca().add_patch(clad)
plt.gca().add_patch(fuel)
plt.title("LEU Fuel")
plt.axis('equal')
plt.legend()
plt.axis([-clad_radius,clad_radius,-clad_radius,clad_radius])
plt.show();
```

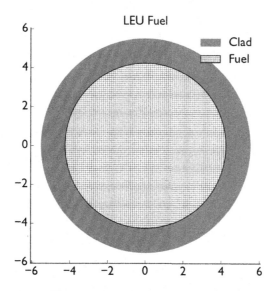

Given that we have a dictionary describing each type of fuel, we can define a lattice of fuel elements. We will make a 10 by 10 lattice of fuel with a 0.5 cm spacing between elements. Also, we will make every third element HEU and the rest LEU. The code below creates this lattice and then plots it.

```
In [10]: fuel_placements = {}
         #10 x 10 lattice with 0.5 cm spacing
         #every third pin is heu
         x = np.arange(6.,120,12)
         y = np.arange(6.,120,12)
         fig = plt.figure(figsize=(8,6), dpi=600)
         count = 1 #set up counting variable
         for i in x:
             for j in y:
                 if not(count % 3): #if count mod 3 is 0, then heu
                     pin_type = "heu"
                     hatch = "/"
                 else: #else leu
                     pin_type = "leu"
                     hatch = "+"
                 fuel_radius = fuel_types[pin_type]["fuel"]["thickness"]
                 clad_radius = fuel_radius +
                         fuel_types[pin_type]["clad"]["thickness"]
```

```
        fuel = plt.Circle((i,j),fuel_radius,facecolor="white",
                          edgecolor="black",hatch=hatch,
                          label="Fuel")
        clad = plt.Circle((i,j),clad_radius,color='gray',label="Clad")
        plt.gca().add_patch(clad)
        plt.gca().add_patch(fuel)
        count += 1 #increment count
plt.title("Fuel Lattice")
plt.axis('equal')
plt.axis([0,120,0,120])
plt.show();
```

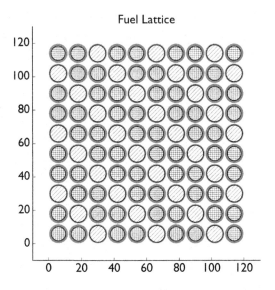

With this lattice, if we knew the shape of the fundamental mode of the scalar flux, we could compute the fission neutron production rate density at each point of the reactor. We will assume a simple scalar flux shape and multiply the flux by the value of $\nu \Sigma_f$, that is the product of the average number of fission neutrons produced, times the macroscopic fission cross-section, to get the fission neutron production rate density. The following code computes this quantity by evaluating $\nu \Sigma_f$ at the middle of each fuel element.

```
In [11]: fuel_placements = {}
         #10 x 10 lattice with 0.5 cm spacing
         #every third pin is heu
         x = np.arange(6.,120,12)
         y = np.arange(6.,120,12)
         fig = plt.figure(figsize=(8,6), dpi=600)
         X =  np.zeros((x.size, x.size))
         Y = X.copy()
         Z = X.copy()
         Zflux = Z.copy()
         row = 0
         col = 0
         count = 1
```

```
for i in x:
    for j in y:
        if not(count % 3): #if count mod 3 is 0, then heu
            pin_type = "heu"
        else: #else leu
            pin_type = "leu"
        nusigf = fuel_types[pin_type]["fuel"]["nu sigma_f"]
        X[row,col] = i
        Y[row,col] = j
        Z[row,col]=nusigf*np.sin(i*np.pi/120)*np.sin(j*np.pi/120)
        Zflux[row,col]=np.sin(i*np.pi/120)*np.sin(j*np.pi/120)
        row += 1 #increment row
        count += 1 #increment count
    col += 1 #increment column
    row = 0

CS = plt.contour(X,Y,Z, colors='k')
plt.clabel(CS, fontsize=9, inline=1)
plt.xlabel("x (cm)");
plt.ylabel("y (cm)");
plt.title("fission neutron production rate (neutrons/cm$^3$/s)");
plt.show();
CS = plt.contour(X,Y,Zflux, colors='k')
plt.clabel(CS, fontsize=9, inline=1)
plt.xlabel("x (cm)");
plt.ylabel("y (cm)");
plt.title("fundamental mode scalar flux");
plt.show();
```

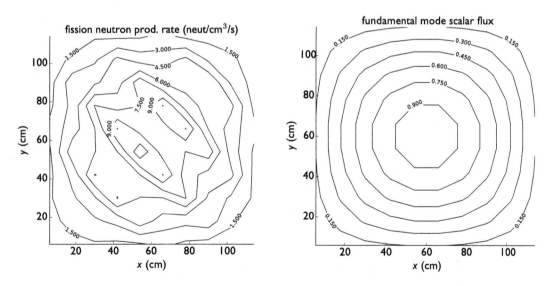

When we solve the diffusion equation for the scalar flux of neutrons in a system, we will revisit these techniques for calculating the fission neutron production rate, and other quantities.

5.2 FUNCTIONS PASSED TO FUNCTIONS

In Python it is possible to pass the name of an existing function as a parameter to another function. This name can then be used to execute the function and manipulate the results. This ability to call a generic function results in very powerful codes as we will see.

BOX 5.2 PYTHON PRINCIPLE

In Python a function can take the name of an arbitrary function as a parameter. In the generic example

```
def new_function(f,x):
    return f(x)
```

The function `new_function` returns the value of function f called as `f(x)`. In this example, the function name that is passed to `new_function` must be a single parameter and x must be of the correct type.

A salient example of this a numerical integration function. If we wanted to write a function to apply an integration formula, we would not want to have to write a new integration routine for each integrand. Instead, we can make the function that forms the integrand a parameter, and then call that function every time we want to evaluate the integrand. The example below uses the midpoint rule to integrate a generic integrand, $f(x)$ between points a and b.

```
In [12]: def midpoint_rule(f,a,b,num_intervals):
             """integrate function f using the midpoint rule

             Args:
                 f: function to be integrated, it must take 1 argument
                 a: lower bound of integral range
                 b: upper bound of integral range
                 num_intervals: the number of intervals in [a,b]
             Returns:
                 estimate of the integral
             """
             L = (b-a) #how big is the range
             dx = L/num_intervals #how big is each interval
             #midpoints are a+dx/2, a+3dx/2, ..., b-dx/2
             midpoints = np.arange(num_intervals)*dx+0.5*dx+a
             integral = 0
             for point in midpoints:
                 integral = integral + f(point)
             return integral*dx
```

To integrate $\sin x$ from 0 to π we pass in the name of NumPy sine function, `np.sin`. The exact value of this integral is 2. With 10 intervals, we get a pretty good answer.

```
In [13]: print(midpoint_rule(np.sin,0,np.pi,10))
```

```
2.00824840791
```

We can see how the numerical integration technique converges to the exact answer as a function of the number of intervals by calling the midpoint rule function with several different values of `num_intervals`. In the next code snippet we do this using a `for` loop to compute the integral using 10^i intervals for $i = 0, 1, 2, \ldots, 8$. Then we plot the error as a function of the number of intervals on a log-log scale. Later, when we study numerical integration in more detail this type of plot will be important.

```
In [14]: num_intervals = 8 #number of interval sizes
         #run several different intervals
         intervals = 10**np.arange(num_intervals)
         ierror = np.zeros(num_intervals)
         fig = plt.figure(figsize=(8,6), dpi=600)
         count = 0
         a = 0
         b = np.pi
         for interval in intervals:
             error[count] = np.fabs(midpoint_rule(np.sin,a,b,interval)-2)
             count += 1
         plt.loglog(intervals,error,marker="o",
                     markersize = 10,linewidth=2);
         plt.xlabel("# of intervals")
         plt.ylabel("Error in midpoint rule")
         plt.show()
```

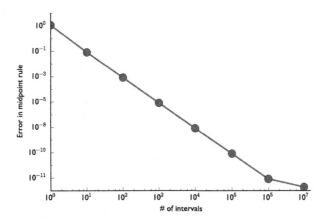

We are not limited to integrating only the sine function. We can define our own functions and pass them to the midpoint rule function we defined. In other words, the midpoint rule can approximate any 1-D definite integral. For example, we can use our midpoint rule function to compute an estimate of the exponential integral function,

$$E_n(x) = \int_1^\infty dt\, \frac{e^{-xt}}{t^n}.$$

Because this is an improper integral, we have to introduce a finite upper bound, and this is another approximation.

```
In [15]: def exp_int_argument(t,n=1,x=1):
             return np.exp(-x*t)/t**n
         num_points = 10**6
         upper_bound = 1000
         print("Exact answer is 0.2193839343")
         print("Our approximation with upper bound",upper_bound,
              "and",num_points,
              "points is",
              midpoint_rule(exp_int_argument,1,upper_bound,num_points))
```

```
Exact answer is 0.2193839343
Our approximation with upper bound 1000 and 1000000 points is 0.2193839038
```

Using Matplotlib we can add the functionality to the integration function to graphically show how the midpoint rule estimates the integral. The function below draws the areas that comprise the integral estimate. It should give the same answer as the previous midpoint rule function, but with pretty graphics. Notice in the docstring for the function, the fact that the function produces a plot is listed as a side effect. A side effect is something that the function does other than return a value. In this case it makes a plot, but a side effect could be printing something to the screen, writing to file, or modifying a NumPy array that was passed to the function.

```
In [16]: def midpoint_rule_graphical(f,a,b,num_intervals):
             """integrate function f using the midpoint rule

             Args:
                 f: function to be integrated, it must take one argument
                 a: lower bound of integral range
                 b: upper bound of integral range
                 num_intervals: the number of intervals to break [a,b] into
             Returns:
                 estimate of the integral
             Side Effect:
                 Plots intervals and areas of midpoint rule
             """
             fig = plt.figure()
             ax = plt.subplot(111)
             L = (b-a) #how big is the range
             dx = L/num_intervals #how big is each interval
             midpoints = np.arange(num_intervals)*dx+0.5*dx+a
             x = midpoints
             y = np.zeros(num_intervals)
             integral = 0
             count = 0
             for point in midpoints:
                 y[count] = f(point)
                 integral = integral + f(point)
                 verts = [(point-dx/2,0)] + [(point-dx/2,f(point))]
                 verts += [(point+dx/2,f(point))] + [(point+dx/2,0)]
                 poly = plt.Polygon(verts, facecolor='0.8', edgecolor='k')
                 ax.add_patch(poly)
                 count += 1
```

```
    y = f(x)
    smooth_x = np.linspace(a,b,10000)
    smooth_y = f(smooth_x)
    plt.plot(smooth_x, smooth_y, linewidth=1)
    plt.xlabel("x")
    plt.ylabel("f(x)")
    plt.title("Integral Estimate is " + str(integral*dx))
    plt.show()
    return integral*dx
midpoint_rule_graphical(np.sin,0,2*np.pi,10)
```

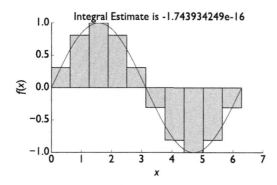

BOX 5.3 PYTHON PRINCIPLE

A side effect of a function is something that the function does that affects or interacts with the function caller in some way. This can be printing to the screen, making a graph, or changing something in a mutable variable (e.g., a NumPy vector) passed to the function. It is important to document side effects so that the code calling the function can handle them.

We can use the same function to compute the exponential integral and visualize how well the approximation of the midpoint rule is doing.

```
In [17]: num_points = 20
         upper_bound = 5
         print("Answer is 0.2193839343")
         print("Our approximation with upper bound",upper_bound,
             "and",num_points,"points is",
               midpoint_rule_graphical(exp_int_argument,1,upper_bound,num_points))


Answer is 0.2193839343
```

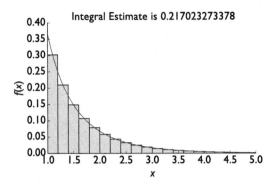

```
Our approximation with upper bound 5 and 20 points is 0.217023273378
```

It appears that the area is better approximated near $x = 5$ than near $x = 1$. Later, we will see other approaches to estimating integrals that give better approximations than rectangles.

5.3 LAMBDA FUNCTIONS

Python also allows you to define simple, one line functions called lambda functions. One of the benefits of a lambda function is that they are very easy to define. Because of this, they are especially good if we want to combine previously defined functions in a simple manner.

Lambda functions also have different scope rules than standard functions. Lambda functions give you access to all the variables available in the scope they are defined. This means that lambda functions do not have their own variable scope.

BOX 5.4 PYTHON PRINCIPLE

Lambda functions are short functions that you define in a single line. The syntax to define a lambda function named `lambda_func` is

```
lambda_func = lambda [parameter list]:
              expression
```

where `[parameter list]` is a comma-separated list of the input parameters to the

function and the result of evaluating `expression` is what the function returns.

Lambda functions have the same scope as the scope in which they are defined. For example, if you define a lambda function inside of a function, it has the same scope as that function.

The following example uses a lambda function to define a line.

```
In [17]: simple_line = lambda x: 2.0*x + 1.0
         print("The line at x = 0 is", simple_line(0))
         print("The line at x = 1 is", simple_line(1))
         print("The line at x = 2 is", simple_line(2))
```

```
x = np.linspace(0,6,50)
y = simple_line(x)
plt.plot(x,y)
plt.ylabel("y")
plt.xlabel("x")
plt.show()
```

```
The line at x = 0 is 1.0
The line at x = 1 is 3.0
The line at x = 2 is 5.0
```

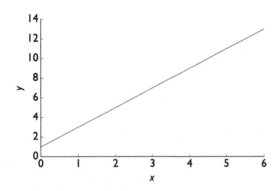

We can use lambda functions in our midpoint integration routine as well. Here we define the probability density function of a Gaussian as the integrand. A unit variance and zero mean Gaussian is given by

$$f(x) = \frac{e^{-x^2}}{\sqrt{\pi}}.$$

The integral over $x \in [-\infty, \infty]$ should be 1. Lambda functions are very useful in this context because the exponential is already defined; we just want to integrate it with a particular form of argument and multiply it by a constant.

```
In [18]: #function to compute gaussian
         gaussian = lambda x: np.exp(-x**2)/np.sqrt(np.pi)
         midpoint_rule_graphical(gaussian,-3,3,20)
```

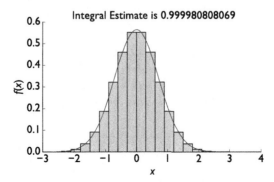

We can use the fact that lambda functions have the same scope as where they are defined to make our midpoint rule integrate two-dimensional functions by defining two lambda functions. In effect, what this code does is treat a 2-D integral as a 1-D integral:

$$\int_{a_y}^{b_y} dy \int_{a_x}^{b_x} dx \, f(x, y) = \int_{a_y}^{b_y} dy \, g(y),$$

where

$$g(y) = \int_{a_x}^{b_x} dx \, f(x, y).$$

This is possible to define because we can define the $g(y)$ using lambda functions. For a test we will estimate the integral

$$\int_0^\pi dy \int_0^\pi dx \, \sin(x)\sin(y) = 4.$$

```
In [19]: def midpoint_2D(f,ax,bx,ay,by,num_intervals_x,num_intervals_y):
             """integrate function f(x,y) using the midpoint rule
             Args:
                 f: function to be integrated, it must take 2 arguments
                 ax: lower bound of integral range in x
                 bx: upper bound of integral range in x
                 ay: lower bound of integral range in y
                 by: upper bound of integral range in y
                 num_intervals_x: the number of intervals in x
                 num_intervals_y: the number of intervals in y
             Returns:
                 estimate of the integral
             """
             g = lambda y: midpoint_rule(lambda x: f(x,y),ax,bx,num_intervals_x)
             return midpoint_rule(g,ay,by,num_intervals_y)
         sin2 = lambda x,y:np.sin(x)*np.sin(y)
         print("Estimate of the integral of sin(x)sin(y), over [0,pi] x [0,pi] is",
             midpoint_2D(sin2,0,np.pi,0,np.pi,1000,1000))
```

```
Estimate of the integral of sin(x)sin(y), over [0,pi] x [0,pi] is 4.00000328987
```

The lambda functions in this example tell Python to treat only a single variable as the function parameter, and evaluate everything else based on the current scope. This means that the code `lambda x: f(x,y)` evaluates `f(x,y)` using whatever the current value of y is and the parameter x that the function was passed. The value of y is supplied by a parameter to the lambda function g.

The intricacies of defining lambda functions inside other lambda functions can get a bit abstract, but remembering that they treat all variables in an expression that is not specified in the parameter list as "known", is the key to understanding how they function.

PROBLEMS

Short Exercises

5.1. Write a Python dictionary that contains the key:value pairs that have for a key the name of the common subatomic particles (i.e., proton, neutron, and electron) and the value the mass of the particle in kilograms.

5.2. Using the `midpoint_rule` function defined above, compute the integral of $\sin^2 x$, over the range $[0, 2\pi]$ with 10, 100, and 1000 intervals.

5.3. Estimate π to five digits of accuracy by computing the integral of $f(x) = 4\sqrt{1 - x^2}$ for $x \in [0, 1]$.

5.4. Integrate the function $f(x, y, z) = \exp(-z^2) \sin(x) \sin(y)$ over the (x, y, z) range $[0, \pi] \times [0, \pi] \times [-4, 4]$ using 10, 100, and 1000 intervals.

Programming Projects

1. Plutonium Decay Chain

Consider the plutonium decay chain from ^{239}Pu to stable ^{207}Pb, as shown below. Construct a dictionary with the keys are A-X where A is the mass number of the nuclide and X is the atomic symbol for the nuclide. For example one key is 239-Pu. The value for the aforementioned keys should be a dictionary with key:value pairs given by:

- key: half-life, value: the half-life of the decay in seconds,
- key: decay_mode, value: the decay mode (i.e., alpha, beta, or stable in this case), and
- key: mass, value: the mass of the nuclide.

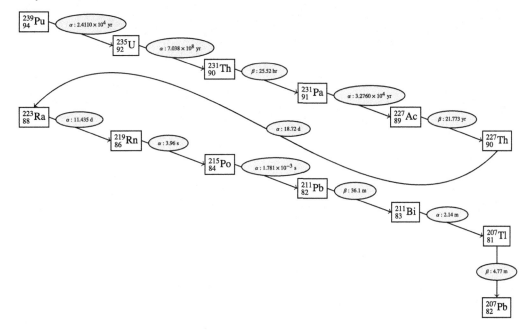

Your code should use the dictionary to print out

- All the nuclides that decay by alpha decay
- The activity of 1 gram of each nuclide that is a beta emitter.

2. *Simple Cryptographic Cipher*

To transmit a message you desire to encrypt it. In the terminology of cryptography the original message is the plain text and the encrypted message is called the cipher text. The means of encrypting the message is called a cipher. A simple method is the ROT-13 cipher, which is an example of the Caesar cipher. In this cipher the letter is replaced by a letter 13 places away in the alphabet. This can be encoded easily in a dictionary:

```
cipherDict = {a:"n", b:"o",...,m:"z",n:"a",...}
```

Write a function called `cipher` that takes in a string and returns an encrypted cipher text using the ROT-13 cipher, your function must also take in cipher text and return plain text. Your code must have the following behavior:

- Handle lower case and capital letters,
- Not do anything to characters that are not alphabet characters, e.g., numbers, punctuation, and other characters, and
- Only take a single parameter as an argument to the function. This argument will be a string containing the plain text or cipher text.

Test your code on the following cipher text "Gur bayl rzcrebe vf gur rzcrebe bs vpr-pernz." Show that it can give the correct plain text, and that it can recover the cipher text by applying the function to the plain text.

6

Testing and Debugging

Als Gregor Samsa eines Morgens aus unruhigen Träumen erwachte, fand er sich in seinem Bett zu einem ungeheuren Ungeziefer verwandelt.

One morning, as Gregor Samsa was waking up from anxious dreams, he discovered that in bed he had been changed into a monstrous verminous bug.

–Franz Kafka, The Metamorphosis, *as translated by Ian Johnston*

CHAPTER POINTS

- Strong testing of code is as important as the ability to write codes.

- Finding problems inside a code, i.e., debugging, is an exercise in intuition as well as inference from how the code fails its tests.

- Assertions can help isolate problems inside of large codes, especially when functions call other functions.

- It is possible to catch errors and handle them in Python using `try-except` blocks.

6.1 TESTING YOUR CODE

For any code it is incumbent on the programmer to make sure that code accomplishes the desired task. For instance, if a function is supposed to return the largest item in a list, the programmer should not deliver that function to a user unless it has been shown, for a variety of test cases, that it indeed gives the correct answer. This needs to be done anytime one writes code where any the following apply:

1. The code will be used by somebody else;
2. The code will be used to be input for another piece of code (e.g., a function that calls another function);
3. The code will be used to make a decisions, or
4. The code will be turned in as a class assignment.

Writing tests for a code, and demonstrating that the code passes the tests, is essential to giving your code credibility. Consider the NumPy dot function. We have not questioned if it will give us the correct answer. To some degree we just believed that the super smart people at NumPy Inc. would not release faulty code. This implicit belief is correct, but not because of the corporate imprimatur of the developers. In fact NumPy is the product of a community of hundreds of developers, and millions of users. The developers have built tests into the source code that take the known value of the dot product, and compare it to the value returned by np.dot. Users of the code can view these tests inside the NumPy source code (i.e., the code that gets executed when NumPy is used). This test, and thousands like it, are run on a regular basis to make sure the code is behaving properly. Furthermore, the user community acts as a secondary test bed. Users of NumPy can report problems with the code to the developers, along with a minimal example demonstrating the flaw, and these will be resolved.

Taken together, we do not believe that np.dot works because some imperious entity tells us; it is because there is evidence of testing, and a user community to help identify errors that might have been missed. The ability to see which tests are used is a clear benefit of an open-source product such as NumPy. Were we not able to view the source code, we would have to rely on the quality of the software vendor, which can vary wildly from vendor to vendor.

In many of our cases, we will need to demonstrate that a piece of code we develop to make a numerical evaluation is giving us the correct answer. To do this we can often use exact solutions to simple problems, the known limits of a system, or look at the overall behavior of the solution compared with some theoretical behavior. It is important that we have more than a single test because a single test could be passed for the wrong reasons.

A simple example of a test being passed for the wrong reasons is a function that is supposed to compute the probability for an event. If the function returns a value of 0.5, no matter the inputs, it would pass a test checking whether the probability is in the interval [0, 1]. Furthermore, it would pass a test designed to check if the function correctly computes a value of 0.5 for a particular set of inputs. Clearly, there are many tests that this function would not pass, but selecting just a few tests may not cover enough of the function's intended use.

A more detailed example of testing, and why multiple tests are needed is presented below. Consider the following code to compute the multiplication factor (k-effective, k_{eff}) for a system of where nuclear fission is present, under the approximation of a one-group, bare reactor [7]. The formula for the multiplication factor is

$$k_{\text{eff}} = \frac{k_\infty}{1 + L^2 B_g^2},$$

where

$$k_\infty = \frac{\nu \Sigma_{\text{f}}}{\Sigma_{\text{a}}},$$

and

$$L^2 = \frac{D}{\Sigma_a} = \frac{1}{3\Sigma_{tr}\Sigma_a},$$

and if the reactor is a slab we have

$$B_g^2 = \left(\frac{\pi}{X}\right)^2.$$

The notation here is standard: Σ_a, Σ_{tr}, and Σ_f are the macroscopic absorption, transport, and fission cross-sections with units of inverse length, ν is the mean number of neutrons per fission, and D is the diffusion coefficient with units of length. The thickness of the slab is X.

The code to compute k-effective is below:

```
In [1]: import numpy as np
        def k_effective(slab_length, nuSigma_f, Sigma_a, Diff_coef):
            """ Computes the eigenvalue
                (k-effective) for a slab reactor

            Args:
                slab_length: the length of the slab
                nuSigma_f: value of nu * the macro. fission x-section
                Sigma_a: value of the macro. absorption x-section
                Diff_coef: the diffusion coefficient

            Returns:
                The value of k-effective
            """
            k_infinity = nuSigma_f / Sigma_a #k-infinity
            L = Diff_coef/Sigma_a #Diffusion length
            B = np.pi/slab_length #geometric buckling.
            k = k_infinity/(1+L**2 * B**2)
            return k
```

To make sure that our function is correct, there are several tests we could run to make sure that the function behaves the way that it should. A simple one would check that $k \to k_\infty$ as the slab size goes to infinity. The code below does this.

```
In [2]: import matplotlib.pyplot as plt
        #20 points from 10^0 to 10^3
        lengths = np.logspace(0,3,20)
        nuSigma_f = 1.1
        Sigma_a = 1.0
        Diff_coef = 1.0
        k_vector = k_effective(lengths, nuSigma_f,
                               Sigma_a, Diff_coef)
        plt.semilogx(lengths,k_vector,'o-')
        plt.semilogx(lengths,nuSigma_f/Sigma_a*np.ones(20))
        plt.xlabel('Slab Width (cm)')
        plt.ylabel('$k_\mathrm{eff}$')
        plt.show()
```

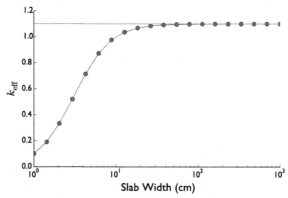

Viewing the resulting figure, we can see that the result goes to k_∞ as the slab size goes to infinity. Next, we test a particular case where we know the answer. If $\Sigma_a = D$ and $X = \pi$, the system will have k-effective equal to $k_\infty/2$. Here is that test:

```
In [3]: test_k = k_effective(slab_length = np.pi, nuSigma_f = 1,
                             Sigma_a = 1, Diff_coef = 1)
        if np.fabs(test_k - 0.5) < 1.0e-8:
            print("The test passed, k =",test_k)
        else:
            print("Test failed, you should probably",
                  "fix the code, k =", test_k)

The test passed, k = 0.5
```

You might be tempted to think that our code is fine, but it turns out that there is a problem (or bug) in the code. Let's try the test again, this time with $D = 2\Sigma_a$. The answer should be one-third of k_∞.

```
In [4]: test_k = k_effective(slab_length = np.pi, nuSigma_f = 1,
                             Sigma_a = 1, Diff_coef = 2)
        kinf = 1.0
        if np.fabs(test_k - 1.0/3.0) < 1.0e-8:
            print("The test passed, k =",test_k)
        else:
            print("Test failed, you should probably",
                  "fix the code, k =", test_k)

Test failed, you should probably fix the code, k = 0.2
```

It is clear that our code has managed to pass some tests, and has failed the third test we tried. Now begins our hunt to find the bug that is giving us the error.

6.2 DEBUGGING

Now that we have identified that our code has a bug, the process of finding that error is called debugging. Given that we ran three tests, and two of them passed we can use how the tests are different to identify where the bug might be.

The first test checked that as `slab_length` went to infinity, $k \to k_\infty$. Looking at the equations we can see that as the slab length goes to infinity, B_g goes to 0. Therefore, this test just checks that k_∞ is calculated properly, and indeed it is. The second test was designed so that B_g would be 1 and L^2 would be 1. In this instance things behaved like they should. When we switched things so that L^2 would be 2 and B_g would be 1, we had a failure. Therefore, the bug does not appear when $L^2 = 1$.

This is a covert bug (in that it is not apparent without a test). Take a moment to look at the function definition, and the equations it was trying to implement to see if you can identify the bug before moving on.

The code below has the bug fixed. It is a very small change, but often the hardest bugs to find are the ones that require just a small correction.

```
In [5]: import numpy as np
        def k_effective(slab_length, nuSigma_f, Sigma_a, Diff_coef):
            """ Computes the eigenvalue
                (k-effective) for a slab reactor

            Args:
                slab_length: the length of the slab
                nuSigma_f: value of nu * the macro. fission x-section
                Sigma_a: value of the macro. absorption x-section
                Diff_coef: the diffusion coefficient

            Returns:
                The value of k-effective
            """
            k_infinity = nuSigma_f / Sigma_a #k-infinity
            L2 = Diff_coef/Sigma_a #Diffusion length squared
            B = np.pi/slab_length #geometric buckling.
            k = k_infinity/(1+L2 * B**2)
            return k
```

Then, re-running the test from before, we get the correct answer.

```
In [6]: test_k = k_effective(slab_length = np.pi, nuSigma_f = 1,
                             Sigma_a = 1, Diff_coef = 2)
        kinf = 1.0
        if np.fabs(test_k - 1.0/3.0) < 1.0e-8:
            print("The test passed, k =",test_k)
        else:
            print("Test failed, you should probably",
                  "fix the code, k =", test_k)
```

```
The test passed, k = 0.3333333333333333
```

The bug in the code was that L^2 and L were mixed up. That is why the error did not show up when $L^2 = 1$ because in that case the values are the same.

In this example we see how tests can help us debug. By running several tests we can hone in on the error in the code: the first test demonstrated that k_∞ was correctly calculated, the second test showed that when $L = 1$, the code was correct. Note also that only one test is not sufficient, as we saw. This is why you want multiple tests for your code, you want to do more than just make sure the code works in one limit or one case, you need to test several possible cases and see that the code works in all of them.

Debugging is a mindset. When looking at your code you need to be thinking critically about how the code works, and what the value of a variable is after each line, etc. Then you want to compare that with what the code is supposed to do. This is one way that comments can be important: they indicate the intension of the programmer and what the code *should* be doing. When comparing code to what it should be doing, a keen attention to detail is necessary because the error could be as small as a single character.

Unfortunately, there is no simple recipe for debugging code. Experience is an important ingredient because the errors that appear in code often repeat, and having made a mistake once will make it possible for the programmer to look for that mistake again in a similar piece of code. The value of experience becomes apparent when a novice programmer presents code with an error to an instructor or another expert. Sometimes without even looking at the code, the expert can identify the problem based on the described behavior. Such an occurrence can cause the novice programmer to despair that he or she "will never be that good", or some other self-defeating watchword. Typically, the expert can diagnose the problem so quickly because of that expert's past mistakes.

Despite not having a simple recipe, when debugging it can be useful to ask the following questions:

- How is the code failing?
- What is the code doing correctly?
- What pieces of the code are most likely to have an error?

Moreover, your best tool for debugging is the `print` function. When in doubt have the code print out what happens after every line and check that with your intuition/expectation. For example, if there is a mistake in a formula you could print out the result and compare it with a hand calculation.

BOX 6.1 LESSON LEARNED

The best debugging tool is the `print` function. When you want to figure out what is going on in your code, pepper it with print statements, like a chef preparing steak au poivre, to see what each line is doing.

There are other tools for debugging. If you use IDLE, Anaconda, or another integrated development environment (such as Visual Studio or XCode), there is a built-in debugger. One of the features a debugger allows you to do is set breakpoints. A breakpoint is a point in the code where execution stops, and you can enter commands interactively to see what the value of different variables are, for example. Also, once the code stops at a breakpoint you can step through the code line by line and see how the code is actually executing, e.g., is a particular `if` statement evaluating to true or false, or how many times does a loop execute. For more information about the debugger check your development environment's documentation.

6.3 ASSERTIONS

In life it pays to be assertive. The same is true in programming. With the `assert` statement, you can check assumptions that are embedded in your code using an `assert` statement. The `assert` statement takes an expression as input. If the expression evaluates to true, then the `assert` does nothing. However, if the expression evaluates to false, an error is thrown and the code stops executing. The benefit of the `assert` statement is that you can make the code stop dead in its tracks, if some assumption made is violated. Then, Python will tell you where the code stopped.

BOX 6.2 PYTHON PRINCIPLE

The `assert` statement is called using the syntax

```
assert expression
```

where `expression` is a python expression that evaluates to true or false. If `expression` evaluates to true, nothing happens. When `expression` evaluates to false, an error occurs and code execution stops. Python will indicate where in the code the `assert` statement failed. This type of error is called an `AssertionError`.

As an example, we use our `k_effective` function from before. Physically, all of the cross-sections in the model we implemented should be non-negative. However, the code will work even if the inputs are negative:

```
In [7]:  test_k = k_effective(slab_length = np.pi, nuSigma_f = 1,
                              Sigma_a = -2, Diff_coef = 1)
         print("With negative Sigma_a, k =", test_k)

With negative Sigma_a, k = -1.0
```

The code worked, but the answer is non-sensical. We would like to tell the user that `k_effective` was called with an improper value. We could just change the help text to indicate that each input needs to be greater than zero. However, if we do this it will not prevent the user from running the code with negative cross-sections. Moreover, if the user does not respect the instruction for the function inputs, the function will behave strangely or give an error

```
In [8]:  test_k = k_effective(slab_length = np.pi, nuSigma_f = 1,
                              Sigma_a = -1, Diff_coef = 1)
         print("With negative Sigma_a, k =", test_k)


-----------------------------------------------------------------
ZeroDivisionError Traceback (most recent call last)

<ipython-input-23-a6b005d701c7> in <module>()
```

```
----> 1 test_k = k_effective(slab_length = np.pi, nuSigma_f = 1,
                             Sigma_a = -1, Diff_coef = 1)
      2 print("With negative Sigma_a, k =", test_k)


<ipython-input-21-2a945842eba5> in
k_effective(slab_length, nuSigma_f, Sigma_a, Diff_coef)
   15     L2 = Diff_coef/Sigma_a #Diffusion length
   16     B = np.pi/slab_length #geometric buckling.
--->17     k = k_infinity/(1+L2 * B**2)
   18     return k


ZeroDivisionError: float division by zero
```

In such a case even though the function caller made a mistake, by passing in a negative cross-section, because the code failed in the function k_effective, the programmer of that function is likely to be blamed.

This is where the assert statement comes in. It can assure that the arguments to the function are what they should be. Therefore, the function can throw an error when it is called improperly. Below, we do this for k_effective; the docstring is not shown for brevity.

```
In [9]: import numpy as np
        def k_effective(slab_length, nuSigma_f, Sigma_a, Diff_coef):
            assert (slab_length > 0)
            assert (nuSigma_f > 0)
            assert (Sigma_a > 0)
            assert (Diff_coef > 0)
            k_infinity = nuSigma_f / Sigma_a #k-infinity
            L2 = Diff_coef/Sigma_a #Diffusion length
            B = np.pi/slab_length #geometric buckling.
            k = k_infinity/(1+L2 * B**2)
            return k
```

If we call this function with invalid parameters, it will give an error and indicate where the error occurred.

```
In [10]: test_k = k_effective(slab_length = np.pi, nuSigma_f = 1,
                              Sigma_a = -1, Diff_coef = 1)
         print("With negative Sigma_a, k =", test_k)


-----------------------------------------------------------------
    AssertionError Traceback (most recent call last)

    <ipython-input-10-a6b005d701c7> in <module>()
----> 1 test_k = k_effective(slab_length = np.pi, nuSigma_f = 1,
                             Sigma_a = -1, Diff_coef = 1)
      2 print("With negative Sigma_a, k =", test_k)
```

```
<ipython-input-9-b097a84a2036> in
k_effective(slab_length, nuSigma_f, Sigma_a, Diff_coef)
     14      assert (slab_length > 0)
     15      assert (nuSigma_f > 0)
---> 16      assert (Sigma_a > 0)
     17      assert (Diff_coef > 0)
     18      k_infinity = nuSigma_f / Sigma_a #k-infinity


AssertionError:
```

Notice that Python tells us which assertion failed so we know that the function call had a bad value of Sigma_a. The program still fails, but it tells us exactly why. Additionally, it indicates that the error is not in the function itself, but with the input parameters.

We can also use assertions to test that the code behaves the way we expect. In the case of this function, we know that k is in $[0, k_\infty]$. Therefore, we can embed that check in two uses of assert.

```
In [11]: import numpy as np
         def k_effective(slab_length, nuSigma_f, Sigma_a, Diff_coef):
             assert (slab_length > 0)
             assert (nuSigma_f > 0)
             assert (Sigma_a > 0)
             assert (Diff_coef > 0)
             k_infinity = nuSigma_f / Sigma_a #k-infinity
             L2 = Diff_coef/Sigma_a #Diffusion length
             B = np.pi/slab_length #geometric buckling.
             k = k_infinity/(1+L2 * B**2)
             assert k >= 0
             assert k <= k_infinity
             return k
```

These types of assert statements can help you debug later on down the road. This is especially true when you have functions being called by other functions. In these situations assert statements can assure that functions were called properly and help isolate where any problems lie.

6.4 ERROR HANDLING

There are times when you want to handle an error so that either the program can continue or print a useful error message before exiting. The complete topic of error handling is outside the scope of this work, and in the purview of writing production software, which we are not tackling. Nevertheless, error handling can make debugging and finding errors in code easier. Additionally, we can use error handling to help us execute tests of our code.

Error handling is often called exception handling. When a program is running, if it encounters an error it can raise an exception. The exception will give some indication of what the error is. In Python, you can place some code in a special block of code called a try block.

TABLE 6.1   Common Exceptions in Python

Exception	Meaning
AssertionError	The argument passed to an assert was False
FloatingPointError	An error happened in a floating point calculation
KeyError	A dictionary key that was not valid was used
KeyboardInterrupt	The user pressed ctrl-c to exit
NameError	An undefined variable was used to exit
OverflowError	A number too large was created/used
RecursionError	Function called itself too many times
TypeError	The wrong type was used in an expression
ZeroDivisionError	Attempted to divide by zero

After the try block, the types of exception to handle are listed using except blocks. Only the exceptions that are explicitly handled are caught.

One type of exception is the ZeroDivisionError that is raised when a number is divided by zero. First, we will look at an uncaught exception:

```
In [12]: z = 10.5/0
```

```
-----------------------------------------------------------------------

        ZeroDivisionError Traceback (most recent call last)

        <ipython-input-12-011b064d3b54> in <module>()
----> 1 z = 10.5/0


        ZeroDivisionError: float division by zero
```

To catch this exception and proceed in the code, we use a try block and except block as

```
In [13]: try:
             z = 10.5/0
         except ZeroDivisionError:
             print("You cannot divide by 0")
You cannot divide by 0
```

Notice that the except block takes the name of the exception as an argument (in this case ZeroDivisionError). A list of common exception types, and what they mean, are given in Table 6.1.

One thing that happens when you catch a raised exception, is that the program will continue on. This can be a useful feature, but often times you want to catch an exception, print a useful error message, and then have the program end. This can be done by adding a raise statement to the end of the except block. The raise statement tells Python to still fail due to the exception, despite the fact that we caught it. This changes the previous example by one line, but changes the output and forces the program to quit:

```
In [14]: try:
             z = 10.5/0
         except ZeroDivisionError:
             print("You cannot divide by 0, exiting")
             raise

You cannot divide by 0, exiting

-----------------------------------------------------------------

         ZeroDivisionError Traceback (most recent call last)

         <ipython-input-14-525c0fef7adc> in <module>()
             1 try:
         ----> 2     z = 10.5/0
             3 except ZeroDivisionError:
             4     print("You cannot divide by 0, exiting")
             5     raise

         ZeroDivisionError: float division by zero
```

BOX 6.3 PYTHON PRINCIPLE

To handle errors, use a try block combined with except blocks. These have the form

```
try:
    [SomeCode]
except ExceptionA:
    [ExcepCodeA]
except ExceptionB:
    [ExcepCodeB]
...
except:
    [CatchAll]
```

The code in the try block, [SomeCode], will be executed. If there is an exception raised while running the code in the try block that matches one of the parameters in the except blocks below (in this example, ExpectionA,

and ExceptionB), the code in that except block will be executed and the code will continue. See Table 6.1 for a list of common exceptions. After executing the except block, the code will continue as normal.

The last except block may not have an exception type. This block will catch all other exceptions and execute its code, in this case [CatchAll]. This should be used with caution as unexpected exceptions may arise that causes errors later in the program.

The raise statement may be used to raise an exception, and if called inside an except block will raise the current exception type. This can be used to make Python quit the program due to an error.

For a more in depth example of error handling we return to the function k_effective. Previously, we added assert statements to make sure that the user gave the function all positive inputs. It would be more useful to the user, and for debugging, to have the code output what the values of the inputs were, if one of the assert calls fails. We do this in the code below (again without the docstring for brevity):

```
In [15]: import numpy as np
         def k_effective(slab_length, nuSigma_f, Sigma_a, Diff_coef):
             try:
                 assert (slab_length >0)
                 assert (nuSigma_f > 0)
                 assert (Sigma_a >0)
                 assert (Diff_coef > 0)
             except AssertionError:
                 print("Input Parameters are not all positive.")
                 print("slab_length =",slab_length)
                 print("nuSigma_f =",nuSigma_f)
                 print("Sigma_a =",Sigma_a)
                 print("Diff_coef =",Diff_coef)
                 raise
             except:
                 print("An unexpected error occurred when",
                       "checking the function parameters")
                 raise

             k_infinity = nuSigma_f / Sigma_a #k-infinity
             L2 = Diff_coef/Sigma_a #Diffusion length
             B = np.pi/slab_length #geometric buckling.
             k = k_infinity/(1+L2 * B**2)
             assert k >= 0
             assert k <= k_infinity
             return k
```

With this function, if it is passed a negative value, will raise an `AssertionError`. The code catches this error, prints out the input parameters to the user, and then exits by raising the exception. This functionality is demonstrated below.

```
In [16]: test_k = k_effective(slab_length = np.pi, nuSigma_f = 1,
                              Sigma_a = -1, Diff_coef = 1)

Input Parameters are not all positive.
slab_length = 3.141592653589793
nuSigma_f = 1
Sigma_a = -1
Diff_coef = 1

-------------------------------------------------------------

         AssertionError Traceback (most recent call last)

         <ipython-input-18-ee866f2992f9> in <module>()
           1 test_k = k_effective(slab_length = np.pi, nuSigma_f = 1,
       ----> 2                           Sigma_a = -1, Diff_coef = 1)


         <ipython-input-17-83b9328123be> in
         k_effective(slab_length, nuSigma_f, Sigma_a, Diff_coef)
           4          assert (slab_length >0)
           5          assert (nuSigma_f > 0)
```

```
----> 6            assert (Sigma_a >0)
      7            assert (Diff_coef > 0)
      8       except AssertionError:


   AssertionError:
```

Also, the function has a generic `except` statement that will catch any other errors in the `try` block. Because the `try` block involves comparison of numbers, if we pass a `string` as a parameter, there will be an error when checking if that parameter is greater than 0. This exception will be caught by the generic `except` statement, and the code prints out an error message, and then quits:

```
In [17]: test_k = k_effective(slab_length = "Pi", nuSigma_f = 1,
                            Sigma_a = -1, Diff_coef = 1)

An unexpected error occurred when checking the function parameters

-----------------------------------------------------------

    TypeError Traceback (most recent call last)

    <ipython-input-22-6c226c0ff76e> in <module>()
      1 test_k = k_effective(slab_length = "Pi", nuSigma_f = 1,
----> 2                        Sigma_a = -1, Diff_coef = 1)


    <ipython-input-21-72247d51776d> in
    k_effective(slab_length, nuSigma_f, Sigma_a, Diff_coef)
      2 def k_effective(slab_length, nuSigma_f, Sigma_a, Diff_coef):
      3     try:
----> 4          assert (slab_length >0)
      5          assert (nuSigma_f > 0)
      6          assert (Sigma_a >0)


    TypeError: unorderable types: str() > int()
```

The exception is a `TypeError`, and was not anticipated, so the program tells the user something unexpected happened.

The `try-except` structure is a useful tool when writing functions that will be called by other functions, or when there is a chance for input parameters to have the wrong form. They go beyond `assert` statements to give the programmer the ability to tell the user what went wrong, and possibly fix the error, before either continuing or quitting.

The concepts of testing, debugging, assertions, and exceptions can be combined to make a capability to readily test and find bugs in code. In the programming exercises below, exceptions and assertions are combined to run a variety of tests on a piece of code, and to raise assertions when tests fail.

FURTHER READING

There are not many books the discuss the process of debugging, but a book by Butcher [8] does discuss how to build the mindset of a master debugger. For a more detailed, though perhaps too detailed for novices, description of exception handling in Python see the Python tutorial https://docs.python.org/3/tutorial/errors.html.

PROBLEMS

Short Exercises

6.1. Write an assert statement that guarantees a variable n is less than 100 and is positive.

6.2. Debug the following code:

```
for i in list_variable:
    list_variable = ['one',2,'III','quatro']
    print(i)
```

6.3. Write a function called soft_equivalence that takes as input 3 parameters: a, b, and tol. The function should return True if the absolute difference between a and b is less than tol, and return False otherwise. The parameter tol should be an optional parameter with a default value of 10^{-6}.

6.4. Create a list of tests you would perform to test a function solve(A,b) that returns the solution to the linear system $\mathbf{Ax} = \mathbf{b}$. You do not need to write any code, just describe the tests, either in equations and/or words.

Programming Projects

1. Test Function for k-Eigenvalue

In this problem you will create a test function that performs all of the tests we performed above on the k-eigenvalue function k_effective. Create a function named test, that takes no input parameters. This function needs to execute the three tests the we developed for the k_effective function: the infinite medium case and the two particular value cases. Each test should raise an assertion error if the test fails to be within some tolerance, appropriately defined by you, of the correct value. If an assertion error is raised, the error needs to be caught, and which test(s) failed should be printed to the screen. The function should return either True, if all of the tests passed, and False, if any test failed.

Demonstrate that you test function works on the k_effective as we defined it correctly above, and demonstrate that it will catch an error in the calculation, if a bug is inserted.

NUMERICAL METHODS

7

Gaussian Elimination

Vaughn's been working on a couple of new pitches, the Eliminator and the Humilator, to complement his fastball, the Terminator.

"Harry Doyle" in the movie Major League II

CHAPTER POINTS

- The most natural way to solve systems of linear equations by hand can be generalized into the algorithm known as Gaussian elimination.

- This algorithm works well for almost all systems, if we allow the order of the equations to be rearranged.

- The number of operations needed to perform Gaussian elimination scales as the number of equations cubed, $O(n^3)$, though on moderately-sized matrices we observe slightly lower growth in the time to solution.

7.1 A MOTIVATING EXAMPLE

In this chapter we will be interested in computing the solution to a system of linear, algebraic equations. The solution of such systems is the basis for many other numerical techniques. For example, the solution of nonlinear systems is often reduced to the solution of successive linear systems that approximate the nonlinear system.

We desire to write a generic algorithm for solving linear systems. We want the systems to have arbitrary size, and arbitrary values for the coefficients. To develop this algorithm, we will start with a concrete example, and then generalize our approach.

We will begin with the following system:

$$3x_1 + 2x_2 + x_3 = 6, \tag{7.1a}$$
$$-x_1 + 4x_2 + 5x_3 = 8, \tag{7.1b}$$
$$2x_1 - 8x_2 + 10x_3 = 4. \tag{7.1c}$$

In matrix form this looks like

$$\begin{pmatrix} 3 & 2 & 1 \\ -1 & 4 & 5 \\ 2 & -8 & 10 \end{pmatrix} \begin{pmatrix} x_1 \\ x_2 \\ x_3 \end{pmatrix} = \begin{pmatrix} 6 \\ 8 \\ 4 \end{pmatrix}.$$

Another way to write this system is a notational shorthand called an augmented matrix, where we put the righthand side into the matrix separated by a vertical line:

$$\left(\begin{array}{ccc|c} 3 & 2 & 1 & 6 \\ -1 & 4 & 5 & 8 \\ 2 & -8 & 10 & 4 \end{array} \right).$$

We will store this matrix in Python for use later:

```
In [1]: import numpy as np
        aug_matrix = np.matrix([(3.0,2,1,6),(-1,4,5,8),(2,-8,10,4)])
        print(aug_matrix)

[[ 3.    2.    1.    6.]
 [ -1.    4.    5.    8.]
 [  2.   -8.   10.    4.]]
```

A straightforward way to solve this system is to use the tools of elementary algebra and try to eliminate variables by adding and subtracting equations from each other. We could do this by taking the second row, and adding to it 1/3 times the first row. Here's how that's done in Python:

```
In [2]: #add row 2 to 1/3 times row 1
        row13 = aug_matrix[0]/3 #row1 * 1/3
        new_row2 = aug_matrix[1] + row13 #add 1/3 row 1 to row 2
        #replace row 2
        aug_matrix[1,:] = new_row2
        print("New matrix =\n",aug_matrix)

New matrix =
 [[ 3.           2.           1.          6.        ]
 [ 0.           4.66666667   5.33333333  10.        ]
 [ 2.          -8.          10.          4.        ]]
```

This eliminated x_1 from the second equation. The next step would be to eliminate x_1 from the third equation by adding $-2/3$ times row 1 to row 3:

```
In [3]: #add row 3 to -2/3 times row 1
        row23 = -2*aug_matrix[0]/3 #row1 * -2/3
        new_row3 = aug_matrix[2] + row23 #add -2/3 row 1 to row 3
        #replace row 3
        aug_matrix[2] = new_row3
        print("New matrix =\n",aug_matrix)

New matrix =
 [[  3.           2.           1.           6.        ]
 [  0.           4.66666667   5.33333333  10.        ]
 [  0.          -9.33333333   9.33333333   0.        ]]
```

The x_2 term from the third equation can be removed by adding to row 3 the quantity $9\frac{1}{3}/4\frac{2}{3}$ times row 2:

```
In [4]: #add row 3 to 9.33333/4.66666 times row 2
        modrow = (9+1./3)/(4+2./3)*aug_matrix[1] #row2 * (-9+1./3)/(4+2.0/3)
        new_row3 = aug_matrix[2] + modrow #add -2/3 row 1 to row 3
        #replace row 3
        aug_matrix[2] = new_row3.copy()
        print("New matrix =\n",aug_matrix)

New matrix =
 [[  3.           2.           1.           6.        ]
 [  0.           4.66666667   5.33333333  10.        ]
 [  0.           0.          20.          20.        ]]
```

Notice that we have manipulated our original system into the equivalent system

$$\left(\begin{array}{ccc|c} 3 & 2 & 1 & 6 \\ 0 & 4\frac{2}{3} & 5\frac{1}{3} & 10 \\ 0 & 0 & 20 & 20 \end{array} \right),$$

or

$$\left(\begin{array}{ccc} 3 & 2 & 1 \\ 0 & 4\frac{2}{3} & 5\frac{1}{3} \\ 0 & 0 & 20 \end{array} \right) \left(\begin{array}{c} x_1 \\ x_2 \\ x_3 \end{array} \right) = \left(\begin{array}{c} 6 \\ 10 \\ 20 \end{array} \right).$$

We can easily solve this via system using a process called "back substitution". In back substitution we start at the last row and solve each equation in succession. In this particular example, take the last equation and solve for x_3, then plug the value of x_3 into the second equation, solve for x_2, and then plug both into the first equation and solve for x_1.

```
In [5]: #backsubstitution
        x3 = aug_matrix[2,3]/aug_matrix[2,2] #solve for x3
        print("x3 =",x3)
        #now solve for x2
        x2 = (aug_matrix[1,3] - x3*aug_matrix[1,2])/aug_matrix[1,1]
```

```
print("x2 =",x2)
#now solve for x1
x1 = (aug_matrix[0,3] - x3*aug_matrix[0,2]-
      x2*aug_matrix[0,1])/aug_matrix[0,0]
print("x1 =",x1)
```

```
x3 = 1.0
x2 = 1.0
x1 = 1.0
```

Therefore, the solution we get is

$$x_1 = 1, \qquad x_2 = 1, \qquad x_3 = 1.$$

We can check this solution by multiplying it by the original coefficient matrix, and showing that the result is equal to the system's righthand side:

```
In [6]: A = np.matrix([(3.0,2,1),(-1,4,5),(2,-8,10)])
        x = np.array([1,1,1])
        b = np.array([6,8,4])
        print(np.dot(A,x),"-",b,"=",np.dot(A,x)-b)
```

```
[[ 6.  8.  4.]] - [6 8 4] = [[ 0.  0.  0.]]
```

Our solution does indeed satisfy the original system. The method we used to solve this system goes by the name Gaussian elimination. The basic idea is to march through the system and eliminate variables one by one until the final equation of the system has only a single unknown. Then back substitution is used to solve for each variable, starting with the last one.

The formula for back-substitution can be written succinctly by noticing that each x_i is the sum of solutions of the unknowns further down the vector divided by the diagonal element of the modified matrix:

$$x_i = \frac{1}{A_{ii}} \left(b_i - \sum_{j=i+1}^{3} A_{ij} x_i \right), \qquad i = 1, 2, 3 \tag{7.2}$$

with **A** being the matrix after applying Gaussian elimination.

7.2 A FUNCTION FOR SOLVING 3 × 3 SYSTEMS

The procedure we used to solve the system given in Eq. (7.1) can be generalized to any coefficients and righthand side. We could just copy and paste the code above, and then change the numbers if we wanted to solve another system, but that would be prone to errors. In the next code block, we define a function that mimics our algorithm above to change the augmented matrix into a form ready for back substitution. The function is then tested on the system in Eq. (7.1).

```
In [7]: def GaussElim33(A,b):
            """create a Gaussian elimination matrix for a 3x3 system

            Args:
                A: 3 by 3 array
                b: array of length 3
            Returns:
                augmented matrix ready for back substitution
            """
            #create augmented matrix
            aug_matrix = np.zeros((3,4))
            aug_matrix[0:3,0:3] = A
            aug_matrix[:,3] = b
            #augmented matrix is created
            for column in range(0,3):
                for row in range(column+1,3):
                    mod_row = aug_matrix[row,:]
                    mod_row -= (mod_row[column]/aug_matrix[column,column]*
                                aug_matrix[column,:])
                    aug_matrix[row] = mod_row
            return aug_matrix
        #test function on the problem above
        aug = GaussElim33(A,b)
        print(aug)

[[  3.          2.          1.          6.        ]
 [  0.          4.66666667  5.33333333  10.       ]
 [  0.          0.          20.         20.       ]]
```

As we can see, this gives us the same result we obtained from our hand calculation. We should test this function more thoroughly, but we will return to that later. The next step is to write a back substitution function that implements Eq. (7.4)

```
In [8]: def BackSub33(aug_matrix,x):
            """back substitute a 3 by 3 system after Gaussian elimination

            Args:
                aug_matrix: augmented matrix with zeros below the diagonal
                x: length 3 vector to hold solution
            Returns:
                nothing
            Side Effect:
                x now contains solution
            """
            #start at the end
            for row in [2,1,0]:
                RHS = aug_matrix[row,3]
                for column in range(row+1,3):
                    RHS -= x[column]*aug_matrix[row,column]
                x[row] = RHS/aug_matrix[row,row]
            return
        x = np.zeros(3)
        BackSub33(aug,x)
        print("The solution is ", x)
```

```
The solution is  [ 1.  1.  1.]
```

The back substitution function has the first example of using a side effect to change an input parameter that we have seen up to this point. To illustrate why we do this, we will revisit the topic of mutable types being passed to a function. Recall, the way that Python and NumPy work: when a NumPy array is passed to a function, the function does not copy the array into local scope. Rather, the function can modify the original values of the array. This is accomplished through a construct called pass by reference. What this means is that the function gets a reference to the memory where the original array lives, and can therefore modify that memory. The function does not make a copy of the NumPy array because the array could easily have millions of elements, and it is a bad idea to be heedlessly copying these large data structures because the computer could easily run out of memory. The upshot of pass by reference in the code above is that when we pass x into the function BackSub33, the function puts the solution into x. This changing of memory outside the function is a side effect because it is a way that a function interacts with the rest of the code not through the mechanism of returning information from a function.

NumPy arrays are not the only data structures that are passed by reference to a function. Any data type that can have its size modified is called a mutable type, and these are passed by reference to functions. We have already encountered two other mutable types in python: the list and dictionary. This means if you pass either of these two types to a function, that function can change the list or dictionary. It is good programming practice to make explicit in the comments and docstring what the side-effects are of a function so that the user knows when a function is called the original data may be changed.

BOX 7.1 PYTHON PRINCIPLE

Mutable data types such as lists, dictionaries, and NumPy arrays, are passed by reference to functions. This means that a function can change the values of a mutable data type when it is the parameter of the function. When a function changes the input data, this is a side effect, and it should be noted in the function's docstring.

We need more vigorous testing of our function above to convince ourselves that it is a tool for solving a general 3×3 system. To make this a general test, we will select a random matrix (using np.random.random) and then multiply it by a random vector to get the righthand side. Then when we solve the system the solution should be the original vector. It would be best to do this many times using a different matrix each time. The code below performs 100 such tests and asserts that the maximum absolute difference between the computed solution and the true solution should be less than 10^{-12}. An assert checks this, and error handling is used to give information about a failed test.

```
In [9]: tests = 100
        for i in range(tests):
            x = np.random.rand(3)
            A = np.random.rand(3,3)
            b = np.dot(A,x)
```

```
        aug = GaussElim33(A,b)
        sol = np.zeros(3)
        BackSub33(aug,sol)
        diff = np.abs(sol-x)
        try:
            assert(np.max(diff) < 1.0e-12)
        except AssertionError:
            print("Test failed with")
            print("A = ",A)
            print("x = ",x)
            raise
    print("All Tests Passed!")
```

```
All Tests Passed!
```

These tests pass, and we have some confidence that our algorithm is working. Later we will see how it might be improved to tackle some pathological cases.

7.3 GAUSSIAN ELIMINATION FOR A GENERAL SYSTEM

The code we wrote above only solved 3×3 systems, that is a useful tool, but we would not want to write a different function for every size system we want to solve. That is, we do not want to have to write a 4×4 solver and a 5×5 solver, etc. Therefore, we need to have a general Gaussian elimination code to solve the system

$$\mathbf{Ax} = \mathbf{b}, \tag{7.3}$$

where $\mathbf{x}$ and $\mathbf{b}$ are vectors of length N, and $\mathbf{A}$ is an $N \times N$ matrix.

Due to the way we structured our 3×3 code, the main change we need to make to adapt the function systems of size N is in the `for` loops. On the whole the code is unchanged: all we must do is replace the 3's in the function with N's. The next code block has this function, and a 4×4 test that has solution $\mathbf{x} = (1, 2, 3, 4)$.

```
In [10]: def GaussElim(A,b):
             """create a Gaussian elimination matrix for a system

             Args:
                 A: N by N array
                 b: array of length N
             Returns:
                 augmented matrix ready for back substitution
             """
             [Nrow, Ncol] = A.shape
             assert Nrow == Ncol
             N = Nrow
             assert b.size == N
             #create augmented matrix
             aug_matrix = np.zeros((N,N+1))
             aug_matrix[0:N,0:N] = A
```

```
        aug_matrix[:,N] = b
        #augmented matrix is created
        for column in range(0,N):
            for row in range(column+1,N):
                mod_row = aug_matrix[row,:]
                mod_row -= (mod_row[column]/aug_matrix[column,column]*
                            aug_matrix[column,:])
                aug_matrix[row] = mod_row
        return aug_matrix

#let's try it on a 4 x 4 to start
A = np.array([(3.0,2,1,1),(-1,4,5,-2),(2,-8,10,-3),(2,3,4,5)])
answer = np.arange(4)+1.0 #1,2,3,4
b = np.dot(A,answer)
aug = GaussElim(A,b)
print(aug)
```

```
[[ 3.          2.          1.          1.          14.         ]
 [ 0.          4.66666667  5.33333333  -1.66666667  18.66666667]
 [ 0.          0.          20.         -7.          32.         ]
 [ 0.          0.          0.          5.42857143  21.71428571]]
```

Notice that we had to determine the size of the system by looking at the size of the matrix and the vector. Also, using `assert` statements, the function checks that the sizes of the input data are compatible. The next step is to define the back substitution function. This function will generalize Eq. (7.4) to have the summation going to N rather than 3, and letting i range from 1 to N:

$$x_i = \frac{1}{A_{ii}} \left(b_i - \sum_{j=i+1}^{N} A_{ij} x_i \right), \qquad i = 1, 2, 3. \tag{7.4}$$

The resulting code is

```
In [11]: def BackSub(aug_matrix,x):
             """back substitute a N by N system after Gauss elimination

             Args:
                 aug_matrix: augmented matrix with zeros below the diagonal
                 x: length N vector to hold solution
             Returns:
                 nothing
             Side Effect:
                 x now contains solution
             """
             N = x.size
             [Nrow, Ncol] = aug_matrix.shape
             assert Nrow + 1 == Ncol
             assert N == Nrow
             for row in range(N-1,-1,-1):
                 RHS = aug_matrix[row,N]
                 for column in range(row+1,N):
                     RHS -= x[column]*aug_matrix[row,column]
```

```
            x[row] = RHS/aug_matrix[row,row]
        return

    x = np.zeros(4)
    BackSub(aug,x)
    print("The solution is ", x)


The solution is  [ 1.  2.  3.  4.]
```

Applying the back substitution function gives us the expected solution, $\mathbf{x} = (1, 2, 3, 4)$.

Now that we have a generic solver, we can make a large matrix that has 2.01 on the diagonal and -1 on the immediate off diagonals, and create a simple righthand side. This code is an example of automatically filling in a matrix. This is an important technique to master because there are many different methods available for solving linear systems, yet filling the matrix can be the hardest part of solving the problem. In the code below we fill the diagonal by setting up a range that runs from 0 to the number of rows in the matrix minus one. We then fill the off diagonals using similar ranges. This type of matrix is called a tridiagonal matrix because it has entries in three diagonal lines in the matrix. We will see these matrices again when we discretize the neutron diffusion equation.

```
In [12]: mat_size = 100
         A = np.zeros((mat_size,mat_size))
         b = np.zeros(mat_size)
         diag = np.arange(mat_size)
         A[diag,diag] = 2.01
         belowDiagRow = np.arange(1,mat_size)
         A[belowDiagRow,belowDiagRow-1] = -1
         aboveDiagRow = np.arange(mat_size-1)
         A[aboveDiagRow,aboveDiagRow+1] = -1
         print(A)

         b[np.floor(mat_size/2)] = 1
         aug_mat = GaussElim(A,b)
         x = np.zeros(mat_size)
         BackSub(aug_mat,x)
         plt.plot(diag,x,color="blue")
         plt.xlabel("Row");
         plt.ylabel("x value");
         plt.show();


[[ 2.01 -1.    0.    ...,  0.    0.    0.   ]
 [-1.    2.01 -1.    ...,  0.    0.    0.   ]
 [ 0.   -1.    2.01  ...,  0.    0.    0.   ]
 ...,
 [ 0.    0.    0.    ...,  2.01 -1.    0.   ]
 [ 0.    0.    0.    ..., -1.    2.01 -1.   ]
 [ 0.    0.    0.    ...,  0.   -1.    2.01]]
```

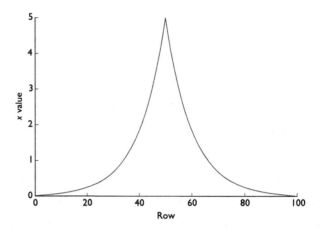

As we will see later, this is related to the solution of the diffusion equation with a Dirac delta function source. The delta function source was specified by having the vector b be zero everywhere except for a single row where it was set to one.

7.4 ROUND OFF AND PIVOTING

For matrices with a large discrepancy in the magnitude of the elements, Gaussian elimination does not perform quite as well. This is due to the fact that the operations used in Gaussian elimination are done with finite precision arithmetic. When large numbers are divided by small numbers, the numerical round off error can build and lead to incorrect answers at the end of Gaussian elimination. An example matrix to demonstrate this is given below.

```
In [13]: epsilon = 1e-14
         A = np.array([(epsilon,-1,1),(-1,2,-1),(2,-1,0)])
         print(A)

[[  1.00000000e-14  -1.00000000e+00   1.00000000e+00]
 [ -1.00000000e+00   2.00000000e+00  -1.00000000e+00]
 [  2.00000000e+00  -1.00000000e+00   0.00000000e+00]]
```

We will set the righthand side of our system to a simple vector:

```
In [14]: b = np.array([0,0,1.0])
         print(b)

[ 0.  0.  1.]
```

Now we solve this system:

```
In [15]: aug_mat = GaussElim(A,b)
         x = np.zeros(3)
         BackSub(aug_mat,x)
         print("The solution is",x)
         print("The residual is",b-np.dot(A,x))
```

```
The solution is [ 0.96589403  0.96969697  0.96969697]
The residual is [ 0.          -0.00380294  0.03790891]
```

As you can see there is a noticeable difference between the calculated solution and the actual solution (as measured by the residual). The residual is the difference between the vector b and the matrix A times the solution. If the solution were exact, the residual would be zero. Part of the reason for this is that we have not used the largest row element as the *pivot* element, or the element we divide by when doing the elimination. In this example, the pivot element is near zero and finite precision error is magnified when we divide by this small number.

To correct this issue we could rearrange the rows so that we divide by larger elements. This is called *row pivoting*.

BOX 7.2 NUMERICAL PRINCIPLE

The error in an approximation is the difference between the true solution and the approximate solution, i.e., for an approximate solution $\hat{x}$ to the vector x the error is defined as

$$e = x - \hat{x}.$$

The residual is equal to the discrepancy between the righthand and lefthand sides of the original equations when the approximate solution is substituted. For a linear system of equations given by $Ax = b$, the residual, r, for an approximation $\hat{x}$ is

$$r = b - A\hat{x}.$$

If one does not know the true solution, the residual can give a measure of the error because the exact solution has a residual of zero. Also, the residual and the error are related via the equation,

$$Ae = r.$$

To further demonstrate the need for row pivoting, consider the system:

```
In [16]: A = np.array([(1.0,0),(0,1.0)])
         b = np.array([2,6.0])
         print("A=\n",A)
         print("b=",b)
```

```
A=
 [[ 1.  0.]
 [ 0.  1.]]
b= [ 2.  6.]
```

The solution to this system is

```
In [17]: aug_mat = GaussElim(A,b)
         x = np.zeros(2)
         BackSub(aug_mat,x)
         print(x)
```

```
[ 2.  6.]
```

This is as expected because **A** is an identity matrix. Now what if we switch the rows in **A** and try the same thing:

```
In [18]: A = np.array([(0,1.0),(1.0,0)])
         b = np.array([6.,2.0])
         print("A =\n",A)
         print("b =",b)
         aug_mat = GaussElim(A,b)
         x = np.zeros(2)
         BackSub(aug_mat,x)
         print("x =",x)

A =
 [[ 0.  1.]
 [ 1.  0.]]
b = [ 6.  2.]
x = [ nan  nan]

-c:21: RuntimeWarning: divide by zero encountered in double_scalars
-c:21: RuntimeWarning: invalid value encountered in multiply
-c:17: RuntimeWarning: invalid value encountered in double_scalars
```

We do not get a solution because the diagonal element in the first row was zero. Therefore, we divided by zero during the Gaussian elimination process.

We can correct both of these issues by checking before we eliminate and rearranging the equations so that we divide by the largest possible entry. This is just rearranging the order that we solve the equations, and will just involve switching rows in the augmented matrix.

It is worth mentioning that largest element is relative to the size of the other entries in the row. If one row is $(10, 11, 10)$ and another is $(2, 1, 1)$ the 2 is actually a better pivot because it is the largest element in its row.

To do pivoting we will need a means to swap rows in our augmented matrix. The following function will modify the augmented matrix by swapping rows.

```
In [19]: def swap_rows(A, a, b):
             """Swap two rows in a matrix: switch row a with row b

             args:
             A: matrix to perform row swaps on
             a: row index of matrix
             b: row index of matrix

             returns: nothing

             side effects:
             changes A to have rows a and b swapped
             """
             assert (a>=0) and (b>=0)
             N = A.shape[0] #number of rows
             assert (a<N) and (b<N) #less than because 0-based indexing
             temp = A[a,:].copy()
             A[a,:] = A[b,:].copy()
             A[b,:] = temp.copy()
```

```
        print("Before swap, A =\n",A)
        swap_rows(A,0,1)
        print("After swapping 0 and 1, A =\n",A)
```
```
Before swap, A =
 [[ 0.  1.]
 [ 1.  0.]]
After swapping 0 and 1, A =
 [[ 1.  0.]
 [ 0.  1.]]
```

The next step is figuring out how to swap during each step. At the beginning of the solve we want find the maximum row element magnitude for each row and store it as a vector. The following code is an example of this:

```
In [20]: N = 5
         A = np.random.rand(N,N)
         print("A =\n",A)
         s = np.zeros(N)
         count = 0
         for row in A:
             s[count] = np.max(np.fabs(row))
             count += 1
         print("s =",s)
```

```
A =
 [[ 0.52115101  0.46323356  0.68539875  0.56665694  0.1093434 ]
 [ 0.00451899  0.29089231  0.9037581   0.83739104  0.90189019]
 [ 0.15378442  0.77964023  0.51662888  0.03317186  0.35942844]
 [ 0.24128529  0.03615319  0.43274014  0.68011597  0.42847979]
 [ 0.96757013  0.94404531  0.56392054  0.32454444  0.97512228]]
s = [ 0.68539875  0.9037581   0.77964023  0.68011597  0.97512228]
```

Then we will have to figure out which row has the largest scaled element in the pivot position and call the `swap_rows` function. We will use the `argmax` function which returns the index of the largest element in a vector.

```
In [21]: pivot_column = 2
         largest_pos = np.argmax(np.fabs(A[:,pivot_column]/s))
         print("Largest scaled element in column",
               pivot_column,"is in row",largest_pos)
```

```
Largest scaled element in column 2 is in row 1
```

Now can put this all together in a new version of Gaussian elimination. This function will find the largest elements in each row of the matrix, then proceed with Gaussian elimination where the pivot element is the element that is the largest element in remaining in the row.

```
In [22]: def GaussElimPivotSolve(A,b,LOUD=0):
             """create a Gaussian elimination with pivoting matrix for a system
```

```python
Args:
    A: N by N array
    b: array of length N
Returns:
    solution vector in the original order
"""
[Nrow, Ncol] = A.shape
assert Nrow == Ncol
N = Nrow
#create augmented matrix
aug_matrix = np.zeros((N,N+1))
aug_matrix[0:N,0:N] = A
aug_matrix[:,N] = b
#augmented matrix is created

#create scale factors
s = np.zeros(N)
count = 0
for row in aug_matrix[:,0:N]: #don't include b
    s[count] = np.max(np.fabs(row))
    count += 1
if LOUD:
    print("s =",s)
if LOUD:
    print("Original Augmented Matrix is\n",aug_matrix)
#perform elimination
for column in range(0,N):

    #swap rows if needed
    largest_pos =(np.argmax(np.fabs(aug_matrix[column:N,column:N]/
                    s[column]))  + column)
    if (largest_pos != column):
        if (LOUD):
            print("Swapping row",column,"with row",largest_pos)
            print("Pre swap\n",aug_matrix)
        swap_rows(aug_matrix,column,largest_pos)
        #re-order s
        tmp = s[column]
        s[column] = s[largest_pos]
        s[largest_pos] = tmp
        if (LOUD):
            print("A =\n",aug_matrix)
    #finish off the row
    for row in range(column+1,N):
        mod_row = aug_matrix[row,:]
        mod_row -= (mod_row[column]/
                    aug_matrix[column,column]*aug_matrix[column,:])
        aug_matrix[row] = mod_row
#now back solve
x = b.copy()
if LOUD:
    print("Final aug_matrix is\n",aug_matrix)
BackSub(aug_matrix,x)
return x
```

As a basic test we will solve a 3×3 system:

```
In [23]: #let's try it on a 3 x 3 to start
         A = np.array([(3.0,2,1),(-1,4,5),(2,-8,10)])
         answer = np.arange(3)+1.0 #1,2,3
         b = np.dot(A,answer)
         x = GaussElimPivotSolve(A,b,LOUD=True)
         print("The solution is",x)
         print("The residual (errors) are",np.dot(A,x)-b)

s = [  3.   5.  10.]
Original Augmented Matrix is
 [[  3.   2.   1.  10.]
  [ -1.   4.   5.  22.]
  [  2.  -8.  10.  16.]]
Swapping row 1 with row 2
Pre swap
 [[  3.          2.          1.         10.        ]
  [  0.          4.66666667  5.33333333 25.33333333]
  [  0.         -9.33333333  9.33333333  9.33333333]]
A =
 [[  3.          2.          1.         10.        ]
  [  0.         -9.33333333  9.33333333  9.33333333]
  [  0.          4.66666667  5.33333333 25.33333333]]
Final aug_matrix is
 [[  3.          2.          1.         10.        ]
  [  0.         -9.33333333  9.33333333  9.33333333]
  [  0.          0.         10.         30.        ]]
The solution is [ 1.  2.  3.]
The residual (errors) are [  0.00000000e+00   0.00000000e+00   3.55271368e-15]
```

Based on the residual, and the fact the we engineered the answer to be known, we can see that the solution is correct. To see the efficacy of the row pivoting we will try our function on the systems that did not work so well before. First we try the rearranged identity matrix:

```
In [24]: A = np.array([(0,1.0),(1.0,0)])
         b = np.array([6.,2.0])
         print("A =\n",A)
         print("b =",b)
         x = GaussElimPivotSolve(A,b,LOUD=False)
         print("x =",x)
         print("The residual (errors) are",np.dot(A,x)-b)

x = [ 2.  6.]
The residual (errors) are [ 0.  0.]
```

The answer is correct. Finally, we try the matrix that had a large discrepancy between the element sizes:

```
In [25]: epsilon = 1e-14
         A = np.array([(epsilon,-1,1),(-1,2,-1),(2,-1,0)])
         print(A)          b = np.array([0,0,1.0])
         x = GaussElimPivotSolve(A,b,LOUD=False)
         print("x =",x)
         print("The residual is",np.dot(A,x)-b)
```

```
x = [ 1.   1.   1.]
The residual is [  0.00000000e+00   -2.22044605e-16   0.00000000e+00]
```

Now we get the correct answer and the residual is effectively zero.

7.5 TIME TO SOLUTION FOR GAUSSIAN ELIMINATION

We will now discuss how long we should expect Gaussian elimination to take. We will do this through counting the number of floating point operations required to perform Gaussian elimination on an $n \times n$ matrix. A floating point operation is either addition, subtraction, multiplication, or division of floating point numbers. The counting of these operations for Gaussian elimination is straightforward, but tedious. In our derivation of the operation count we will ignore the extra column we added to get the augmented matrix because we care about how the operation count scales for large matrices. If the matrix is large, the addition of a single column is a small perturbation.

At the start of the algorithm we have to eliminate the first column from the $(n-1)$ rows that are not the first row. We first need to compute the elimination factors which involve dividing the row 1, column 1 element by the appropriate pivot element, requiring $(n-1)$ divisions because this must be done for each row that is not the first. Next, we multiply each row by the elimination factor. This involves $(n-1)^2$ multiplications because we multiply every element of the matrix, besides the first row and the first column, by the appropriate factor to eliminate the first column below the first row. Then we need to add $(n-1)$ elements from row 1, columns 2 through n to the corresponding elements in all the other rows: this is $(n-1)^2$ additions in all. Therefore after eliminating the first column we have performed $2(n-1)^2 + (n-1)$ floating point operations.

After eliminating the first column below the diagonal, we now have an $(n-1)$ by $(n-1)$ matrix that we need to eliminate a column from. This will require $2(n-2)^2 + (n-2)$ floating point operations. The column after that requires $2(n-3)^2 + (n-3)$, and so on until we get to the last column. Using summation notation we can write the total number of floating point operations as

$$\text{Total Floating Point Operations} = \sum_{l=1}^{n} \left[2(n-l)^2 + (n-l) \right].$$

We can factor this expression to get a simpler form:

$$\sum_{\ell=1}^{n} \left[2(n-\ell)^2 + (n-\ell) \right] = \sum_{\ell=1}^{n} \left[2n^2 - 4n\ell + 2\ell^2 + n - \ell \right]$$

$$= 2n^3 + n^2 - \sum_{\ell=1}^{n} \left[4n\ell - 2\ell^2 + \ell \right]$$

$$= 2n^3 + n^2 - 2n^2(n+1) - \frac{n(n+1)}{2} + \frac{n(n+1)(2n+1)}{3}$$

$$= \frac{2}{3}n^3 - \frac{1}{2}n^2 - \frac{1}{6}n,$$

where we have used the identities

$$\sum_{\ell=1}^{n} \ell = \frac{n(n+1)}{2}, \qquad \sum_{\ell=1}^{n} \ell^2 = \frac{n(n+1)(2n+1)}{6}.$$

Therefore, the total number of floating point operations in Gaussian elimination for a matrix of size n by n is

$$\text{Total Floating Point Operations} = \frac{2}{3}n^3 - \frac{1}{2}n^2 - \frac{1}{6}n.$$

Often we are concerned with how the number of floating point operations scales as n increases. Therefore, we want to know how Gaussian elimination trends as n is large. For large n, $n^3 \gg n^2 \gg n$. For this reason we say that Gaussian elimination is an $O(n^3)$, or order n cubed, algorithm because the leading term when n is large is the n^3 term. What this means in practice is that if we double the number of rows in my matrix, i.e., n goes to $2n$, we should expect the code to take $2^3 = 8$ times longer. Nevertheless, this is only for large n as will see next.

BOX 7.3 NUMERICAL PRINCIPLE

When talking about the number of operations for an algorithm, we often use Big-O notation to describe how the leading order term behaves as the problem gets bigger. The leading-order growth of Gaussian elimi-nation scales as the number of equations, n, to the third power. Therefore, we say that this algorithm is $O(n^3)$. As an example, an algo-rithm that increased linearly in the problem size would be an $O(n)$ algorithm.

We can use the time module that Python provides to time how long it takes to run our Gaussian elimination code as we increase n.

```
In [26]: import time
         num_tests = 13
         N = 2**np.arange(num_tests)
         times = np.zeros(num_tests)
         for test in range(1,num_tests):
             A = np.random.rand(N[test],N[test])
             b = np.dot(A,np.ones(N[test]))
             x = np.zeros(N[test])
             start = time.clock()
             aug = GaussElim(A,b)
             BackSub(aug,x)
             end = time.clock()
             times[test] = end-start
         plt.loglog(N,times,'o-')
         y_comp = times.copy()/4
```

```
for comp_place in range(num_tests-1,0,-1):
    #because x goes up by factor 2 each time, time should go up by 8
    y_comp[comp_place -1] = np.exp(np.log(y_comp[test])-3*(
                              np.log(N[test])-np.log(N[comp_place-1])))
plt.loglog(N[4:num_tests],y_comp[4:num_tests],'r',linewidth=4)
#make a comparison line with slope 2
y_comp = times.copy()*4
for comp_place in range(num_tests-1,0,-1):
    y_comp[comp_place -1] = np.exp(np.log(y_comp[test])-2*(
                              np.log(N[test])-np.log(N[comp_place-1])))
plt.loglog(N[4:num_tests],y_comp[4:num_tests],'g-',linewidth=4)
plt.xlabel("Number of Equations")
plt.ylabel("Seconds to complete solve")
plt.show()
```

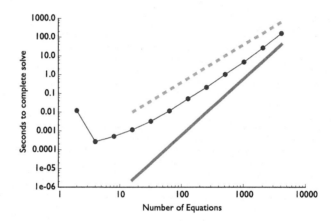

In this figure the solid line is a slope of 3 on a log-log scale (which corresponds time growing as the number of equations to the third power). If the time to solve a system was consistent with the theory, then this dots would be parallel to the red line. That assumes that the time it takes to execute the code is linearly related to the number of floating point operations. Starting around tens of equations, the slope is closer to 2, the slope of the dashed line. This implies that initially the time to solution is not growing as fast as the theory. The reason for this is that Python is not giving the CPU work to do fast enough so that the floating point operations are not the bottleneck to the calculation, rather, feeding instructions to the CPU is taking up the bulk of the time. As the matrices get bigger, each instruction takes longer and the overhead of feeding instructions is a smaller fraction of the execution time. We see this in the figure: as n grows, the slope is getting steeper: when we have thousands of equations the slope is around 2.6. This indicates that if we ran a large enough problem, and had enough patience and computer memory, we would see the predicted result.

FURTHER READING

Gaussian elimination is a classical method for solving linear systems, and there are many possible references for it. The book on numerical linear algebra by Trefethen and Bau [9] provides an in-depth discussion of the stability of Gaussian elimination with row pivoting.

PROBLEMS

Short Exercises

7.1. Solve the linear system $\mathbf{A}\mathbf{x} = \mathbf{b}$ for $\mathbf{x}$ where

$$\mathbf{A} = \begin{pmatrix} -1 & 2 & 3 \\ 4 & -5 & 6 \\ 7 & 8 & -9 \end{pmatrix} \qquad \mathbf{b} = (12.9, -5.1, 10.7)^{\mathrm{T}}.$$

7.2. Consider the n by n matrix defined by

$$\mathbf{A} = \begin{pmatrix} 1 & \frac{1}{2} & \frac{1}{3} & \cdots & \frac{1}{n} \\ \frac{1}{2} & \frac{1}{3} & \frac{1}{4} & \cdots & \frac{1}{n+1} \\ \vdots & & & & \\ \frac{1}{n} & \frac{1}{n+1} & \frac{1}{n+2} & \cdots & \frac{1}{2n-1} \end{pmatrix},$$

and the vector

$$\mathbf{b} = (1, -1, 1, -1, \ldots)^{\mathrm{T}}.$$

Write a function to generate $\mathbf{A}$ and $\mathbf{b}$ as a function of n. Use Gaussian elimination to solve the system $\mathbf{A}\mathbf{x} = \mathbf{b}$ for $n = 2, 4, 8, 16$. What do you notice about your answers? Are the answers that you get from Gaussian elimination correct?

Programming Projects

1. Xenon Poisoning

^{135}Xe is produced in a nuclear reactor through two mechanisms:

- The production of ^{135}Te from fission, which then decays to ^{135}I via β decay with a half-life of 19 s, which then decays to ^{135}Xe via β decay with a half-life of 6.6 h.
- The production of ^{135}Xe directly from fission.

^{135}Xe is important because it has a very large capture cross-section of 2.6×10^6 barns. ^{135}Xe is also radioactive, and decays with a half-life of 9.1 h.

For a reactor with an average scalar flux, ϕ in units [neutrons/cm^2/s], the equations for the production of ^{135}Xe are (using the abbreviation number density [nuclei/cm^3] of ^{135}Xe$\rightarrow X$, ^{135}I$\rightarrow I$, ^{135}Te$\rightarrow T$),

$$\frac{d}{dt}T = \gamma_T \, \Sigma_f \phi - \lambda_T T,$$

$$\frac{d}{dt}I = \lambda_T T - \lambda_I I,$$

$$\frac{d}{dt}X = \lambda_I I + \gamma_X \, \Sigma_f \phi - (\lambda_X + \sigma_a^X \phi)X.$$

If we consider the steady-state limit, i.e., $\frac{d}{dt} = 0$, we have an equation for the equilibrium concentration of all three nuclides. For a ^{235}U fueled reactor, the fission yields for Te and Xe are $\gamma_T = 0.061$, $\gamma_X = 0.003$, and the macroscopic fission cross-section is $\Sigma_f = 0.07136$ cm^{-1}. Compute the equilibrium concentrations of these three nuclides at power densities of 5, 50, and 100 W/cm^3, using the energy per fission of 200 MeV/fission. Also, compute the absorption rate density of neutrons in ^{135}Xe, given by $\sigma_a^X \phi X$.

2. Flux Capacitor Waste

In the movie *Back to the Future*, the time machine is fueled by plutonium. Assuming that the time machine is refueled with 1 mg of pure ^{239}Pu every year and the old fuel, still with the same mass, is placed in a disposal site. The stable lead is recovered from the site at a rate of 50% a year. What are the equilibrium mass of ^{239}Pu and its daughter products after this refueling cycle goes on for a long time. Also, compute the alpha particle and beta particle production rate. The decay chain for ^{239}Pu is given on page 92.

3. Four-Group Reactor Theory

A large subcritical system with a source has its scalar flux in each of four groups described by these four equations,

$$\Sigma_{a,1}\phi_1 + \Sigma_{s,1\to2}\phi_1 = \chi_1 \sum_{g=1}^{4} \nu\Sigma_{f,g}\phi_g + Q_1$$

$$\Sigma_{a,2}\phi_2 + \Sigma_{s,2\to3}\phi_2 = \chi_2 \sum_{g=1}^{4} \nu\Sigma_{f,g}\phi_g + \Sigma_{s,1\to2}\phi_1 + Q_2$$

$$\Sigma_{a,3}\phi_3 + \Sigma_{s,3\to4}\phi_3 = \chi_3 \sum_{g=1}^{4} \nu\Sigma_{f,g}\phi_g + \Sigma_{s,2\to3}\phi_2 + Q_3$$

$$\Sigma_{a,4}\phi_4 = \chi_4 \sum_{g=1}^{4} \Sigma_{f,g}\phi_g + \Sigma_{s,3\to4}\phi_3 + Q_4.$$

For source strength, $Q = [10^{12}, 0, 0, 0]$ n/cm^3/s and the cross-sections given below in Table 7.1 [10] for the group bounds, [1.35 MeV, 10 MeV], [9.1 keV, 1.35 MeV], [0.4 eV, 9.1 keV], and [0.0 eV, 0.4 eV], compute the scalar flux in each group in system where neutrons can only scatter within group or to the energy group below.

TABLE 7.1 Four Group Constants

	Group 1	Group 2	Group 3	Group 4
χ_g	0.57500	0.42500	0.00000	0.00000
$\nu\Sigma_{fg}$ (cm^{-1})	0.00480	0.00060	0.00885	0.09255
Σ_{ag} (cm^{-1})	0.00490	0.00280	0.03050	0.12100
$\Sigma_{sg\to g+1}$ (cm^{-1})	0.08310	0.05850	0.06510	
D_g (cm)	2.16200	1.08700	0.63200	0.35400

4. Matrix Inverse

The inverse of a matrix **A** is defined as the matrix such that

$$\mathbf{AA}^{-1} = \begin{pmatrix} 1 & 0 & 0 & \cdots & 0 \\ 0 & 1 & 0 & \cdots & 0 \\ 0 & 0 & 1 & \cdots & 0 \\ \vdots & \vdots & \vdots & \ddots & \vdots \\ 0 & 0 & 0 & \cdots & 1 \end{pmatrix}.$$

Given this fact, we can compute the inverse of an n by n matrix, column by column, by solving the following set of problems:

$$\mathbf{Ax}_i = \mathbf{b}_i \qquad i = 1, \ldots, n,$$

where $\mathbf{b}_i$ is a vector of zeros except at position i where it has 1, and $\mathbf{x}_i$ is the ith column of the inverse. Using this formulation write a Python function that computes the inverse of a general $n \times n$ matrix. There are two ways of doing this, one way computes Gaussian elimination n times, the other more sophisticated way has the augmented matrix have n extra columns corresponding to all the $\mathbf{b}_i$. Test your function on the matrix from Short Exercise 7.1 of this chapter and demonstrate that the inverse matrix you compute gives the identity when multiplied by **A**.

LU Factorization and Banded Matrices

Lu Lu Lu, I've got some apples. Lu Lu Lu you've got some too.

–**"Butters Stotch" in the television series South Park**

CHAPTER POINTS

- In the case where one wants to solve several systems with changing righthand sides, LU factorization is an efficient solution method.

- Banded matrices have non-zero entries only on or near the diagonal, and the banded structure is preserved by LU factorization.

- With our methods we are able to solve realistic neutron diffusion problems.

In practice it is often the case that we want to solve the same linear system multiple times, just with different right-hand sides. One example in nuclear engineering involves a subcritical nuclear system with a source. We can write the neutron diffusion equation for this system as

$$-\nabla \cdot D(\mathbf{r})\nabla\phi(\mathbf{r}) + \Sigma_a\phi(\mathbf{r}) = \nu\Sigma_f\phi(\mathbf{r}) + Q(\mathbf{r}),$$

where the energy-integrated scalar flux is $\phi(\mathbf{r})$, the diffusion coefficient is $D(\mathbf{r})$, the absorption macroscopic cross-section is Σ_a, and the source rate density is $Q(\mathbf{r})$. As we will see later, this

131

equation can, after numerical discretization, be written as

$$A\phi = q,$$

where the entries in ϕ and q represent a discrete value of the scalar flux and source rate density, respectively, and the matrix A is a discrete version of the diffusion operator.

From the previous chapter, we know that solving this system with Gaussian elimination is possible. Nevertheless, if you we going to analyze the system with several different source configurations (e.g., strength and position of the source) you would have to perform Gaussian elimination on each source configuration. If you have the intuition that the solution of one of these systems should make it possible to solve all of them, you are on the right track. To put it another way, besides the manipulations to the right-hand side of the equation, Gaussian elimination manipulates the matrix to get it in a form where the solution is reduced to back-substitution. We would like to do this work only once.

It is possible to do the work of Gaussian elimination to put the matrix in a form to make the solution with any right-hand side simple. By simple we mean that it involves one back substitution and one forward substitution, that is back substitution in the opposite direction (from top to bottom).

8.1 LU FACTORIZATION

LU factorization writes a matrix A as the product of a lower triangular matrix (that is a matrix with nonzero elements on the diagonal and below) times an upper triangular matrix (that is a matrix with nonzero elements on or above the diagonal). In other words,

$$A = LU,$$

where

$$L = \begin{pmatrix} * & 0 & \cdots & & \\ * & * & 0 & \cdots & \\ * & * & * & 0 & \cdots \\ \vdots & \vdots & \vdots & \ddots & \\ * & * & * & \cdots & * \end{pmatrix},$$

and

$$U = \begin{pmatrix} * & * & * & \cdots & * \\ 0 & * & * & \cdots & * \\ 0 & 0 & \ddots & \cdots & * \\ \vdots & \vdots & \vdots & \ddots & \\ 0 & \cdots & & 0 & * \end{pmatrix},$$

where the $*$ denote a potentially nonzero matrix element.

The question remains how we find the entries in the matrices in the LU factorization. To see this, consider a matrix we used in the last chapter:

$$\begin{pmatrix} 3 & 2 & 1 \\ -1 & 4 & 5 \\ 2 & -8 & 10 \end{pmatrix}.$$

For this matrix, after Gaussian elimination, the matrix became

$$\begin{pmatrix} 3 & 2 & 1 \\ 0 & 4\frac{2}{3} & 5\frac{1}{3} \\ 0 & 0 & 20 \end{pmatrix}.$$

Notice that this matrix is upper triangular.

We want to find the matrices **L** and **U** such that

$$\begin{pmatrix} l_{11} & 0 & 0 \\ l_{21} & l_{22} & 0 \\ l_{31} & l_{32} & l_{33} \end{pmatrix} \begin{pmatrix} u_{11} & u_{12} & u_{13} \\ 0 & u_{22} & u_{23} \\ 0 & 0 & u_{33} \end{pmatrix} = \begin{pmatrix} 3 & 2 & 1 \\ -1 & 4 & 5 \\ 2 & -8 & 10 \end{pmatrix}.$$

We have to make a choice at this point. The choice we have is which diagonal, either **L** or **U**, we want to be all ones. We will choose to have all the $l_{ii} = 1$, that is the **L** matrix has a diagonal of all 1's (this choice makes our algorithm equivalent to an algorithm called Doolittle's method). Performing the matrix multiplication we get 9 equations and 9 unknowns for the factorization. The product of the first column of **U** by the first column of **L** gives

$$u_{11} = 3$$
$$l_{21}u_{11} = -1$$
$$l_{31}u_{11} = 2.$$

Which simplifies to,

$$u_{11} = 3,$$
$$l_{21} = -\frac{1}{3},$$
$$l_{31} = \frac{2}{3}.$$

Notice that the solutions below the diagonal, the l_{21} and l_{31}, are the factors we used in our Gaussian elimination algorithm to remove the first column in the second and third equations (see Section 7.1). Also, the u_{11} value is that found in the same position in our Gaussian-eliminated matrix.

Continuing on, the equations from the second column are

$$u_{12} = 2,$$
$$u_{12}l_{21} + u_{22} = 4,$$

$$u_{12}l_{31} + u_{22}l_{32} = -8,$$

the solutions to these equations are

$$u_{12} = 2,$$

$$u_{22} = 4\frac{2}{3},$$

$$l_{32} = -2.$$

Looking back to Section 7.1 of the previous chapter, we see that the value of l_{32} is the value of the factor used to eliminate the second column below the diagonal in the Gaussian elimination example from before. Finally, the last three equations are

$$u_{13} = 1,$$

$$u_{13}l_{21} + u_{23} = 5,$$

$$u_{13}l_{31} + u_{23}l_{32} + u_{33} = 10.$$

The solutions to these equations are

$$u_{13} = 1,$$

$$u_{23} = 5\frac{1}{3},$$

$$u_{33} = 20.$$

These are precisely the values of the third column in our Gaussian elimination example.

In summary, our factorization looks like:

$$
\begin{pmatrix} 1 & 0 & 0 \\ -\frac{1}{3} & 1 & 0 \\ \frac{2}{3} & -2 & 1 \end{pmatrix}
\begin{pmatrix} 3 & 2 & 1 \\ 0 & 4\frac{2}{3} & 5\frac{1}{3} \\ 0 & 0 & 20 \end{pmatrix}
=
\begin{pmatrix} 3 & 2 & 1 \\ -1 & 4 & 5 \\ 2 & -8 & 10 \end{pmatrix}.
$$

Some observations are in order:

1. The upper triangular matrix is the same as the matrix we received after doing Gaussian elimination.
2. The non-zero and non-diagonal elements of the lower triangular matrix are the factors we used to arrive at our Gaussian matrix.

This suggests that we can reformulate our Gaussian elimination example to give us an LU factorization. What we will have to do is

1. Store the factors used to eliminate matrix elements in the appropriate place to get the **L** matrix, and
2. Perform Gaussian elimination as usual.

The code below modifies our Gauss elimination function to do just this. Also, we apply the function to our example matrix.

```
In [1]: import numpy as np
        def LU_factor(A):
            """Factor in place A in L*U=A. The lower triangular parts of A
            are the L matrix.  The L has implied ones on the diagonal.
            Args:
                A: N by N array
            Side Effects:
                A is factored in place.
            """
            [Nrow, Ncol] = A.shape
            assert Nrow == Ncol
            N = Nrow
            for column in range(0,N):
                for row in range(column+1,N):
                    mod_row = A[row]
                    factor = mod_row[column]/A[column,column]
                    mod_row = mod_row - factor*A[column,:]
                    #put the factor in the correct place in the modified row
                    mod_row[column] = factor
                    #only take the part of the modified row we need
                    mod_row = mod_row[column:N]
                    A[row,column:N] = mod_row
            return

        #let's try it on a 3 x 3 to start
        A = np.array([(3.0,2,1),(-1,4,5),(2,-8,10)])
        print("The original matrix is\n",A)
        LU_factor(A)
        print("The LU factored A is\n",A)
The original matrix is
 [[  3.    2.    1.]
 [ -1.    4.    5.]
 [  2.   -8.   10.]]
The LU factored A is
 [[  3.           2.           1.         ]
 [ -0.33333333   4.66666667   5.33333333]
 [  0.66666667  -2.          20.         ]]
```

This algorithm gives us the same thing our onerous, by-hand procedure did. Notice that we did not return an **L** and a **U** matrix, rather we factored in place the original matrix **A** by overwriting it. This is a reasonable thing to do because **A** could be a very large system and we do not want to duplicate that memory unnecessarily.

8.1.1 Forward and Backward Substitution

Now that we have discussed how to do LU factorization, the question we have not answered is how we can take an LU factored matrix and get the solution to the system $\mathbf{Ax} = \mathbf{b}$. To do this, we note that we can easily solve systems that only involve a lower (or upper) triangular matrix using back (or forward) substitution. For example, both

$$\mathbf{Ly} = \mathbf{b},$$

and

$$Ux = y,$$

are easy to solve. In fact that is exactly what we want to do, notice that if we solve

$$Ly = b,$$

that implies we know

$$y = L^{-1}b.$$

Therefore, if we take the original system

$$Ax = LUx = b,$$

and operate by L^{-1}, we get

$$Ux = L^{-1}b.$$

Therefore, given an LU factorization of a matrix **A** we want to solve

$$Ly = b,$$

then

$$Ux = y.$$

The following code does just this.

```
In [2]: def LU_solve(A,b):
            """Take a LU factorized matrix and solve it for RHS b
            Args:
                A: N by N array that has been LU factored with
                assumed 1's on the diagonal of the L matrix
                b: N by 1 array of righthand side
            Returns:
                x: N by 1 array of solutions
            """
            [Nrow, Ncol] = A.shape
            assert Nrow == Ncol
            N = Nrow
            x = np.zeros(N)
            #temporary vector for L^-1 b
            y = np.zeros(N)
            #forward solve
            for row in range(N):
                RHS = b[row]
                for column in range(0,row):
                    RHS -= y[column]*A[row,column]
                y[row] = RHS
            #back solve
            for row in range(N-1,-1,-1):
                RHS = y[row]
```

```
            for column in range(row+1,N):
                RHS -= x[column]*A[row,column]
            x[row] = RHS/A[row,row]
        return x
```

```
In [3]: #let's try it on a 3 x 3 to start
        A = np.array([(3.0,2,1),(-1,4,5),(2,-8,10)])
        LU_factor(A)
        b = np.array([6,8,4])
        x = LU_solve(A,b)
        print("The solution to the system is",x)
```

```
The solution to the system is [ 1.  1.  1.]
```

This code gives the correct answer, a fact that is easily checked as you see that the sum of each row of this matrix equals the corresponding row on the right-hand side.

8.2 LU WITH PIVOTING, AND OTHER SUPPOSEDLY FUN THINGS

One thing we have not discussed is pivoting. Pivoting with LU is needed in the same cases as it is in Gaussian elimination. Nevertheless, when we switch equations around, we need to make sure that we keep track of where we have switched rows, so we can do the same to the right-hand side when we do our forward and back solves. This complicates things considerably, but only because we have to add that bookkeeping to our algorithm.

First, we will need to be able to swap rows in the matrix:

```
In [4]: def swap_rows(A, a, b):
            """Rows two rows in a matrix, switch row a with row b
            args:
            A: matrix to perform row swaps on
            a: row index of matrix
            b: row index of matrix

            returns:
            nothing

            side effects:
            changes A to have rows a and b swapped
            """
            assert (a>=0) and (b>=0)
            N = A.shape[0] #number of rows
            assert (a<N) and (b<N) #less than because 0-based indexing
            temp = A[a,:].copy()
            A[a,:] = A[b,:].copy()
            A[b,:] = temp.copy()
```

Next, we will change our algorithm to handle pivoting and keep track of the pivots:

```
In [5]: def LU_factor(A,LOUD=True):
            """Factor in place A in L*U=A. The lower triangular parts of A
```

```
are the L matrix.   The L has implied ones on the diagonal.

Args:
    A: N by N array
Returns:
    a vector holding the order of the rows,
    relative to the original order
Side Effects:
    A is factored in place.
"""
[Nrow, Ncol] = A.shape
assert Nrow == Ncol
N = Nrow
#create scale factors
s = np.zeros(N)
count = 0
row_order = np.arange(N)
for row in A:
    s[count] = np.max(np.fabs(row))
    count += 1
if LOUD:
    print("s =",s)
if LOUD:
    print("Original Matrix is\n",A)
for column in range(0,N):
    #swap rows if needed
    largest_pos = np.argmax(np.fabs(A[column:N,column]/s[column]))
            + column
    if (largest_pos != column):
        if (LOUD):
            print("Swapping row",column,"with row",largest_pos)
            print("Pre swap\n",A)
        swap_rows(A,column,largest_pos)
        #keep track of changes to RHS
        tmp = row_order[column]
        row_order[column] = row_order[largest_pos]
        row_order[largest_pos] = tmp
        #re-order s
        tmp = s[column]
        s[column] = s[largest_pos]
        s[largest_pos] = tmp
        if (LOUD):
            print("A =\n",A)
    for row in range(column+1,N):
        mod_row = A[row]
        factor = mod_row[column]/A[column,column]
        mod_row = mod_row - factor*A[column,:]
        #put the factor in the correct place in the modified row
        mod_row[column] = factor
        #only take the part of the modified row we need
        mod_row = mod_row[column:N]
        A[row,column:N] = mod_row
return row_order
#let's try it on a 4 x 4 to start
```

```
A = np.array([(3.0,2,1,-2),(-1,4,5,4),(2,-8,10,3),(-2,-8,10,0.1)])
answer = np.ones(4)
b = np.dot(A,answer)
print(b)
row_order = LU_factor(A)
print(A)
print("The new row order is",row_order)
```

```
[  4.   12.    7.    0.1]
s = [  3.    5.   10.   10.]
Original Matrix is
 [[  3.    2.    1.   -2. ]
 [ -1.    4.    5.    4. ]
 [  2.   -8.   10.    3. ]
 [ -2.   -8.   10.    0.1]]
Swapping row 1 with row 2
Pre swap
 [[  3.           2.           1.          -2.        ]
 [ -0.33333333   4.66666667   5.33333333   3.33333333]
 [  0.66666667  -9.33333333   9.33333333   4.33333333]
 [ -0.66666667  -6.66666667  10.66666667  -1.23333333]]
A =
 [[  3.           2.           1.          -2.        ]
 [  0.66666667  -9.33333333   9.33333333   4.33333333]
 [ -0.33333333   4.66666667   5.33333333   3.33333333]
 [ -0.66666667  -6.66666667  10.66666667  -1.23333333]]
[[  3.           2.           1.          -2.        ]
 [  0.66666667  -9.33333333   9.33333333   4.33333333]
 [ -0.33333333  -0.5         10.          5.5        ]
 [ -0.66666667   0.71428571   0.4         -6.52857143]]
The new row order is [0 2 1 3]
```

We need to change the LU_solve function to take advantage of the swapped rows. In function below, we make the necessary modification, and test the solution

```
In [6]: def LU_solve(A,b,row_order):
            """Take a LU factorized matrix and solve it for RHS b

            Args:
                A: N by N array that has been LU factored with
                assumed 1's on the diagonal of the L matrix
                b: N by 1 array of righthand side
                row_order: list giving the re-ordered equations
                from the LU factorization with pivoting
            Returns:
                x: N by 1 array of solutions
            """
            [Nrow, Ncol] = A.shape
            assert Nrow == Ncol
            assert b.size == Ncol
            assert row_order.max() == Ncol-1
            N = Nrow
```

```
#reorder the equations
tmp = b.copy()
for row in range(N):
    b[row_order[row]] = tmp[row]

x = np.zeros(N)
#temporary vector for L^-1 b
y = np.zeros(N)
#forward solve
for row in range(N):
    RHS = b[row]
    for column in range(0,row):
        RHS -= y[column]*A[row,column]
    y[row] = RHS
#back solve
for row in range(N-1,-1,-1):
    RHS = y[row]
    for column in range(row+1,N):
        RHS -= x[column]*A[row,column]
    x[row] = RHS/A[row,row]
return x


print(A,row_order)
x = LU_solve(A,b,row_order)
print("The solution to the system is",x)
```

```
[[  3.          2.          1.          -2.        ]
 [  0.66666667 -9.33333333  9.33333333   4.33333333]
 [ -0.33333333 -0.5        10.           5.5       ]
 [ -0.66666667  0.71428571  0.4         -6.52857143]] [0 2 1 3]
The solution to the system is [ 1.  1.  1.  1.]
```

We now have a fully functional LU factorization algorithm and solving capability. We have not presented an exhaustive testing of the algorithm here, but this should be done before you use it in your own coding.

8.3 BANDED AND SYMMETRIC MATRICES

In many engineering calculations, symmetric and banded matrices arise from the equations. These matrices are often the result of a discretization of a diffusion equation. This is an important topic because, in truth, until about 20 years ago, large matrices that were not banded and symmetric were too difficult for all but the most specialized codes to handle.

For a matrix to be symmetric is means that if I transpose the matrix, the matrix does not change, i.e. if,

$$A_{ij} = A_{ji}$$

then the matrix is symmetric. This is equivalent to saying $\mathbf{A} = \mathbf{A}^T$. One feature of a symmetric matrix is that it is possible to store only the lower or upper triangular part of the matrix because of the symmetry.

Banded matrices take their name from the structure of the matrix. When the non-zeros of the matrix are shown they form bands around the diagonal. A diagonal matrix is a banded matrix of bandwith 1:

$$\begin{pmatrix} * & 0 & \ldots & & & \\ 0 & * & 0 & \ldots & & \\ 0 & 0 & * & 0 & \ldots & \\ \vdots & \vdots & \ddots & \ddots & \ddots & \\ 0 & 0 & \ldots & \ldots & * & \end{pmatrix}$$

The bandwidth is 1 because there is only 1 diagonal or off diagonal that has non-zero elements.

A common banded matrix is the tri-diagonal matrix. This matrix has a bandwidth of 3 and looks like

$$\begin{pmatrix} * & * & 0 & \ldots & & & & \\ * & * & * & 0 & \ldots & & & \\ 0 & * & * & * & 0 & \ldots & & \\ \vdots & \ddots & \ddots & \ddots & \ddots & & & \\ \vdots & & 0 & * & * & * & 0 \\ 0 & \ldots & & & 0 & * & * & * \\ 0 & 0 & \ldots & & & 0 & * & * \end{pmatrix}$$

To solve a tridiagonal system we can make some shortcuts. The primary shortcut is that we do not need to create a matrix at all. We only need to store three vectors, one for the diagonal and two for the off diagonals. We could redefine our algorithms to respect this sparsity of the matrix, i.e., take advantage of all the 0's. We will not do this here because the algorithm gets much more complicated. If we were writing a code to solve very large tridiagonal systems, we would want to take advantage of this efficiency because the amount of memory needed to store three vectors grows linearly with the number of equations, but grows quadratically if we store the whole matrix and all those wasteful zeros.

Pentadiagonal systems also arise in practice, namely 2-D diffusion equations will have 5 non-zero terms in each row. Furthermore, a 3-D diffusion equation will lead to a heptadiagonal matrix or a matrix that has 7 non-zero elements in each row. These matrices typically have a bandwidth much larger than 5 or 7 because the ordering of unknowns often makes it so that there are zeros between the non-zero elements in a row.

FURTHER READING

LU factorization is a common technique for solving linear systems of equations. It is also the basis for the SuperLU package [11], which has a Python interface through the `scipy` package.

PROBLEMS

Short Exercises

8.1. Compute the LU factorization of the matrix

$$\mathbf{A} = \begin{pmatrix} -1 & 2 & 3 \\ 4 & -5 & 6 \\ 7 & 8 & -9 \end{pmatrix}.$$

8.2. Consider the n by n matrix defined by

$$\mathbf{A} = \begin{pmatrix} 1 & \frac{1}{2} & \frac{1}{3} & \cdots & \frac{1}{n} \\ \frac{1}{2} & \frac{1}{3} & \frac{1}{4} & \cdots & \frac{1}{n+1} \\ \vdots & & & & \\ \frac{1}{n} & \frac{1}{n+1} & \frac{1}{n+2} & \cdots & \frac{1}{2n-1} \end{pmatrix},$$

and the vector

$$\mathbf{b} = (1, -1, 1, -1\ldots)^{\mathsf{T}}.$$

Write a function to generate $\mathbf{A}$ and $\mathbf{b}$ as a function of n. Use it to solve the system $\mathbf{A}\mathbf{x} = \mathbf{b}$ for $n = 2, 4, 8, 16$ via LU factorization. What do you notice about your answers and the resulting factorization?

Programming Projects

1. Matrix Inverse via LU Factorization

In the previous chapter we presented an approach to compute the inverse of a matrix. Here is another way to compute the inverse; this time using LU factorization. Compute the LU factorization of **A** and then use this to solve

$$\mathbf{A}\mathbf{x}_i = \mathbf{b}_i \qquad i = 1, \ldots, n,$$

where $\mathbf{b}_i$ is a vector of zeros except at position i where it has 1, and $\mathbf{x}_i$ is the ith column of the inverse. Test your implementation on the matrix from Short Exercise problem 8.1 above. Check that the inverse of **A** times the original matrix gives the identity matrix.

2. Shielding a Radioactive Source

Consider the problem of a slab geometry radiation source surrounded by a shield. You are interested in computing the leakage rate of radiation outside the shield. If the shield is a pure absorber and the source is collimated into a beam, the net neutron current density J [neutrons/cm^2·s] of neutrons can be described by the equation

$$\frac{dJ}{dx} + \Sigma_a J(x) = Q(x), \qquad J(0) = 0,$$

where x is the position in the slab which extends from $x = 0$ to $x = 10$ cm, $Q(x)$ [neutrons/cm^3·s] is the spatially dependent source strength, and Σ_a [cm^{-1}] is the macroscopic absorption cross-section. The leakage rate density out of the shield is $J(10)$. A simple discretization of this equation leads to the system of equations

$$\frac{J_i - J_{i-1}}{\Delta x} + \frac{\Sigma_a}{2}(J_i + J_{i-1}) = Q_i, \qquad i = 0 \ldots I.$$

In this equation $\Delta x = 10/I$, $J_i \approx J(i\Delta x)$, and $Q_i = Q(i\Delta x + \Delta x/2)$. Solve this system of equations and find the net leakage rate density with $I = 100$ and $\Sigma_a = 1$ with several different source configurations:

- A thin source at the left of the slab

$$Q(x) = \begin{cases} 1 & 0 \leq x \leq 1 \\ 0 & \text{otherwise} \end{cases},$$

- A thin source at the middle of the slab

$$Q(x) = \begin{cases} 1 & 4.5 \leq x \leq 5.5 \\ 0 & \text{otherwise} \end{cases},$$

- A thick source at the left of the slab

$$Q(x) = \begin{cases} 1 & 0 \leq x \leq 4 \\ 0 & \text{otherwise} \end{cases},$$

- A thin source at the middle of the slab

$$Q(x) = \begin{cases} 1 & 3 \le x \le 7 \\ 0 & \text{otherwise} \end{cases}.$$

Compare your solution to the analytic solution to problem. The analytic solution is an integral over an exponential function. To compute your numerical solution you will not need Gaussian elimination or LU factorization.

3. LU Factorization of a Tridiagonal System

Before we mentioned that it was possible to LU factorize a tridiagonal matrix. Modify the LU factorization without pivoting function, LU_factor, defined above to work with tridiagonal matrix. Your modified function should take as input three vectors, representing the main diagonal and two off diagonals. The function should return the three vectors that yield the LU factorization. Check your algorithm on one of the tridiagonal matrices used in this chapter. Also, use this function to see how large of a tridiagonal system you can solve on your computer.

Iterative Methods for Linear Systems

These evils thou repeat'st upon thyself
Have banish'd me from Scotland. O my breast,
Thy hope ends here!

–"MacDuff" in the play **Macbeth** *by William Shakespeare*

CHAPTER POINTS

- Iterative methods are an alternative to direct methods such as Gaussian Elimination and LU factorization.

- These methods can be faster than direct methods.

- The number of iterations required to compute a solution may be less important than how long it takes each iteration to complete.

The methods we have talked about up to this point are called "direct" methods because they pass through the matrix a fixed number of times to get the answer. For example, Gaussian elimination makes one pass through the augmented matrix to get it in upper triangular form, and then back solves to get the answers. You can say ahead of time exactly how many operations it will take to solve the system (modulo any pivoting). One thing that is also true is that no matter the system, the amount of work is the same if the number of equations is the same.

If we have a large system, it may be possible to solve the linear system in a faster way by making a guess at the solution and refining that guess until we are "close enough". These methods are called iterative methods, and most of the methods we cover require diagonal dominance. Diagonal dominance means that for each row i in the matrix $\mathbf{A}$ of size I by I

$$|A_{ii}| > \sum_{j=1, j\neq i}^{n} |A_{ij}|, \qquad i = 1 \ldots n$$

or in words, the diagonal is larger than the sum of the magnitudes all the off-diagonal elements of each row.

BOX 9.1 NUMERICAL PRINCIPLE

When the absolute value of the each element of the main diagonal in the matrix is larger than the sum of the absolute values of the off diagonal terms in each row, the matrix is said to be diagonally dominant. In equation form an $n \times n$ matrix is diagonally dominant if

$$|A_{ii}| > \sum_{j=1, j\neq i}^{n} |A_{ij}|, \qquad i = 1 \ldots n.$$

9.1 JACOBI ITERATION

The first iterative method we will meet is the simplest, and it is called the Jacobi method. For a system $\mathbf{Ax} = \mathbf{b}$, what the Jacobi method does is start with a guess $\mathbf{x}^{(0)}$ and then solve each row of the system by evaluating the diagonal term at the $\ell + 1$ iteration and the off diagonal terms at the ℓ iteration. It then does this over and over until either the change $\|\mathbf{x}^{(\ell+1)} - \mathbf{x}^{(\ell)}\|$ is small, the residual $\|\mathbf{Ax}^{(\ell+1)} - \mathbf{b}\|$ is small, or the maximum number of iterations is met.

We will demonstrate how the Jacobi method works with an example. Consider the linear system of a 4×4 tri-diagonal matrix

$$\begin{pmatrix} 2.5 & -1 & 0 & 0 \\ -1 & 2.5 & -1 & 0 \\ 0 & -1 & 2.5 & -1 \\ 0 & 0 & -1 & 2.5 \end{pmatrix} \begin{pmatrix} x_1 \\ x_2 \\ x_3 \\ x_4 \end{pmatrix} = \begin{pmatrix} 1 \\ 1 \\ 1 \\ 1 \end{pmatrix}.$$

Take an initial guess of $\mathbf{x}^{(0)} = [0, 0.5, 0.5, 0]^T$. The first iteration looks like:

$$2.5x_1^{(1)} - x_2^{(0)} = 1$$
$$-x_1^{(0)} + 2.5x_2^{(1)} - x_3^{(0)} = 1$$
$$-x_2^{(0)} + 2.5x_3^{(1)} - x_4^{(0)} = 1$$
$$-x_3^{(0)} + 2.5x_4^{(1)} = 1.$$

Each of these equations is independent of the other because we know the value of $x_i^{(0)}$ from our initial guess. Therefore, we can solve to get

$$x_1^{(1)} = 0.6, \ x_2^{(1)} = 0.6, \ x_3^{(1)} = 0.6, \ x_4^{(1)} = 0.6.$$

We can solve for $\mathbf{x}^{(2)}$ from the system

$$2.5x_1^{(2)} - x_2^{(1)} = 1,$$
$$-x_1^{(1)} + 2.5x_2^{(2)} - x_3^{(1)} = 1,$$
$$-x_2^{(1)} + 2.5x_3^{(2)} - x_4^{(1)} = 1,$$
$$-x_3^{(1)} + 2.5x_4^{(2)} = 1,$$

for $x_i^{(2)}$ and repeat the procedure until we are happy with the answer.

The general idea behind Jacobi iteration, is that given an initial guess $\mathbf{x}^{(0)}$, we compute

$$x_i^{(\ell+1)} = \frac{1}{A_{ii}} \left(b_i - \sum_{j=1, i \neq j}^{I} A_{ij} x_j^{(\ell)} \right), \qquad \text{for } i = 1 \dots I, \ \ell = 0 \dots .$$

We stop when the following criteria is met:

$$\frac{\|\mathbf{x}^{(\ell+1)} - \mathbf{x}^{(\ell)}\|_2}{\|\mathbf{x}^{(\ell+1)}\|_2} < \epsilon,$$

where ϵ is a user-defined tolerance, and the 2-norm is

$$\|\mathbf{y}\|_2 = \sqrt{\sum_{i=1}^{I} y_i^2}.$$

Now we will do these Jacobi iterations in Python on the system from above.

```
In [1]:   import numpy as np
          A = np.array([(2.5,-1,0,0),(-1,2.5,-1,0),(0,-1,2.5,-1),(0,0,-1,2.5)])
          print("Our matrix is\n",A)
          b = np.array([1.0,1,1,1])
          x = np.array([0,0.5,0.5,0])
          print("Our RHS is",b)
          print("The initial guess is",x)
```

```
Our matrix is
[[ 2.5 -1.   0.   0. ]
 [-1.   2.5 -1.   0. ]
 [ 0.  -1.   2.5 -1. ]
 [ 0.   0.  -1.   2.5]]
Our RHS is [ 1.  1.  1.  1.]
The initial guess is [ 0.   0.5  0.5  0. ]
```

```
In [2]:   #first Jacobi iteration
          x_new = np.zeros(4)
          for row in range(4):
              x_new[row] = b[row]
              for column in range(4):
                  if column != row:
                      x_new[row] -= A[row,column]*x[column]
              x_new[row] /= A[row,row]
          #check difference
          relative_change = np.linalg.norm(x_new-x)/np.linalg.norm(x_new)
          print("New guess is",x_new)
          print("Norm of change is",relative_change)
```

```
New guess is [ 0.6  0.6  0.6  0.6]
Norm of change is 0.71686043892
```

```
In [3]:   #Second Jacobi Iteration
          x = x_new.copy() #replace old value
          x_new *= 0 #reset x_new
          for row in range(4):
              x_new[row] = b[row]
              for column in range(4):
                  if column != row:
                      x_new[row] -= A[row,column]*x[column]
              x_new[row] /= A[row,row]
          #check difference
          relative_change = np.linalg.norm(x_new-x)/np.linalg.norm(x_new)
          print("New guess is",x_new)
          print("Norm of change is",relative_change)
```

```
New guess is [ 0.64  0.88  0.88  0.64]
Norm of change is 0.259937622455
```

Notice that the change is going down. We will perform one more iteration in this example.

```
In [4]:   #Third Jacobi Iteration
          x = x_new.copy() #replace old value
          x_new *= 0 #reset x_new
          for row in range(4):
              x_new[row] = b[row]
              for column in range(4):
                  if column != row:
                      x_new[row] -= A[row,column]*x[column]
              x_new[row] /= A[row,row]
          #check difference
          relative_change = np.linalg.norm(x_new-x)/np.linalg.norm(x_new)
          print("New guess is",x_new)
          print("Norm of change is",relative_change)
```

Data: Matrix **A**, vector **b**, vector $\mathbf{x}_{\text{old}}$ (initial guess), `tolerance`
Result: The solution **x** to $\mathbf{Ax} = \mathbf{b}$
$\mathbf{x} = 0$;
while $\|\mathbf{x} - \mathbf{x}_{\text{old}}\|_2 >$`tolerance` **do**
$\quad$ **for** $i \in [0,$ *number of rows of* **A**$]$ **do**
$\quad\quad$ **for** $j \in [0,$ *number of columns of* **A**$]$ **do**
$\quad\quad\quad$ **if** $i \neq j$ **then**
$\quad\quad\quad\quad$ $x_i = x_i - A_{ij}x_{\text{old }j}$;
$\quad\quad\quad$ **end**
$\quad\quad$ **end**
$\quad\quad$ $x_i = x_i/A_{ii}$;
$\quad$ **end**
$\quad$ Compute the change from $\mathbf{x}_{\text{old}}$ to $\mathbf{x}$;
$\quad$ $\mathbf{x}_{\text{old}} = \mathbf{x}$;
end

Algorithm 9.1. Jacobi Iterations

```
New guess is [ 0.752  1.008  1.008  0.752]
Norm of change is 0.13524314758
```

We could continue by hand, but it makes sense at this point to write out the general algorithm. The Jacobi method is written in pseudocode in Algorithm 9.1.

This pseudocode is converted to a Python implementation in the code below. One change between the pseudocode and the Python implementation is that the user is not required to input an initial guess: if the user does not provide an initial guess, the initial guess is set to a random vector. Also, in practice it may be the case that the method does not converge because, for instance, the matrix is not diagonally dominant. To make sure that the function returns something in this case, we set a maximum number of iterations as a user input with a default value of 100.

```python
In [5]: def JacobiSolve(A,b,x0=np.array([]),tol=1.0e-6,
                   max_iterations=100,LOUD=False,):
           """Solve a linear system by Jacobi iteration.
           Note: system must be diagonally dominant
           Args:
               A: N by N array
               b: array of length N
               x0: initial guess (if none given will be random)
               tol: Relative L2 norm tolerance for convergence
               max_iterations: maximum number of iterations
           Returns:
               The approximate solution to the linear system
           """
           [Nrow, Ncol] = A.shape
           assert Nrow == Ncol
           N = Nrow
           converged = False
```

```
            iteration = 1
            if (x0.size==0):
                #random initial guess
                x0 = np.random.rand(N)
            x = x0.copy()
            while not(converged):
                #replace old value
                x0 = x.copy()
                for row in range(N):
                    x[row] = b[row]
                    for column in range(N):
                        if column != row:
                            x[row] -= A[row,column]*x0[column]
                    x[row] /= A[row,row]
                relative_change = np.linalg.norm(x-x0)/np.linalg.norm(x)
                if (LOUD):
                    print("Iteration",iteration,
                        ": Relative Change =",relative_change)
                if (relative_change < tol) or (iteration >= max_iterations):
                    converged = True
                iteration += 1
            return x
```

We will try this function on our system from above. Only a portion of the output is shown because it takes tens of iterations to converge to the solution.

```
In [6]:  A = np.array([(2.5,-1,0,0),(-1,2.5,-1,0),(0,-1,2.5,-1),(0,0,-1,2.5)])
         print("Our matrix is\n",A)
         b = np.array([1.0,1,1,1])
         x = JacobiSolve(A,b,tol=1.0e-6,LOUD=True)
         print(x)

Our matrix is
 [[ 2.5 -1.   0.   0. ]
 [-1.   2.5 -1.   0. ]
 [ 0.  -1.   2.5 -1. ]
 [ 0.   0.  -1.   2.5]]
Iteration 1 : Relative Change = 1.0
Iteration 2 : Relative Change = 0.392232270276
Iteration 3 : Relative Change = 0.201990875657
...
Iteration 30 : Relative Change = 1.16661625519e-06
Iteration 31 : Relative Change = 7.55049332263e-07
[ 0.90908977  1.27272543  1.27272543  0.90908977]
```

Note that the iterations stopped when the change in the iterate was smaller than the tolerance 10^{-6}. It is a good idea to check the answer by comparing $\mathbf{Ax}$ to $\mathbf{b}$.

```
In [7]:  #Check the answer
         print("Ax =",np.dot(A,x))
         print("b =",b)

Ax = [ 0.999999    0.99999837  0.99999837  0.999999  ]
b = [ 1.  1.  1.  1.]
```

We see that the solution does indeed have a small residual. Our Jacobi method could be used on a much larger matrix, as in the next example:

```
In [8]:  N = 100
         A = np.zeros((N,N))
         b = np.ones(N)
         #same structure as before
         for i in range(N):
             A[i,i] = 2.5
             if (i>0):
                 A[i,i-1] = -1
             if (i < N-1):
                 A[i,i+1] = -1
         x100 = JacobiSolve(A,b,tol=1.0e-6,LOUD=True)
         print(x100)

Iteration 1 : Relative Change = 1.0
Iteration 2 : Relative Change = 0.44285152295
Iteration 3 : Relative Change = 0.260809758969
...
Iteration 54 : Relative Change = 1.36670150228e-06
Iteration 55 : Relative Change = 1.09236419158e-06
Iteration 56 : Relative Change = 8.73100528097e-07
[ 0.99999923  1.49999848  1.74999775  1.87499706  1.93749641  1.96874581
  1.98437027  1.99218229  1.99608811  1.99804087  1.99901712  1.99950515
...
  1.99804087  1.99608811  1.99218229  1.98437027  1.96874581  1.93749641
  1.87499706  1.74999775  1.49999848  0.99999923]
```

9.1.1 Convergence of the Jacobi Method

We can demonstrate graphically the convergence of the Jacobi method for a 2 by 2 system. This will help us compare different methods and how they converge. For Jacobi, and subsequent methods, we will solve the system

$$\begin{pmatrix} 2 & 1.9 \\ 1.9 & 4 \end{pmatrix} \begin{pmatrix} x_1 \\ x_2 \end{pmatrix} = \begin{pmatrix} 0.2 \\ -4.2 \end{pmatrix}.$$

The solution to this system is $x_1 = 2$ and $x_2 = -2$. We will start the Jacobi method with an initial guess of $\mathbf{x}^{(0)} = (-1, 0.5)^{\mathrm{T}}$. This time we will accompany the solution with a graph showing how the approximate solution for Jacobi changes with each iteration.

```
In [9]:  A = np.array([(2,1.9),(1.9,4)])
         solution = np.ones(2)*2
         solution[1] = -2
         b = np.dot(A,solution)
         x0 = np.array([-1,.5])
         xp, yp = JacobiSolve(A,b,x0=x0, LOUD=True)
         plt.plot(xp,yp,'o-',label='Jacobi')
         xpoint = np.linspace(-5,5,100)
         line1 = lambda x: (0.2 - 2*x)/1.9
```

```
line2 = lambda x: (-4.2 - 1.9*x)/4
plt.plot(xpoint,line1(xpoint),'-',color='black')
plt.plot(xpoint,line2(xpoint),'-',color='black')
plt.axis([-1.2,3,-3,2])
plt.legend(); plt.xlabel('x$_1$'); plt.ylabel('x$_2$')
plt.show()
```

```
Iteration 1 : Relative Change = 1.0
Iteration 2 : Relative Change = 0.643406164463
Iteration 3 : Relative Change = 0.244980153199
...
Iteration 33 : Relative Change = 1.10225320325e-06
Iteration 34 : Relative Change = 1.04360408096e-06
Iteration 35 : Relative Change = 4.97391183094e-07
```

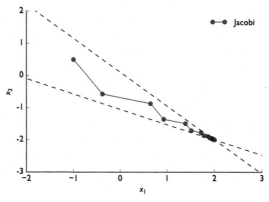

In this figure, the dashed lines are the two lines that represent our system of equations

$$y = -\frac{2}{1.9}x + \frac{0.2}{1.9},$$
$$y = -\frac{1.9}{4}x - \frac{4.2}{4}.$$

Where these lines intersect is the solution to our system. In this figure, there is one striking phenomenon we can see. The Jacobi solution bounces around in the area between the two lines. Because the approximation is always in this area and moves toward the solution, it will eventually get to the solution. Also, note that the initial steps are large, and the step size shrinks as the approximate solutions gets closer to the true answer. This is a common phenomenon for iterative methods. The behavior of Jacobi is similar when there are more equations than two, it just becomes harder to visualize the convergence.

9.1.2 Time to Solution for Jacobi Method

As we did with Gaussian elimination, we can time how long it takes to get a solution and see how that time scales with the number of unknowns by plotting the time to solution versus the number of equations on a log-log scale. The slope of this line (on the log-log scale) gives us the leading order behavior of the Jacobi method as the number of equations gets large.

Recall, for Gaussian elimination we observed a slope between 2 and 3.

```
In [10]: import time
         num_tests = 10
         I = 2**np.arange(num_tests)
         times = np.zeros(num_tests)
         for test in range(num_tests):
             N = I[test]
             A = np.zeros((N,N))
             b = np.ones(N)
             #same structure as before
             for i in range(N):
                 A[i,i] = 2.5
                 if (i>0):
                     A[i,i-1] = -1
                 if (i < N-1):
                     A[i,i+1] = -1
             start = time.clock()
             x = JacobiSolve(A,b)
             end = time.clock()
             times[test] = end-start
         plt.loglog(I,times,'o-')
         plt.xlabel("Number of Equations")
         plt.ylabel("Seconds to complete solve")
         plt.show()
         print("Approximate growth rate is n^",
             (np.log(times[test])-np.log(times[test-1]))/
             (np.log(I[test])-np.log(I[test-1])))
```

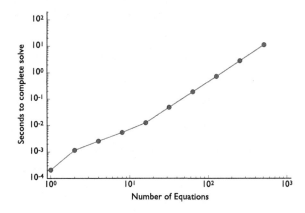

```
Approximate growth rate is n^ 2.00477104343
```

The slope of the line is 2, so that we can conclude that our implementation of Jacobi method requires $O(n^2)$ operations where n is the number of equations. This means the time to solution is growing more slowly than we saw for Gauss elimination, which has a theoretical growth rate of three or $O(n^3)$. This means that for a large enough system, we will expect Jacobi be faster than Gaussian elimination. However, Jacobi can only be applied to diagonally dominant systems and, therefore, is not as general purpose as Gaussian elimination. Another

Data: Matrix **A**, vector **b**, vector $\mathbf{x}_{\text{old}}$ (initial guess), `tolerance`
Result: The solution **x** to $\mathbf{Ax} = \mathbf{b}$
$\mathbf{x} = 0$;
while $|\mathbf{x} - \mathbf{x}_{\text{old}}|_2 >$ `tolerance` **do**
$\quad \mathbf{x}_{\text{old}} = \mathbf{x}$;
$\quad \mathbf{x} = (\mathbf{b} - \mathbf{Ax} + \text{diag}(\mathbf{A}) \cdot \mathbf{x})$;
end

Algorithm 9.2. Jacobi Algorithm without any `for` loops

difference between Jacobi and Gaussian elimination is that we cannot guarantee the number of iterations in a Jacobi solve, whereas with Gaussian elimination we know it takes one pass through the system to get a upper triangular form plus a back solve.

Just saying that Gaussian elimination is $O(n^3)$ and Jacobi is $O(n^2)$ does not mean that for a given problem Jacobi will be faster. The constant that multiples the leading order term could be large in one method making the comparison only valid as $n \to \infty$. For example, $n^3 < 1000n^2$ for $n < 1000$. Therefore, to compare two methods at a given system size requires knowing more than just the scaling. We will see this phenomenon concretely in the next section where we speed up Jacobi, but do not change its scaling as $n \to \infty$.

9.2 A FASTER JACOBI METHOD

With NumPy we can make a simpler Jacobi iteration that should also be faster by removing the inner two `for` loops. We can do this by using the fact that we can represent the Jacobi method using matrix-vector products, vector addition, and scalar division. To do this we notice that the inner update of a Jacobi iteration can be written as

$$x_i^{(\ell+1)} = \frac{1}{A_{ii}} \left(b_i - \sum_{j=1, i \neq j}^{I} A_{ij} x_j^{(\ell)} \right) = \frac{1}{A_{ii}} \left(b_i - \mathbf{a}_i \cdot \mathbf{x}^{(\ell)} + A_{ii} x_i^{(\ell)} \right),$$

where the ith row of the matrix is denoted as $\mathbf{a}_i$. Furthermore, we can write the entire update as a matrix vector product

$$\mathbf{x}^{(\ell+1)} = \frac{1}{\text{Diag}(\mathbf{A})} \left(\mathbf{b} - \mathbf{A} \cdot \mathbf{x}^{(\ell)} + \text{Diag}(\mathbf{A}) \cdot \mathbf{x}^{(\ell)} \right),$$

where the diagonal of the matrix is denoted as $\text{Diag}(\mathbf{A})$, and division by a vector is understood to be elementwise.

Therefore, instead of writing a `for` loop to perform the sum in the update and then using another `for` loop to look over the rows, we can use the optimized, built-in NumPy routine for a matrix vector product. The resulting function is not quite as easy to read, but the algorithm runs faster. We implement Algorithm 9.2 and test it out.

```
In [11]: def JacobiSolve_Short(A,b,x0=np.array([]),tol=1.0e-6,
                    max_iterations=100,LOUD=False):
             """Solve a linear system by Jacobi iteration.
             This implementation removes the for loops to make it faster
```

```
Note: system must be diagonally dominant
Args:
    A: N by N array
    b: array of length N
    tol: Relative L2 norm tolerance for convergence
    max_iterations: maximum number of iterations
Returns:
    The approximate solution to the linear system
"""
[Nrow, Ncol] = A.shape
assert Nrow == Ncol
N = Nrow
converged = False
iteration = 1
if (x0.size==0):
    #random initial guess
    x0 = np.random.rand(N)
x = x0.copy()
while not(converged):
    x0 = x.copy() #replace old value
    #update is (b - whole row * x + diagonal part * x)/diagonal
    x = (b - np.dot(A,x0)+ A.diagonal()*x0)/A.diagonal()
    relative_change = np.linalg.norm(x-x0)/np.linalg.norm(x)
    if (LOUD):
        print("Iteration", iteration,
                ": Relative Change =",relative_change)
    if (relative_change < tol) or (iteration >= max_iterations):
        converged = True
    iteration += 1
return x
```

A reasonable first check is that the new algorithm gives the same answer as before.

```
In [12]: N = 100
         A = np.zeros((N,N))
         b = np.ones(N)
         #same structure as before
         for i in range(N):
             A[i,i] = 2.5
             if (i>0):
                 A[i,i-1] = -1
             if (i < N-1):
                 A[i,i+1] = -1
         x100_Short = JacobiSolve_Short(A,b,tol=1.0e-6,LOUD=False)
         print(x100_Short-x100)
```

```
[  0.00000000e+00   2.22044605e-16   0.00000000e+00   4.44089210e-16
   0.00000000e+00  -4.44089210e-16  -4.44089210e-16   0.00000000e+00
 ...
   0.00000000e+00  -4.44089210e-16   0.00000000e+00   0.00000000e+00
   0.00000000e+00   0.00000000e+00   0.00000000e+00   0.00000000e+00]
```

Those are almost identical. We can also demonstrate that this algorithm is much faster than the previous one.

Below, we show results from the same timing study that we did for plain Jacobi. The results in the figure show that we can solve more than 1000 equations in less than 1 second. The calculated growth rate of the time to complete the solve is $n^{2.12}$, basically quadratic scaling in the number of equations.

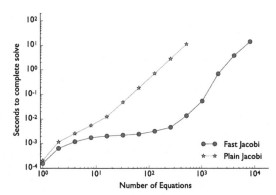

This curve is typical of the performance of an algorithm that is initially not filling up all the resources of the computer (fast memory called cache, for example). The bulk of the computation is spent on setting up the problem or moving memory around, two things that are not dependent on the number of floating point operations required. Eventually, once this set up time is small compared to the overall computation, we see that the scaling is about quadratic, which is better than the theoretical growth of Gaussian elimination. Despite having the same growth rate as the previous implementation, the absolute speed is much better.

One way to understand why the second implementation is faster is because the second has fewer `for` loops. When I have a `for` loop, the computer has to wait for each pass through the loop to complete before beginning the loop body again. With my "fast" implementation, all the work that needs to be done for an iteration is specified in a few lines using NumPy functions. This gives Python and NumPy the ability to keep the CPU of my computer completely filled up with tasks to do, instead of waiting for other tasks to finish and for memory to move around. As a result the time to solution is basically constant at less than 0.01 seconds until we get to about 256 equations.

9.3 GAUSS-SEIDEL

The Gauss-Seidel (GS) method is a twist on Jacobi iteration. The difference between the Jacobi and Gauss-Seidel method is that GS uses the most up to date information when performing an iteration. In equation form, Gauss-Seidel computes

$$x_i^{(\ell+1)} = \frac{1}{A_{ii}} \left(b_i - \sum_{j=1}^{i-1} A_{ij} x_j^{(\ell+1)} - \sum_{j=i+1}^{I} A_{ij} x_j^{(\ell)} \right), \qquad \text{for } i = 1 \dots I, \ \ell = 0 \dots. \qquad (9.1)$$

In particular, when the method is updating x_i, it can use the updates already computed for the previous equations to get a better approximation: these are the $j = 1 \dots i - 1$ terms on the right-hand side of Eq. (9.1). Because Gauss-Seidel is using the most up to date information,

Data: Matrix **A**, vector **b**, vector **x** (initial guess), tolerance

Result: The solution **x** to **Ax** = **b**

x_{old} = **0**;

while $\|x - x_{old}\|_2$ > tolerance **do**

    $x_i = b_i$;

    for $i \in [0, \text{number of rows of } \mathbf{A}]$ **do**

        for $j \in [0, \text{number of columns of } \mathbf{A}]$ **do**

            if $i \neq j$ **then**

                $x_i = x_i - A_{ij}x_j$;

            end

        end

        $x_i = x_i / A_{ii}$;

    end

    Compute the change from x_{old} to **x**;

    x_{old} = **x**;

end

Algorithm 9.3. Gauss-Seidel Iterations

by the time it gets to the last equation it is basically using information all at iteration $\ell + 1$. Therefore, we might expect Gauss-Seidel to converge in fewer iterations than Jacobi. That being said, it may not be faster than Jacobi if the iterations take longer to compute than a Jacobi iteration.

The algorithm for Gauss-Seidel is given in Algorithm 9.3. Notice that the only change is that x_{old} is only used to compute the change during an iteration and is not otherwise used in the update.

Below, we implement a standard Gauss-Seidel algorithm in Python:

```python
In [14]: def Gauss_Seidel_Solve(A,b,x0=np.array([]),tol=1.0e-6,
                                max_iterations=100,LOUD=False):
             """Solve a linear system by Gauss-Seidel iteration.
             Note: system must be diagonally dominant
             Args:
                 A: N by N array
                 b: array of length N
                 x0: initial guess (if none given will be random)
                 tol: Relative L2 norm tolerance for convergence
                 max_iterations: maximum number of iterations
             Returns:
                 The approximate solution to the linear system
             """
             [Nrow, Ncol] = A.shape
             assert Nrow == Ncol
             N = Nrow
             converged = False
             iteration = 1
             if (x0.size==0):
                 x0 = np.random.rand(N) #random initial guess
             x = x0.copy()
             while not(converged):
                 x0 = x.copy() #replace old value
```

```
    for row in range(N):
        x[row] = b[row]
        for column in range(N):
            if column != row:
                #use x in update
                x[row] -= A[row,column]*x[column]
        x[row] /= A[row,row]
    relative_change = np.linalg.norm(x-x0)/np.linalg.norm(x)
    if (LOUD):
        print("Iteration",iteration,
            ": Relative Change =",relative_change)
    if (relative_change < tol) or (iteration >= max_iterations):
        converged = True
    iteration += 1
return x
```

We now repeat the solution of a system that we previously solved using the Jacobi method, where it took 56 iterations. The code below performs the same test using Gauss-Seidel.

```
In [15]: N = 100
         A = np.zeros((N,N))
         b = np.ones(N)
         #same structure as before
         for i in range(N):
             A[i,i] = 2.5
             if (i>0):
                 A[i,i-1] = -1
             if (i < N-1):
                 A[i,i+1] = -1
         x100_GS = Gauss_Seidel_Solve(A,b,LOUD=True)
         print(x100_GS)
Iteration 1 : Relative Change = 1.0
Iteration 2 : Relative Change = 0.398660303888
Iteration 3 : Relative Change = 0.209236888361
...
Iteration 31 : Relative Change = 1.61012686252e-06
Iteration 32 : Relative Change = 1.07080293175e-06
Iteration 33 : Relative Change = 7.12128943502e-07
[ 0.99999896  1.49999827  1.74999781  1.8749975   1.9374973   1.96874716
  1.98437208  1.99218452  1.99609073  1.99804383  1.99902038  1.99950865
 ...
  1.99804605  1.99609309  1.99218699  1.98437462  1.96874972  1.93749981
  1.87499987  1.74999992  1.49999996  0.99999998]
```

The 33 iterations Gauss-Seidel required is quite a bit fewer than the 56 required by Jacobi (about 41% less). Using the most up-to-date information nearly cut the number of iterations in half. Compared to the Jacobi method, the complexity of the algorithm should be roughly the same per iteration because each iteration is solving a simple one-equation system for the next iteration's value for each unknown.

9.3.1 Convergence of Gauss-Seidel

We can compare the convergence between Jacobi and Gauss-Seidel graphically on the same 2 by 2 system as before and visually see how the iterations proceed. The figure below shows different behavior than the Jacobi method.

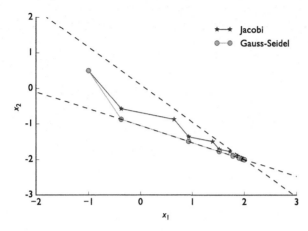

When we observe the convergence of Gauss-Seidel, we see that unlike Jacobi, it does not bounce between the lines. Rather it goes to one of the lines and then moves along that line to the solution. We can also see that Gauss-Seidel is "skipping" some of the Jacobi iterates, but getting to what looks like nearly the same points. For example, we can see that the third point for Gauss-Seidel is very near the fourth Jacobi point.

9.3.2 Time to Solution for Gauss-Seidel

We can do the same timing study we did for Jacobi. We expect Gauss-Seidel to be faster than the plain Jacobi (our first implementation), but probably slower than the fast Jacobi. We expect it to be faster than our simple Jacobi because it should take fewer iterations, and slower than fast Jacobi because there is not a way to implement Gauss-Seidel using matrix vector products, like we did in our fast Jacobi implementation and the reduction in iterations is not sufficient to overcome the additional cost per iteration. This is a visual demonstration of the benefit using up-to-date information in the iteration. In this test we observe a quadratic growth rate again.

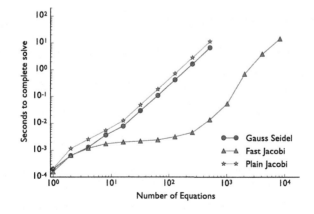

Gauss-Seidel, like our two Jacobi implementations, appears to be an $O(n^2)$ algorithm. Due to the fewer number of iterations, the absolute time to solution is faster than plain Jacobi, but it is slower than fast Jacobi because each iteration is more expensive.

Before moving on, it is worthwhile to discuss in more detail why Gauss-Seidel cannot be done in the same fast manner as Jacobi. This is because each equation in Jacobi is independent of another: to update $x_{10}^{(\ell)}$ does not depend on the update to $x_9^{(\ell)}$, for example. This independence in updates is not the case for Gauss-Seidel iterations: $x_{10}^{(\ell)}$ cannot be updated until $x_9^{(\ell)}$ is. This means that I cannot write the algorithm as briefly as I could Jacobi. Gauss-Seidel has an intra-iteration dependence that makes each row's unknown dependent on those that come before it. This dependence is the fundamental reason we do not have a fast Gauss-Seidel implementation.

All hope is not lost for Gauss-Seidel, however. One can do something fancy called red-black Gauss-Seidel where all the even unknowns (called the red unknowns) are updated, and then all the odd unknowns (the black unknowns) are updated. Given the structure of the tridiagonal matrix above, to do a Gauss-Seidel update on an even unknown only requires knowledge of odd unknowns, and vice-versa. The updates for red and black unknowns can be done independently. This can make for a fast algorithm, however it gets more complicated because you need to split your system into "red" and "black" parts. As a general algorithm, this can be much trickier to code. For a tridiagonal system, the coding is pretty straightforward and comprises an exercise at the end of the chapter.

9.4 SUCCESSIVE OVER-RELAXATION

Though it sounds like an injury you can get sitting in your easy chair, over-relaxation takes the update from a Gauss-Seidel iteration and moves the solution further in that direction, i.e., it over relaxes it. The basic idea is to combine the current iterate with the Gauss-Seidel calculation of the next iterate using a factor ω:

$$x_i^{(\ell+1)} = (1-\omega)x_i^{(\ell)} + \frac{\omega}{A_{ii}}\left(b_i - \sum_{j=1}^{i-1} A_{ij}x_j^{(\ell+1)} - \sum_{j=i+1}^{I} A_{ij}x_j^{(\ell)}\right), \quad \text{for } i = 1\ldots I, \ \ell = 0\ldots. \quad (9.2)$$

Clearly, if $\omega = 1$ then nothing has changed about the update. On the other hand, making $\omega > 1$ over-relaxes the system. In a sense it tries to take the correction to the current iteration and move the solution farther in that direction. The algorithm hardly changes at all. All we do is compute the update, and then compute the linear combination in Eq. (9.2).

```
In [18]: def SOR_Solve(A,b, x0= np.array([]),tol=1.0e-6,
                    omega=1,max_iterations=100,LOUD=False):
             """Solve a linear system by Gauss-Seidel iteration with SOR.
             Note: system must be diagonally dominant
             Args:
                 A: N by N array
                 b: array of length N
                 x0: initial guess (if none given will be random)
                 tol: Relative L2 norm tolerance for convergence
```

```
    omega: the over-relaxation parameter
    max_iterations: maximum number of iterations
Returns:
    The approximate solution to the linear system
"""
[Nrow, Ncol] = A.shape
assert Nrow == Ncol
N = Nrow
converged = False
iteration = 1
if (x0.size==0):
    #random initial guess
    x0 = np.random.rand(N)
x = x0.copy()
while not(converged):
    x0 = x.copy() #replace old value
    for row in range(N):
        x[row] = b[row]
        for column in range(N):
            if column != row:
                x[row] -= A[row,column]*x[column]
        x[row] /= A[row,row]
        x[row] = (1.0-omega) * x0[row] + omega*x[row]
    relative_change = np.linalg.norm(x-x0)/np.linalg.norm(x)
    if (LOUD):
        print("Iteration",iteration,
            ": Relative Change =",relative_change)
    if (relative_change < tol) or (iteration >= max_iterations):
        converged = True
    iteration += 1
return x
```

Using our standard example of 100 tri-diagonal equations, we can get a reduction in the number of iterations using $\omega = 1.2$:

```
In [19]: N = 100
         A = np.zeros((N,N))
         b = np.ones(N)
         #same structure as before
         for i in range(N):
             A[i,i] = 2.5
             if (i>0):
                 A[i,i-1] = -1
             if (i < N-1):
                 A[i,i+1] = -1
         x100_GS11 = SOR_Solve(A,b,omega=1.2,LOUD=True)

Iteration 1 : Relative Change = 1.0
Iteration 2 : Relative Change = 0.348994473141
Iteration 3 : Relative Change = 0.15758242743
...
Iteration 20 : Relative Change = 3.29823212222e-06
Iteration 21 : Relative Change = 1.76671341242e-06
Iteration 22 : Relative Change = 9.46306176483e-07
```

That saved us 11 iterations: Gauss-Seidel required 33 compared with 22 here. If we tweak ω some more, we get that $\omega \approx 1.3$ is the best value of ω, leading to 19 iterations.

9.4.1 Convergence of SOR

On our 2 by 2 system, we expect that the graphical convergence of SOR should look similar to that of Gauss-Seidel, but exaggerated because SOR is taking the Gauss-Seidel update and moving further in that direction. The figure below compares an SOR solution with a tuned value of ω with an SOR solution using too large a value of ω.

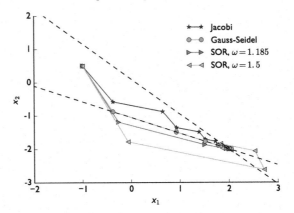

With SOR the convergence with a tuned value of ω, in this case $\omega = 1.185$, is similar to Gauss-Seidel, but faster. This happens because SOR allows the solution to go off the dashed lines during convergence to get a better approximation. Indeed, after the first iteration, the iterates zoom into the solution quickly. There can be too much of good thing, however. When the value of ω is too large, in this case $\omega = 1.5$, the approximation overshoots the true solution, before coming back. This leads to the solution requiring more iterations than Gauss-Seidel.

9.4.2 Time to Solution for SOR

The results from the timing study on SOR is given next. In this figure we observe the same trend as the standard Jacobi, and Gauss-Seidel methods.

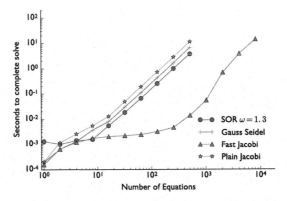

Once again, we get a little faster because we saved on some iterations, also the growth rate shrunk mildly. It is worth noting at this point that the improvement from plain Jacobi to Gauss-Seidel to SOR are all numerical improvements: that is the numerical method improved resulting in fewer iterations. The speed increase going to Fast Jacobi is due to an implementation improvement: the fast implementation uses the computer's resources better.

One drawback with SOR is that it is not generally possible to determine ahead of time what the appropriate value of ω is. Experience on similar systems is generally required to determine the value of ω. As a result, it may be more expensive to determine ω properly, by trial and error, than the savings it gives on a similar system. In the timing study above, the value of ω used was determined using this trial and error approach.

9.5 CONJUGATE GRADIENT

The conjugate gradient method will work on systems where the matrix $\mathbf{A}$ is symmetric, and positive definite (it does not need to be diagonally dominant in this case). Positive definite means that for any $\mathbf{x}$ that is not all zeros,

$$\mathbf{x}^T \mathbf{A} \mathbf{x} > 0.$$

We will need this fact when deriving the method because we will divide by $\mathbf{x}^T \mathbf{A} \mathbf{x}$. We will also need the definition of the residual for iteration ℓ, $\mathbf{r}^{(\ell)}$, given by

$$\mathbf{r}^{(\ell)} = \mathbf{b} - \mathbf{A} \mathbf{x}^{(\ell)}.$$

To derive the method consider the function of $\mathbf{x}$:

$$f(\mathbf{x}) = \frac{1}{2} \mathbf{x}^T \mathbf{A} \mathbf{x} - \mathbf{b}^T \mathbf{x}, \tag{9.3}$$

the minimum of this function occurs when the gradient of $f(\mathbf{x})$ with respect to $\mathbf{x}$ is 0 or when

$$\mathbf{A} \mathbf{x} = \mathbf{b}.$$

Therefore, we can derive an iterative method by computing $f(\mathbf{x}^{(0)})$ for an initial guess and refining the solution by decreasing the value of $f(\mathbf{x}^{(1)})$ and continuing on. In particular we write

$$\mathbf{x}^{(\ell+1)} = \mathbf{x}^{(\ell)} + \alpha_\ell \mathbf{s}^{(\ell)}.$$

The vector $\mathbf{s}$ is the search direction, and α_ℓ is the step length. All that we have said up to this point is that at each iteration we are moving the solution in a particular direction $\mathbf{s}^{(\ell)}$ by an amount α_ℓ. We would like to pick α_ℓ so that the error after step $\ell + 1$ the function is at a minimum along direction $\mathbf{s}^{(\ell)}$. To accomplish this we take the derivative of $f(\mathbf{x}^{(\ell+1)})$ with respect to α_ℓ and set it to zero,

$$\frac{\partial}{\partial \alpha_\ell} f(\mathbf{x}^{(\ell+1)}) = \left(\frac{\partial}{\partial \mathbf{x}^{(\ell+1)}} f(\mathbf{x}^{(\ell+1)}) \right)^T \frac{\partial}{\partial \alpha_\ell} (\mathbf{x}^{(\ell+1)})$$

$$= \left(\mathbf{r}^{(\ell+1)}\right)^{\mathrm{T}} \mathbf{s}^{(\ell)}$$

$$= -\left(\mathbf{A}\left(\mathbf{x}^{(\ell)} + \alpha_\ell \mathbf{s}^{(\ell)}\right) - \mathbf{b}\right)^{\mathrm{T}} \mathbf{s}^{(\ell)}$$

$$= \left(\mathbf{r}^{(\ell)} - \mathbf{A}\alpha_\ell \mathbf{s}^{(\ell)}\right)^{\mathrm{T}} \mathbf{s}^{(\ell)}$$

$$= 0.$$

When we solve this for α_ℓ, we get,

$$\alpha_\ell = \frac{\left(\mathbf{r}^{(\ell)}\right)^{\mathrm{T}} \mathbf{s}^{(\ell)}}{\left(\mathbf{s}^{(\ell)}\right)^{\mathrm{T}} \mathbf{A}\mathbf{s}^{(\ell)}}. \tag{9.4}$$

We still have not specified the search direction.

The power of the conjugate gradient method is in how it selects the search direction. To understand this we will need to understand what it means for two vectors to be conjugate. Two vectors, $\mathbf{u}$ and $\mathbf{v}$, are conjugate with respect to a matrix $\mathbf{A}$ if

$$\mathbf{u}^{\mathrm{T}}\mathbf{A}\mathbf{v} = 0.$$

It is true that if $\mathbf{u}$ is conjugate to $\mathbf{v}$, then $\mathbf{v}$ is conjugate to $\mathbf{u}$. The property of conjugacy is related to orthogonality of two vectors. If two vectors are orthogonal then,

$$\mathbf{u} \cdot \mathbf{v} = \mathbf{u}^{\mathrm{T}}\mathbf{v} = 0.$$

When two vectors are conjugate, they are orthogonal after one is multiplied by $\mathbf{A}$.

Conjugate gradient seeks search directions that are conjugate to all the previous search directions. What this means is that every search direction is orthogonal to each of the previous directions when multiplied by $\mathbf{A}$. That is, we do not step in the same direction multiple times. To do this we want to write the search direction for the $\ell + 1$ step as a linear combination of the residual plus a constant times the previous step as

$$\mathbf{s}^{(\ell+1)} = \mathbf{r}^{(\ell+1)} + \beta_l \mathbf{s}^{(\ell)}. \tag{9.5}$$

We want step $\ell + 1$ to be conjugate to the previous step so that $(\mathbf{s}^{(\ell)})^{\mathrm{T}}\mathbf{A}\mathbf{s}^{(\ell+1)} = 0$. We want to pick β_ℓ so that this is the case. We will multiply Eq. (9.5) by $(\mathbf{s}^{(\ell)})^{\mathrm{T}}\mathbf{A}$ and set the result to zero to get

$$\left(\mathbf{s}^{(\ell)}\right)^{\mathrm{T}} \mathbf{A}\mathbf{r}^{(\ell+1)} + \beta_\ell \left(\mathbf{s}^{(\ell)}\right)^{\mathrm{T}} \mathbf{A}\mathbf{s}^{(\ell)} = 0. \tag{9.6}$$

Solving for β_ℓ gives

$$\beta_\ell = -\frac{\left(\mathbf{r}^{(\ell+1)}\right)^{\mathrm{T}} \mathbf{A}\mathbf{s}^{(\ell)}}{\left(\mathbf{s}^{(\ell)}\right)^{\mathrm{T}} \mathbf{A}\mathbf{s}^{(\ell)}}. \tag{9.7}$$

This process that will ensure that every subsequent search direction is conjugate to all previous search directions. To set the initial search direction, we set $\mathbf{s}^{(0)} = \mathbf{r}^{(0)}$. This also bounds

Data: Matrix **A**, vector **b**, vector **x** (initial guess)
Result: The solution **x** to **Ax** = **b**
residual = **b** − **Ax**;
s = residual;
while *change in* **x** *is small* **do**
    | compute α defined by Eq. (9.4);
    | **x** = **x** + α**s**;
    | residual = **b** − **Ax**;
    | compute β defined by Eq. (9.7);
    | compute **s** defined by Eq. (9.5);
end

Algorithm 9.4. Conjugate Gradient Algorithm

the number of iterations in the solution. For an N by N matrix there are at most N mutually conjugate vectors. This means that there are at most N iterations in a CG solve because each step is in a conjugate direction *and* we minimize the error along each step.

We have talked through the mathematics of how conjugate gradient works, we write the method in pseudocode in Algorithm 9.4.

Notice that the CG algorithm is expressed entirely in matrix-vector products and vector addition/subtraction. This algorithm for conjugate gradient is implemented in Python below.

```
In [22]: def CG(A,b, x= np.array([]),tol=1.0e-6,
             max_iterations=100,LOUD=False):
         """Solve a linear system by Conjugate Gradient
         Note: system must be SPD
         Args:
             A: N by N array
             b: array of length N
             x0: initial guess (if none given will be random)
             tol: Relative L2 norm tolerance for convergence
             max_iterations: maximum number of iterations
         Returns:
             The approximate solution to the linear system
         """
         [Nrow, Ncol] = A.shape
         assert Nrow == Ncol
         N = Nrow
         converged = False
         iteration = 1
         if (x.size==0):
             #random initial guess
             x = np.random.rand(N)
         r = b - np.dot(A,x)
         s = r.copy()
         while not(converged):
             denom = np.dot(s, np.dot(A,s))
             alpha = np.dot(s,r)/denom
             x = x + alpha*s
```

```
        r = b - np.dot(A,x)
        beta = - np.dot(r,np.dot(A,s))/denom
        s = r + beta * s
        relative_change = np.linalg.norm(r)
        if (LOUD):
            print("Iteration",iteration,
                  ": Relative Change =",relative_change)
        if (relative_change < tol) or (iteration >= max_iterations):
            converged = True
        iteration += 1
    return x
```

The results from our standard example are below.

```
In [23]: N = 100
         A = np.zeros((N,N))
         b = np.ones(N)
         #same structure as before
         for i in range(N):
             A[i,i] = 2.5
             if (i>0):
                 A[i,i-1] = -1
             if (i < N-1):
                 A[i,i+1] = -1
         x100_CG = CG(A,b,LOUD=True)


Iteration 1 : Relative Change = 8.2630468612
Iteration 2 : Relative Change = 5.11537771495
Iteration 3 : Relative Change = 1.90670704394
...
Iteration 22 : Relative Change = 2.34099376734e-06
Iteration 23 : Relative Change = 1.1809715533e-06
Iteration 24 : Relative Change = 5.42003345606e-07
```

A couple of things to note. The number of iterations of this algorithm is about the same as Gauss-Seidel with SOR, in other words it converges faster than Jacobi. Also, there are no `for` loops in the algorithm, it is all matrix-vector operations. Therefore, it should be competitive with our fast implementation of Jacobi in terms of speed.

9.5.1 Convergence of CG

On our 2 by 2 system, we can compare the graphical convergence of conjugate gradient relative to the other methods. Given that our search directions will be conjugate, we expect that the solution should converge in two iterations because there are not more than two conjugate directions in a two-dimensional space. After the first iteration, the conjugate gradient approximate moves directly to the solution, as predicted.

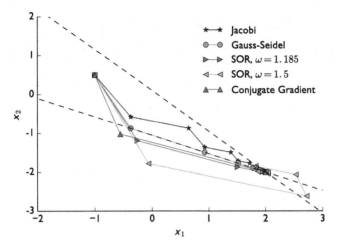

9.5.2 Time to Solution for CG

We now replicate the timing study we did before for Jacobi, Gauss-Seidel, and SOR for CG. Given that the number of iterations was similar to SOR for our $N = 100$ system, we expect the scaling for CG to be similar to SOR. However, the CG algorithm is expressed in terms of matrix-vector product so we expect the time to solution should exhibit the behavior we saw with the fast Jacobi method: for small systems the time to solution will be roughly constant, and as the system gets larger the time to solution should scale as $O(N^2)$.

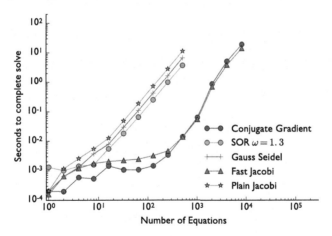

In these results the time to solution for conjugate gradient is roughly constant until there are hundreds of equations. Past this point, the time to solution grows roughly as the number of equations squared. In the end the time to solution is roughly equal to that from Jacobi. Given that the two methods have a similar time to solution, there are still reasons to favor conjugate gradient: the matrix need not be diagonally dominant and it is guaranteed to converge in N iterations.

9.6 TAKING ADVANTAGE OF TRI-DIAGONAL FORM

In the example problems above we have been storing the whole N by N matrix, when really we only need about $3N$ numbers because the rest are zeros. We can do this by reformulating our matrix to be N by 3 and putting the matrix elements in the appropriate place as done here:

```
In [26]:  N = 10
          A_tri = np.zeros((N,3))
          b = np.ones(N)
          #same structure as before
          #but fill it more easily
          A_tri[:,1] = 2.5 #middle column is diagonal
          A_tri[:,0] = -1.0 #left column is left of diagonal
          A_tri[:,2] = -1.0 #right column is right of diagonal
          A_tri[0,0] = 0 #remove left column in first row
          A_tri[N-1,2] = 0 #remove right column in last row
          print("Our matrix is\n",A_tri)
```

```
Our matrix is
 [[ 0.   2.5 -1. ]
 [-1.   2.5 -1. ]
 [-1.   2.5 -1. ]
 [-1.   2.5 -1. ]
 [-1.   2.5 -1. ]
 [-1.   2.5 -1. ]
 [-1.   2.5 -1. ]
 [-1.   2.5 -1. ]
 [-1.   2.5 -1. ]
 [-1.   2.5  0. ]]
```

This is a much smaller matrix, but to use it we need to define special algorithms. We will do this for Jacobi because it is the simplest to show here. The modification for Gauss-Seidel, SOR, or conjugate gradient would also be possible (in fact implementing Gauss-Seidel is an exercise at the end of this chapter). The code below is our Jacobi method for a tri-diagonal system. In the code, the major change is that the main diagonal is always in column 1 of a row and the off diagonals are in columns 0 and 2 respectively.

```
In [27]: def JacobiTri(A,b,tol=1.0e-6,max_iterations=100,LOUD=False):
            """Solve a tridiagonal system by Jacobi iteration.
            Note: system must be diagonally dominant
            Args:
                A: N by 3 array
                b: array of length N
                tol: Relative L2 norm tolerance for convergence
                max_iterations: maximum number of iterations
            Returns:
                The approximate solution to the linear system
            """
            [Nrow, Ncol] = A.shape
            assert 3 == Ncol
```

```
N = Nrow
converged = False
iteration = 1
x0 = np.random.rand(N) #random initial guess
x = np.zeros(N)
while not(converged):
    x0 = x.copy() #replace old value
    for i in range(1,N-1):
        x[i] = (b[i] - A[i,0]*x0[i-1] - A[i,2]*x0[i+1])/A[i,1]
    i = 0
    x[0] = (b[i]  - A[i,2]*x0[i+1])/A[i,1]
    i = N-1
    x[i] = (b[i] - A[i,0]*x0[i-1])/A[i,1]
    relative_change = (np.linalg.norm(x-x0)/
                       np.linalg.norm(x))
    if (LOUD):
        print("Iteration",iteration,
              ": Relative Change =",relative_change)
    if (relative_change < tol) or (iteration >= max_iterations):
        converged = True
    iteration += 1
return x_new
```

The results from our timing study indicate that the rate of growth in the solution time is linear, rather than quadratic. This is because the number of elements in the system is growing linearly in n, whereas the elements in a full matrix grows as the number of equations squared.

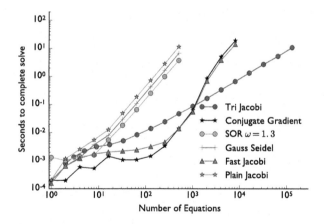

When the solution time is only growing linearly, we can run matrices as large as 100,000 by 100,000 in as little as 10 seconds. This is about 3 orders of magnitude improvement on a simple problem. This last figure tells makes three key points that are common in numerical methods:

1. The number of iterations is not necessarily as important as the speed at which the iterations are performed (compare SOR with fast Jacobi).

2. Eventually, the method that grows more slowly will be more efficient. This is seen in the fact that eventually the tridiagonal Jacobi method is the fastest because it grows linearly in the number of equations whereas the other methods grow quadratically in the number of equations.

This will be our last chapter on linear solvers. In the future we will use them to solve a variety of problems. Linear solvers also form the basis for several other numerical methods: nonlinear solvers, eigenvalue solvers, and discretizations for initial and boundary value problems. The fact that linear solvers are the foundation for so many other topics in scientific computing is one of the motivations for starting with that topic.

FURTHER READING

There are a variety of iterative methods that can be used to solve linear systems that extend beyond those discussed here. The classic reference for these methods is Saad's treatment aimed at *sparse* linear systems [12], i.e., systems where the matrix has many zero entries, such as a tridiagonal matrix. Trefethen and Bau also provide detailed discussion of iterative methods [9].

PROBLEMS

Short Exercises

9.1. Write a Python function called `isSymmetric` which takes a single parameter A and checks if the NumPy matrix A is symmetric. It should return 1 if the matrix is symmetric, and 0 if the matrix is non-symmetric.

9.2. Write a Python function called `isDiagonallyDominant` which takes a single parameter A and checks if the NumPy matrix A is diagonally dominant. It should return 1 if the matrix is diagonally dominant, and 0 if the matrix is not diagonally dominant.

Programming Projects

1. Exiting Gracefully

The Jacobi and Gauss-Seidel implementations in this chapter have a maximum number of iterations before they return a solution. They do not, however, tell the user that the maximum number of iterations was reached. Modify the Jacobi implementation given above to alert the user that the maximum number of iterations has been reached. You can do this by inserting a `print` statement that is executed or by using an `assert` statement.

2. Tri-diagonal Gauss-Seidel

Write a Gauss-Seidel solver for tri-diagonal matrices. The implementation should take as input a tri-diagonal matrix just as the tri-diagonal Jacobi defined previously. Test your implementations on the same timing study performed above in Section 9.6. Comment on the results.

```
import numpy as np
import math
delta = 0.05;
L = 1.0;
k = 0.001;
ndim = round(L/delta)
nCells = ndim*ndim;
A = np.zeros((nCells,nCells));
b = np.zeros(nCells)
#save us some work for later
idelta2 = 1.0/(delta*delta);

#now fill in A and b
for cellVar in range(nCells):
    xCell = cellVar % ndim; #x % y means x modulo y
    yCell = (cellVar-xCell)/ndim;
    xVal = xCell*delta + 0.5*delta;
    yVal = yCell*delta + 0.5*delta;
    #put source only in the middle of the problem
    if ( ( math.fabs(xVal - L*0.5) < .25*L) and
         ( math.fabs(yVal - L*0.5) < .25*L) ):
        b[cellVar] = 1;
    #end if

    A[cellVar,cellVar] = 4.0*k*idelta2;

    if (xCell > 0):
      A[cellVar,ndim*yCell + xCell -1] = -k*idelta2;
    if (xCell < ndim-1):
      A[cellVar,ndim*yCell + xCell + 1] = -k*idelta2;
    if (yCell > 0):
      A[cellVar,ndim*(yCell-1) + xCell] = -k*idelta2;
    if (yCell < ndim-1):
      A[cellVar,ndim*(yCell+1) + xCell] = -k*idelta2;

if (nCells <= 20):
    #print the matrix
    print("The A matrix in Ax = b is\n",A)

    #print the righthand side
    print("The RHS is",b)

x, residual = CG(A,b,LOUD=True,max_iterations=1000)
```

Algorithm 9.5. 2-D Heat Equation Code

3. 2-D Heat Equation

Below is a program in Algorithm 9.5, which builds a matrix and righthand side for a heat conduction problem in 2-D. The discretization of the 2-D heat equation gives a linear system $\mathbf{Ax} = \mathbf{b}$ where the solution vector $\mathbf{x}$ is the temperature at the grid points of the 2-D domain. In particular

1. The 2-D heat equation in a homogeneous material of constant conductivity k with a uniform volumetric heat source q and zero-temperature conditions on the boundary of the rectangular domain of length L_x and width L_y is

$$-k\nabla^2 T = q, \qquad \text{for } x \in [0, L_x] \quad y \in [0, L_y].$$

With the boundary condition

$$T(x, y) = 0 \qquad \text{for } x, y \text{ on the boundary.}$$

2. The 2-D Laplacian is discretized using a mathematical technique known as finite differences (which we will see later on).
3. As a result of the spatial discretization, the heat equation forms a linear system of the form $\mathbf{Ax} = \mathbf{b}$. The code below forms this matrix.
4. The size of this matrix is `nCells` is determined by the value of Δ (the distance between points at which we want to evaluate the temperature), and the size of L_x and L_y. In the code $L_x = L_y$ and this value is called L. The total number of values in each direction is `ndim`$= L/\Delta$ leading to a total number of unknowns is `nCells` $=$ `ndim` $\star$ `ndim`.

Your work:

1. Look at Algorithm 9.5. It is a working program except that the conjugate gradient function has been deleted from it. To make it run you will have to add in a conjugate gradient solver. The conjugate gradient solver given in class should return a vector containing the residual at each iteration.
2. Plot the logarithm of the 2-norm of the residual, $\|\mathbf{Ax}_i - \mathbf{b}\|_2$ error versus the iteration number.
3. The solution vector $\mathbf{x}$ contains the `nCells` $=$ `ndim` $\star$ `ndim` unknowns. The x and y positions corresponding to a given row of the matrix are defined in the lines that assign `xVal` and `yVal`. Plot the 2-D temperature distribution in the two following cases:
   a. A very coarse grid (small number of grid points)
   b. A fine grid (high number of grid points).

Islands in the stream
That is what we are
No one in between
How can we be wrong?

–"Islands in the Stream" by Dolly Parton and Kenny Rogers

CHAPTER POINTS

- Polynomials can be used to approximate an unknown function between known values.

- High-degree polynomial interpolation is subject to oscillations known as the Runge phenomenon.

- Spline interpolants can give accurate and smooth interpolating functions when many points of the function are known.

In this chapter we will look at how we can take functions evaluated at particular points and fit polynomials to them. The points which we know the function could be from a table of values, measurements, etc. The process of creating a function that fits observed data is called inter-

173

polation: the interpolation function will pass through the given points and make it possible to give values between the known data.

Polynomials are useful for interpolation because they are easy to evaluate and can be readily generated. As we will see, high-degree polynomials are not always the best choice, but a set of low-degree polynomials can be quite useful when we have many data points.

10.1 POLYNOMIALS

We denote a polynomial that is of degree n as $P_n(x)$. A polynomial of degree n has $n + 1$ coefficients and is written as

$$P_n(x) = a_0 + a_1 x + a_2 x^2 + \cdots + a_n x^n = \sum_{i=0}^{n} a_i x^i.$$

It is possible to evaluate polynomials efficiently because there is a natural recursion. To compute the value of x^n we need to compute x^{n-1}, x^{n-2}, ..., x. Therefore, if we compute a term at a time, we only have to perform n multiplications to compute the values $\{x^n, x^{n-1}, \ldots, x\}$. Below is a function that evaluates polynomials in this efficient manner.

```
In [1]: def polynomial(a,x):
            """Evaluate a general 1-D polynomial at point
            Args:
                a: array of the n+1 coefficients of a polynomial
                x: the point to evaluate the polynomial at
            Returns:
                Pn(x)
            """
            num_coefficients = a.size
            answer = a[0]
            xpower = 1
            for i in range(1,num_coefficients):
                #the next power of x is x*previous power
                xpower *= x
                answer += a[i]*xpower
            return answer
```

The total number of floating point operations in this calculation is comprised of $2n$ multiplications (n from computing x^n and n more from the coefficient multiplications) and n additions. Testing out this function reveals that in practice there is basically no difference in the time it takes to evaluate a degree 10 polynomial and a degree 10^6 polynomial.

To demonstrate how this function works, we will test it on a simple quadratic polynomial:

$$P_2(x) = (x - 1)(x + 1) = x^2 - 1.$$

```
In [2]: a = np.array([-1,0,1])
        X_points = np.linspace(-3,3,100)
        y = polynomial(a,X_points)
```

```
plt.plot(X_points,y)
plt.xlabel("x")
plt.ylabel("y")
plt.show()
```

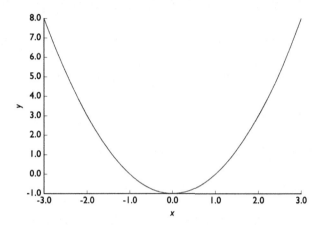

A more complicated polynomial can also be evaluated. Next we evaluate a quintic, or fifth degree, polynomial with random coefficients.

```
In [4]: poly_degree = 5
        a = np.random.uniform(-2,2,poly_degree + 1)#+1 because of 0
        y = polynomial(a,X_points)
        plt.plot(X_points,y)
        plt.xlabel("x")
        plt.ylabel("y")
        plt.show()
```

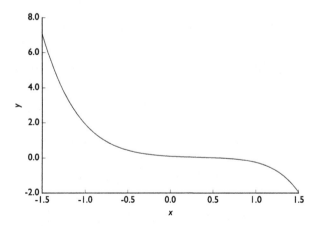

Polynomials have the property that any function can be approximated by a polynomial to any desired degree of accuracy. This result, known as the Weierstrass approximation theorem, is written as

Weierstrass approximation theorem. *For any function $f(x)$ defined on the interval $a \le x \le b$, there exists a degree n polynomial with n finite where*

$$|P_n(x) - f(x)| < \epsilon,$$

for all $x \in [a, b]$ and any $\epsilon > 0$.

Therefore, any function you name, we can approximate it as well as you like with a polynomial inside a given interval.

BOX 10.1 NUMERICAL PRINCIPLE

The Weierstrass approximation theorem states that we can approximate any function well using polynomials. It does not state what degree the polynomial needs to be to make a good approximation.

In addition to the approximation properties of polynomials, these functions are also unique. For a given set of $n + 1$ pairs of points $[x_i, f(x_i)]$, there is only one polynomial of degree n that passes through all those points. This makes sense because a polynomial of degree n has $n + 1$ coefficients. Therefore, the $n + 1$ data points define a unique polynomial. We will now discuss a method to compute the coefficients of polynomials given data.

10.2 LAGRANGE POLYNOMIALS

Lagrange polynomials are the simplest way to interpolate a set of points. This approach is not necessarily the most efficient for generating polynomial interpolating functions, but the difference is minimal for most applications. Regardless of the method used to compute the polynomial, the polynomial coefficients will be the same due to the uniqueness of interpolating polynomials.

The equations to construct a linear Lagrange polynomial are straightforward. Given points a_0 and a_1 and $f(a_0)$ and $f(a_1)$, the linear Lagrange polynomial formula is

$$P_1(x) = \frac{x - a_1}{a_0 - a_1} f(a_0) + \frac{x - a_0}{a_1 - a_0} f(a_1).$$

It is clear that $P_1(a_0) = f(a_0)$ and $P_1(a_1) = f(a_1)$ because in each case one of the numerators goes to 0. Also, it is clear that the function is a linear polynomial. We can translate this formula for linear interpolation into a function for Python:

```
In [5]: def linear_interp(a,f,x):
            """Compute linear interpolant
            Args:
                a: array of the 2 points
                f: array of the value of f(a) at the 2 points
```

```
Returns:
    The value of the linear interpolant at x
"""
assert a.size == f.size
answer = (x-a[1])/(a[0]-a[1])*f[0] + (x-a[0])/(a[1]-a[0])*f[1]
return answer
```

We can check this on an example when the function is $\sin(x)$ and we interpolate $x = 0$ and $x = \pi/2$.

```
In [6]: a = np.array([0,np.pi*0.5])
        f = np.sin(a)
        x = np.linspace(-0.05*np.pi,0.55*np.pi,200)
        y = linear_interp(a,f,x)
        plt.plot(x,y,linestyle="--",label="Linear Interpolant")
        plt.plot(x,np.sin(x),",label="$\sin(x)$")
        plt.scatter(a,f,c="black",label="Data Points")
        plt.xlabel("x")
        plt.ylabel("y")
        plt.legend(loc="best")
        plt.show()
```

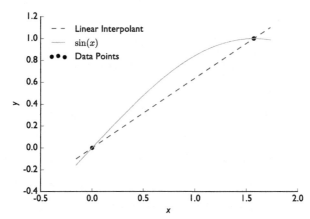

Note how the line passes through the two points we passed to the function. The line captures the overall trend of $\sin(x)$ between the two points, but we might want a better approximation.

The formulation of the quadratic Lagrange polynomial is also straightforward. Given points a_0, a_1, and a_2 and $f(a_0)$, $f(a_1)$, and $f(a_2)$, the quadratic Lagrange polynomial formula is

$$P_2(x) = \frac{(x-a_1)(x-a_2)}{(a_0-a_1)(a_0-a_2)} f(a_0) + \frac{(x-a_0)(x-a_2)}{(a_1-a_0)(a_1-a_2)} f(a_1)$$
$$+ \frac{(x-a_0)(x-a_1)}{(a_2-a_0)(a_2-a_1)} f(a_2).$$

This formula now has three terms that each have the product of two linear polynomials. Similar to the linear Lagrange polynomial, the numerator vanishes on two of the terms when we evaluate the formula at one of the data points. A function to implement the quadratic Lagrange polynomial is given next.

```
In [7]: def quadratic_interp(a,f,x):
            """Compute the quadratic interpolant
            Args:
                a: array of the 3 points
                f: array of the value of f(a) at the 3 points
            Returns:
                The value of the quadratic interpolant at x
            """
            answer = (x-a[1])*(x-a[2])/(a[0]-a[1])/(a[0]-a[2])*f[0]
            answer += (x-a[0])*(x-a[2])/(a[1]-a[0])/(a[1]-a[2])*f[1]
            answer += (x-a[0])*(x-a[1])/(a[2]-a[0])/(a[2]-a[1])*f[2]
            return answer
```

We will do the same test as before on $\sin(x)$, the point we add is in the middle and compare our result to the linear interpolant.

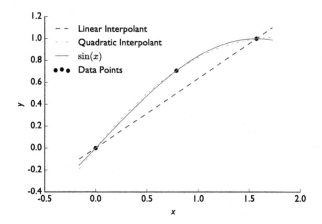

Notice that the quadratic interpolant does a much better job than a linear function. Part of this is due to the fact that by adding the point in the middle we have decreased the distance between known values of the function. Moreover, between the known data points, the curvature of the quadratic interpolant follows the original function more closely.

Comparing the linear and quadratic Lagrange interpolation formulas, we can begin to see a pattern in the Lagrange interpolation formulas. Each term has in the numerator the product of $(x - a_i)$ where the a_i's are different than the point we evaluate $f(x)$ at. Also, the denominator is the product of the a point minus each other a point. This leads us to the general Lagrange interpolation formula:

$$P_n(x) = \sum_{i=0}^{n} f(a_i) \frac{\prod_{j=1, j\neq i}^{n}(x - x_j)}{\prod_{j=1, j\neq i}^{n}(x_i - x_j)},$$

Data: Degree: n, Points: $a_0, a_1, \ldots, a_n$, Function values: $f(a_0), f(a_1), \ldots, f(a_n)$,
     Evaluation Point: x
Result: The value of the nth degree Lagrange interpolant at point x
$\mathsf{answer} = 0$;
for $i \in [0, n]$ **do**
$\quad | \quad \mathsf{product} = 1$;
$\quad |$ **for** $j \in [0, n]$ **do**
$\quad | \quad | \quad$ **if** $i \neq j$ **then**
$\quad | \quad | \quad | \quad \mathsf{product} = (\mathsf{product}) \times \frac{x - a_j}{a_i - a_j}$;
$\quad | \quad | \quad$ **end**
$\quad |$ **end**
$\quad | \quad \mathsf{answer} = \mathsf{answer} + (\mathsf{product}) \times f(a_i)$;
end

Algorithm 10.1: Lagrange Polynomial Interpolation

where we have used the product notation:

$$\prod_{i=1}^{n} a_i = a_1 a_2 \ldots a_n.$$

The general Lagrange polynomial interpolation algorithm is given in pseudocode in Algorithm 10.1.

This algorithm is implemented in Python below.

```
In [9]: def lagrange_interp(a,f,x):
            """Compute a lagrange interpolant
            Args:
                a: array of n points
                f: array of the value of f(a) at the n points
            Returns:
                The value of the Lagrange interpolant at x
            """
            answer = 0
            assert a.size == f.size
            n = a.size
            for i in range(n):
                product = 1
                for j in range(n):
                    if (i != j):
                        product *= (x-a[j])/(a[i]-a[j])
                answer += product*f[i]
            return answer
```

We can use this function to compute the interpolation of $\sin(x)$ using a degree 3, or cubic, polynomial.

```
In [10]: a = np.linspace(0,np.pi*0.5,4)
         f = np.sin(a)
```

```
x = np.linspace(-0.1*np.pi,0.6*np.pi,200)
y = lagrange_interp(a,f,x)
```

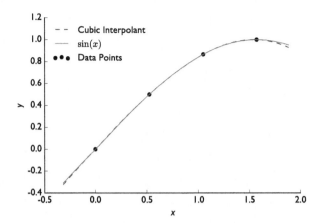

This interpolant is nearly indistinguishable from the original function over this range. With a general Lagrange interpolation routine, we can go even further and create a quintic interpolant. The result of running the code above with 6 points instead of 4 gives the following result:

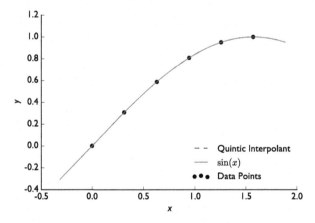

In this case we cannot distinguish between the original function and the interpolating polynomial. These results seem to indicate that higher degree polynomial interpolants are superior to lower degree, but there is a problem with high-degree interpolation called the Runge phenomenon that we discuss later.

10.2.1 Efficiency of Lagrange Polynomials

The Lagrange polynomial formula is not the most efficient way to compute an interpolating polynomial. Nevertheless, for almost any application the speed of modern computers means that our Lagrange polynomial is fast enough. This is especially true because using

NumPy we can compute the value of the polynomial at many x points at the same time for a given set of input data points. As shown in the figure below, it is possible to evaluate degree 512 polynomials at one million points in less than a second.

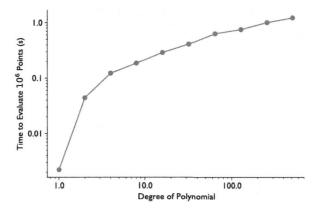

If the Lagrange polynomial formula is too slow for a particular application, there are other approaches discussed at the end this chapter.

10.2.2 The Runge Phenomenon

At degrees higher than 3, polynomial interpolation can be ill-behaved between the interpolation data. Specifically, the interpolating polynomial can have large oscillations. We can see this in a simple example with the function

$$f(x) = \frac{1}{1 + 25x^2}.$$

This function seems to be fairly innocuous, but it tortures polynomial interpolation. We will look at 5th and 7th degree polynomials. Using our Lagrange polynomial function defined above, we produce the interpolating polynomials we get the following result:

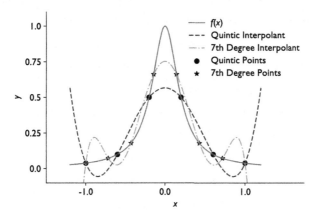

Notice how the polynomial interpolants are very inaccurate near the edges of the domain and the behavior of the interpolating polynomials are completely different than the underlying function. In particular the higher degree polynomial has a larger maximum error and larger average error than the lower degree polynomial. This can be seen in the plot of the absolute value of the interpolation error below:

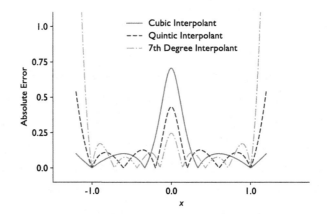

This behavior of high degree interpolants is called the Runge phenomenon: with equally spaced points a polynomial interpolant can have large oscillations going both high and below the actual function. This also makes extrapolating, that is evaluating the function outside the range of the data a dangerous endeavor. In this example, whether the function increases or decreases beyond the data depends on the degree of the interpolating polynomial. This high sensitivity to the choices made in interpolation make extrapolation untrustworthy outside the domain of the data, especially when the polynomial degree is high.

In general, performing high degree polynomial interpolation is a bad idea. In general it is better to fit low-degree polynomials to subsets of the data and then connect those polynomials. This is the idea behind spline interpolation.

10.3 CUBIC SPLINE INTERPOLATION

A cubic spline is a piecewise cubic function that interpolates a set of data points and guarantees smoothness at the data points. Before we discuss cubic splines, we will develop the concept of piecewise linear fits.

If we have several points, but do not want to have a high degree polynomial interpolant because of fear of the Runge phenomenon, we could do linear interpolation between each of the points.

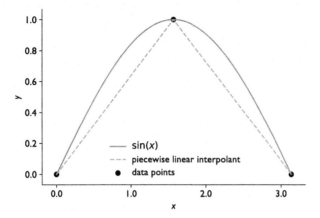

This fit is OK, but it has some problems. The primary one is that it is not smooth at the data points: the function has a discontinuous derivative at some of the points. This would be the case even if we knew that the underlying function should be smooth. Also, a linear interpolant is not a good fit to the function: above we had much better luck with quadratics and cubics.

To guarantee a degree of accuracy, avoid the oscillations we have seen before, and get smooth functions we can, and should, use cubic spline interpolants. Cubic spline interpolants are continuous in the zeroth through second derivatives and pass through all the data points. The name spline comes from thin sticks, called splines, that were used in drafting in the days before computers. One also could imagine that these flexible sticks were used to strike colleagues in moments of merriment or anger.

To set up our algorithm we begin with the $n+1$ data points (x_i, y_i) (also called knot points), this implies n intervals and $n-1$ interior (not at the beginning or end) points. We will denote the cubic on the interval from point $(i-1)$ to i as

$$f_i(x) = a_i + b_i x + c_i x^2 + d_i x^3.$$

The cubics need to match at the knot points so

$$f_1(x_1) = y_1,$$

$$f_i(x_{i+1}) = f_{i+1}(x_{i+1}) = y_{i+1} \qquad 1 \le i < n,$$

$$f_n(x_{n+1}) = y_{n+1},$$

which are $2n$ total conditions, when the end points are included. In equation form these become

$$a_1 + b_1 x_1 + c_1 x_1^2 + d_1 x_1^3 = y_1,$$

$$a_{i+1} + b_{i+1} x_{i+1} + c_{i+1} x_{i+1}^2 + d_{i+1} x_{i+1}^3 = a_i + b_i x_{i+1} + c_i x_{i+1}^2 + d_i x_{i+1}^3 = y_{i+1},$$

$$a_n + b_n x_{n+1} + c_n x_{n+1}^2 + d_n x_{n+1}^3 = y_{n+1}.$$

Also, we need to make the derivatives continuous at the interior knot points,

$$f_i'(x_{i+1}) = f_{i+1}'(x_{i+1}) \qquad 1 \le i < n.$$

The $n - 1$ equations for this are

$$b_{i+1} + 2c_{i+1}x_{i+1} + 3d_{i+1}x_{i+1}^2 = b_{i+1} + 2c_{i+1}x_i + 3d_{i+1}x_i^2 \qquad 1 \le i < n.$$

Finally, we need to make the second derivatives continuous at the interior knot points,

$$f_i''(x_{i+1}) = f_{i+1}''(x_{i+1}) \qquad 1 \le i < n.$$

The $n - 1$ equations for equality are

$$2c_{i+1} + 6d_{i+1}x_{i+1} = 2c_i + 6d_ix_{i+1}.$$

For the n intervals there are $4n$ unknowns (4 coefficients for each cubic). We have $4n - 2$ equations at this point so we need two more equations. The natural choice is to set the second derivative to be zero at the two endpoints:

$$f_1''(x_1) = 0, \qquad f_n''(x_{n+1}) = 0.$$

We now have $4n$ equations and $4n$ unknowns.

We will build a cubic spline for $\sin(x)$ using $x = (0, \pi/2, \pi)$. This spline will have two intervals, meaning that there are 8 cubic coefficients we need to find. We fill a matrix with the equations for matching the function at the knot points first. The code below does this.

```
In [11]: #knot points are sin(x) at 0, pi/2,pi
         n = 2 #2 intervals
         data = np.zeros((n+1,2))
         data[:,0] = a
         data[:,1] = y_a
         coef_matrix = np.zeros((4*n,4*n))
         rhs = np.zeros(4*n)
         #set up the 2n equations that match the data at the knot points
         #first point
         x = data[0,0]
         coef_matrix[0,0:4] = [1,x,x**2,x**3]
         rhs[0] = data[0,1]
         #second point
         x = data[1,0]
         coef_matrix[1,0:4] = [1,x,x**2,x**3]
         rhs[1] = data[1,1]
         x = data[1,0]
         coef_matrix[2,4:8] = [1,x,x**2,x**3]
         rhs[2] = data[1,1]
         #third point
         x = data[2,0]
         coef_matrix[3,4:8] = [1,x,x**2,x**3]
         rhs[3] = data[2,1]
         print(coef_matrix[0:4,:])
```

```
[[ 1.         0.         0.         0.         0.         0.         0.         0.       ]
 [ 1.         1.57079633 2.4674011  3.87578459 0.         0.         0.         0.       ]
 [ 0.         0.         0.         0.         1.         1.57079633 2.4674011  3.87578459]
 [ 0.         0.         0.         0.         1.         3.14159265 9.8696044  31.00627668]]
```

These are the first four rows of the matrix to determine the 8 unknowns in our spline fit.

The next step is defining the equations for the first derivative at the interior point. This adds one more equation.

```
In [12]:  #now the first derivative equations
          #second point
          x = data[1,0]
          coef_matrix[4,0:4] = [0,1,2*x,3*x**2]
          rhs[4] = 0
          coef_matrix[4,4:8] = [0,-1,-2*x,-3*x**2]
```

The last step in the construction of the equations is to create the equations for the second-derivatives at the knot points. One of these equations will be at the middle point, $x = \pi/2$, and the other two specify that the second derivative goes to zero at the endpoints.

```
In [13]:  #now the second derivative equations
          #second point
          x = data[1,0]
          coef_matrix[5,0:4] = [0,0,2,6*x]
          rhs[5] = 0
          coef_matrix[5,4:8] = [0,0,-2,-6*x]
          #set first point to 0
          x = data[0,0]
          coef_matrix[6,0:4] = [0,0,-2,6*x]
          rhs[6] = 0
          #set last point to 0
          x = data[2,0]
          coef_matrix[7,4:8] = [0,0,2,6*x]
          rhs[7] = 0
          print(coef_matrix)
```

```
[[ 1.         0.         0.         0.         0.         0.         0.         0.       ]
 [ 1.         1.57079633 2.4674011  3.87578459 0.         0.
  0.         0.        ]
 [ 0.         0.         0.         0.         1.         1.57079633
   2.4674011  3.87578459]
 [ 0.         0.         0.         0.         1.         3.14159265
   9.8696044  31.00627668]
 [ 0.         1.         3.14159265 7.4022033  0.        -1.
  -3.14159265 -7.4022033 ]
 [ 0.         0.         2.         9.42477796 0.         0.
  -2.        -9.42477796]
 [ 0.         0.        -2.         0.         0.         0.         0.         0.       ]
 [ 0.         0.         0.         0.         0.         0.         2.         18.84955592]]
```

Finally, we solve the system of equations using Gauss elimination (from Chapter 7) and get the coefficients of the cubic spline interpolant.

```
In [14]:  #solve for the cubic coefficients
          coefs = GaussElim(coef_matrix,rhs)
          print(coefs)
```

```
[ 0.           0.95492966         0.          -0.12900614
 -1.           2.86478898        -1.2158542     0.12900614]
```

Therefore, our approximation is

$$f(x) = \begin{cases} 0.95492966x - 0.12900614x^3 & x \leq \frac{\pi}{2} \\ -1 + 2.86478898x - 1.2158542x^2 + 0.12900614x^3 & x \geq \frac{\pi}{2} \end{cases}.$$

The beauty of the cubic spline interpolant is how well it approximates a function. If we look at the sine wave example from before, it is hard to distinguish the cubic spline interpolant from the original function. In the code below we evaluate the cubic spline fit. For a given point we need to determine which spline to use and this logic is expressed in the if-elif-else block.

```
In [15]:  #evaluate function
          points = 200
          X = np.linspace(-0.1,np.pi+0.1,points)
          y_interp = np.zeros(points)
          for i in range(points):
              if (X[i] < np.min(data[:,0])):
                  spline = 0
              elif (X[i] > np.max(data[0:n,0])):
                  spline = n-1
              else:
                  #knot to the left is spline
                  spline = np.argmax(X[i]-data[:,0])
              y_interp[i] = np.sum(coefs[4*spline:(4*spline+4)] *
                            [1,X[i],X[i]**2,X[i]**3])
          plt.plot(X,np.sin(X), label="$\sin(x)$")
          plt.plot(X,y_interp, linestyle="--", label="Cubic Spline")
          plt.scatter(data[:,0],data[:,1],label="knot points")
```

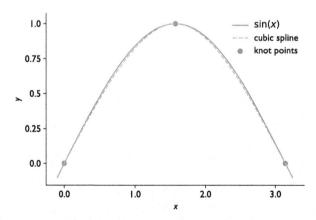

In the following graph, the error, as defined by the difference between the true function and the interpolant, is quantified so that we can see how small it is:

```
In [16]:   plt.plot(X,y_interp-np.sin(X), label="cubic spline $- \sin(x)$")
           plt.scatter(data[:,0],0*data[:,1],label="knot points");
           plt.xlabel("x")
           plt.ylabel("Error")
           plt.legend(loc="best")
```

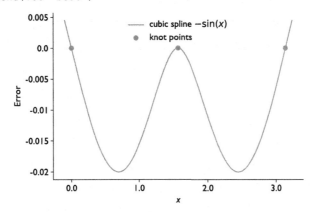

Making the cubic spline was straightforward, but tedious. Additionally, evaluating the splines involved determining which spline to use and then evaluating the function. Thankfully, the package SciPy, a companion package for NumPy that implements many numerical algorithms, has a cubic spline function that we can use.

The function in the interpolate section of SciPy that we want to use is called CubicSpline. It takes in the data points x and y as inputs, and returns a function that we can evaluate. There are several options for the conditions at the beginning and end points. We have already discussed the natural spline conditions where the second derivative of the splines at the first and last knot point are set to zero, i.e.,

$$f_1''(x_1) = f_n''(x_{n+1}) = 0, \qquad \text{natural spline conditions.}$$

Another type of spline is the "clamped" spline where the first-derivative is set to zero at the end points:

$$f_1'(x_1) = f_n'(x_{n+1}) = 0, \qquad \text{clamped spline conditions.}$$

The default condition used by CubicSpline is the "not-a-knot" condition where the third derivative of the first and last splines is fixed so that it matches the third derivative at the nearest interior point:

$$f_1'''(x_2) = f_2'''(x_2), \qquad f_{n-1}'''(x_n) = f_n'''(x_n), \qquad \text{not-a-knot spline conditions.}$$

Plugging in the cubics to this equation we get the conditions

$$a_1 = a_2, \qquad a_{n-1} = a_n.$$

This implies that the spline in the first and nth interval are the same as the splines in the second and $(n-1)$ intervals, respectively. In this sense, the endpoints are not treated as knots, just a point that the interior spline must pass through.

The code below compares these splines on an interpolation problem. This time we apply it to the hyperbolic sine function using five knot points, meaning there are four different splines.

```
In [17]:   from scipy.interpolate import CubicSpline
           #define data
           a = np.linspace(0.2,np.pi*1.8,5)
           data = np.zeros((5,2))
           data[:,0] = a
           data[:,1] = np.sinh(a)
           #define splines
           splineFunction = CubicSpline(data[:,0],data[:,1],bc_type='natural')
           splineFuncClamp = CubicSpline(data[:,0],data[:,1],bc_type='clamped')
           splineFuncNot = CubicSpline(data[:,0],data[:,1],bc_type='not-a-knot')
           #make plot
           points = 200
           X = np.linspace(0,np.pi*2,points)
           plt.plot(X,splineFunction(X),
                   label="Natural Cubic Spline")
           plt.plot(X,splineFuncClamp(X),linestyle="--",
                   label="Clamped Cubic Spline")
           plt.plot(X,splineFuncNot(X),linestyle="-.",
                   label="NaK Cubic Spline")
           plt.plot(X,np.sin(X), label="$\sinh(x)$")
           plt.scatter(data[:,0],data[:,1],label="knot points")
```

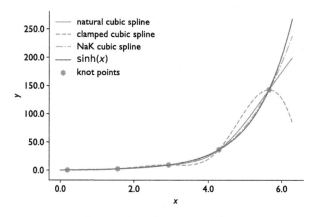

It is apparent that the choice of spline type does have an affect on the interpolation. The clamped splines force a local extreme point to be created at the endpoints because the derivative goes to zero. Similarly, the natural splines create an inflection point (i.e., the second-derivative is zero at the endpoints). For this particular problem the not-a-knot splines work best near the large values of x.

On the Runge phenomenon example from before, cubic spline interpolants perform better than high-degree polynomials as shown in this next figure.

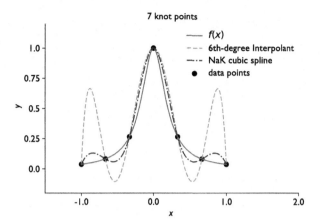

Adding two more knot points makes the spline fit better

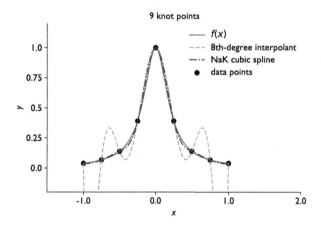

Notice how the splines get better, but the full polynomial gets worse. That is why cubic splines are the go to method for fitting curves through a series of points.

CODA

Next time we will talk about those times when it is not desirable to have the function touch every point. These problems are curve-fitting problems and sometimes called regression. In these problems, as demonstrated below, we do not want the function to interpolate the data but to find some trend in the data. In the following figure a line that has had random noise added to it is shown, along with the original function, a cubic spline interpolant, and a line fitted by linear regression. We will cover this topic in the next chapter.

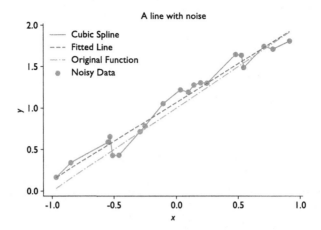

FURTHER READING

As mentioned above, there are other ways to compute interpolating polynomials than the Lagrange polynomial formula. All of the methods give the same the result, but different algorithms to compute them. Neville's algorithm and Newton's divided difference formula are two well-known polynomial construction techniques. Additionally, it is possible to use other functions for interpolation. Rational function interpolation writes the interpolant as the quotient of two polynomials and can give well-behaved interpolants where polynomial interpolation is too oscillatory. For periodic functions, trigonometric interpolation is another approach.

PROBLEMS

Short Exercises

10.1. Compute, by hand, the second-degree Lagrange polynomial for the points, $x = \{0, 2, 4\}$ and $f(x) = \{1, 0.223891, -0.39715\}$. If $f(3) = -0.260052$, how accurate is the interpolation at this point. Plot your interpolating polynomial for $x \in [0, 5]$. The points were generated from the Bessel function of the first kind, $J_0(x)$. Compare your interpolating polynomial to the actual function.

10.2. Repeat the previous exercise with a clamped cubic spline and a not-a-knot cubic spline.

Programming Projects

1. Root-Finding via Interpolation

Given the following function data

x	2.5	2.4	1.2	1	0.5	0.2	0.1
$f(x)$	−0.0600307	−0.0594935	−0.039363	−0.0279285	0.071541	0.43361	0.73361

Use interpolation to find the root of the function, that is find x such that $f(x) = 0$. You can do this by fitting interpolating functions to the data using $\hat{x}_i = f(x_i)$, and $\hat{f}(\hat{x}_i) = x_i$, and then evaluating $\hat{f}(0)$. Find the best approximation you can using an interpolating polynomial of maximum degree, and the three kinds of splines discussed above. The correct answer is $x = 0.75$. Comment on the results.

2. Extrapolation

Consider the following data for an unknown function. In the problems below $\log x$ denotes the natural logarithm of x.

x	1.2	1.525	1.85	2.175	2.5
$f(x)$	5.48481	2.3697	1.62553	1.28695	1.09136

Use interpolating polynomials of different degree, and three kinds of splines to estimate $f(3)$. Repeat the previous procedure after logarithmically transforming the data, i.e., set $\hat{x}_i = \log x_i$, and $\hat{f}(\hat{x}_i) = \log f(x_i)$. The correct answer is $f(3) = 0.910239226627$. Discuss which approaches performed best for both the linear and logarithmic interpolation.

3. Moderator Temperature Coefficient of Reactivity

The change in the reactivity for a nuclear reactor due to changes in the moderator temperature is called the moderator temperature coefficient of α_m is the logarithmic derivative of k_∞ for the reactor as

$$\alpha_m = \frac{1}{k_\infty} \frac{\partial k_\infty}{\partial T_m} = -\beta_m \left(\log \frac{1}{p} - (1 - f) \right),$$

where the subscript m denotes "moderator", p is the resonance escape probability for the reactor, f is the thermal utilization, and β_m is

$$\beta_m = -\frac{1}{N_m} \frac{\partial N_m}{\partial T_m},$$

with N_m the number density of the moderator.

Consider a research reactor that is cooled by natural convection. It has $p = 0.63$ and $f = 0.94$. Plot the moderator temperature coefficient from $T_m = 285$ K to 373 K as a function of temperature for this reactor, using the data below from the National Institute for Standards and Technology (NIST) for T_m in K, and density, ρ_m, in mol/liter:

T_m	ρ_m	$\frac{\partial \rho_m}{\partial T_m}$
289.99	55.442	−0.00769
300.12	55.315	−0.0143
320.53	54.909	−0.0231
346.13	54.178	−0.0308
366	53.475	−0.0385
369.87	53.326	−0.0385

Thus the unfacts, did we possess them, are too imprecisely few to warrant our certitude...

–James Joyce, Finnegan's Wake

CHAPTER POINTS

- Given independent and dependent variables, we can fit a linear model to the data using least squares regression.

- Least squares can handle multiple independent variables and nonlinearity in the independent variables.

- Logarithmic transforms allow us to fit exponential and power law models.

In the previous chapter we investigated methods that exactly interpolated a set of data. That is we derived functions that passed through a set of data points. Sometimes it is better to find a function that does not exactly interpolate the data points. For example, if we did an experiment and we expected the measured value to be linear in some variable x we might write

$$f(x) = a + bx + \epsilon,$$

where $f(x)$ is the measured value, and ϵ is an error term because the experimental data probably does not fall in exactly a straight line. In this case the ϵ contains the measurement error and inherent variability in the system. In such a case, rather than performing linear interpolation, we want to find the values of a and b that best match the measured data, but do not necessarily exactly match the data.

Another reason we might want to find this best match is that there might be another hidden variable we do not know about, but we want to find the best possible model given the variables we do know about. As an example say we want to know the radiation dose rate outside of a shield, but there is an unknown source variability, we could write

$$\log(\text{dose}) = a + b(\text{shield thickness}) + \epsilon,$$

where now ϵ is capturing the error in not including the source variation, as well as the measurement uncertainty.

In a less scientific example, a retail corporation has stores throughout the country and wants to know how the area around the store (sometimes called a trade area) and the size of the store affect the sales at a store. In this case we may write

$$\text{sales} = a + b(\text{Population within 5 minute drive of store}) + c(\text{Size of Store}) + \epsilon.$$

In this case there may be hundred of hidden variables such as the presence of competition in the market, the age of the store, or the time of year.

The question that we are faced with is that we have set of input variables and the value of an output variable at several points, and we want to fit a linear model to this data. By linear model we mean a function that is the sum of contributions from each input or some operation on the inputs. The form of the model could come from theory or we could just be looking for an explanation of the output using the data we have. In this case Lagrange polynomials or cubic splines will not work because we have a model form we want to approximate, and interpolation methods either prescribe a model form or have the form determined by the amount of data. Furthermore, if we do an experiment and measure at the same point multiple times, we will likely get several slightly different values of the outputs. Interpolation methods will not work because they expect the interpolating function to only take on a single value at each input.

11.1 FITTING A SIMPLE LINE

To proceed, we will take some data and try to fit a linear model to the data. Along the way we will see that the problem is not well-posed, and demonstrate a means to give it a unique solution.

Say we are given data for an input x and an output y:

x	y
1	11.94688
2	22.30126
3	32.56929
4	41.65564

and we want to fit a model

$$y = a + bx,$$

which makes sense because y does look roughly linear in x. In the parlance of curve fitting, x is an independent variable, and y is the dependent variable.

What we can do is write this as a linear system of equations, where the unknowns are $\mathbf{u} = (a, b)^\mathrm{T}$. That is, we want to solve for $\mathbf{u}$ in the equation

$$\mathbf{X}\mathbf{u} = \mathbf{y},$$

where $\mathbf{X}$ is the data matrix,

$$\mathbf{X} = \begin{pmatrix} 1 & 1 \\ 1 & 2 \\ 1 & 3 \\ 1 & 4 \end{pmatrix},$$

and $\mathbf{y}$ is the vector of dependent variables

$$\mathbf{y} = (11.94688, 22.30126, 32.56929, 41.65564)^\mathrm{T}.$$

Notice that the data matrix has a column of ones because for each equation in the system the constant a is the same. Putting in our data values our system gives

$$\begin{pmatrix} 1 & 1 \\ 1 & 2 \\ 1 & 3 \\ 1 & 4 \end{pmatrix} \begin{pmatrix} a \\ b \end{pmatrix} = \begin{pmatrix} 11.94688 \\ 22.30126 \\ 32.56929 \\ 41.65564 \end{pmatrix}.$$

I think that the problem here is obvious: we have 4 equations and 2 unknowns. This means that there is not expected to be a vector $\mathbf{u}$ that can satisfy every equation. To make the problem well-posed, we can multiply both sides of the equation by the transpose of the matrix:

$$\begin{pmatrix} 1 & 1 & 1 & 1 \\ 1 & 2 & 3 & 4 \end{pmatrix} \begin{pmatrix} 1 & 1 \\ 1 & 2 \\ 1 & 3 \\ 1 & 4 \end{pmatrix} \begin{pmatrix} a \\ b \end{pmatrix} = \begin{pmatrix} 1 & 1 & 1 & 1 \\ 1 & 2 & 3 & 4 \end{pmatrix} \begin{pmatrix} 11.94688 \\ 22.30126 \\ 32.56929 \\ 41.65564 \end{pmatrix}.$$

We will perform this calculation using Python, rather than by hand:

```
In [1]: import numpy as np
        A = np.array([(1,1),(1,2),(1,3),(1,4)])
        RHS = np.array([11.94688, 22.30126, 32.56929, 41.65564])
        print("The system A x = b has A =\n",A)
        print("And b =",RHS)

The system A x = b has A =
 [[1 1]
 [1 2]
```

```
[1 3]
 [1 4]]
And b = [ 11.94688  22.30126  32.56929  41.65564]

In [2]: AT_times_A = np.dot(A.transpose(),A)
        AT_times_RHS = np.dot(A.transpose(),RHS)
        print("The system after multiplying by A transpose is A^T A =\n",AT_times_A)
        print("A^T b =",AT_times_RHS)

The system after multiplying by A transpose is A^T A =
 [[ 4 10]
 [10 30]]
A^T b = [ 108.47307  320.87983]
```

Now we have a two-by-two system that we can solve:

$$\begin{pmatrix} 4 & 10 \\ 10 & 30 \end{pmatrix} \mathbf{u} = \begin{pmatrix} 108.47307 \\ 320.87983 \end{pmatrix}.$$

This system is known as the "normal equations". We will solve this system with the Gaussian elimination code we wrote earlier, and that I have handily stored in GaussElim.py.

```
In [3]: from GaussElim import *
        ab = GaussElimPivotSolve(AT_times_A,AT_times_RHS)
        print("The constant a =",ab[0]," with a slope of b =",ab[1])

The constant a = 2.26969  with a slope of b = 9.939431
```

The data and the fitted line are shown next.

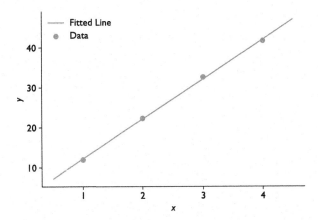

To get a measure of how this model fits the data we look at a quantity called R^2 (pronounced "R-squared"), defined as

$$R^2 = 1 - \frac{\sum_i (\hat{y}_i - y_i)^2}{\sum_i (\bar{y} - y_i)^2},$$

where $\hat{y}_i$ is the ith predicted data point by the model,

$$\hat{y}_i = \mathbf{x}_i \cdot \mathbf{u} = a + bx_i,$$

and $\bar{y}$ is the mean of the data. Notice that in the definition of $\mathbf{x}$ we had to include a 1 to handle the intercept term: $\mathbf{x}_i = (1, x_i)^{\mathrm{T}}$.

For our data we can compute R^2 in Python as

```
In [4]: yhat = ab[0] + ab[1]*A[:,1]
        print("yhat =\t",yhat)
        print("y =\t",RHS)
        residual = yhat - RHS
        print("error =\t",residual)
        r2Num = np.sum(residual**2)
        r2Denom = np.sum((RHS-RHS.mean())**2)
        print("R2 numerator =", r2Num)
        print("R2 denominator =", r2Denom)
        r2 = 1 - r2Num/r2Denom
        print("R2 =", r2)

yhat =    [ 12.209121  22.148552  32.087983  42.027414]
y =  [ 11.94688  22.30126  32.56929  41.65564]
error =   [ 0.262241 -0.152708 -0.481307  0.371774]
R2 numerator = 0.46196241067
R2 denominator = 494.423405429
R2 = 0.999065654244
```

A perfect R^2 is 1. Sometimes R^2 is called the fraction of variance explained by the model. Therefore, 1 or 100% implies that 100% of the variance in the data is explained by the model. A value of 0.999, as in this fit, is very high and typically only appears in contrived data sets used as examples.

11.1.1 Least-Squares Regression

So why did we multiply by the transpose of the matrix? It turns out that the above procedure, besides giving us a system that we can solve, minimizes the error

$$\bar{E} = \sum_{i=1}^{n} (\hat{y}_i - y_i)^2.$$

That is why the above procedure is called least-squares regression because it minimizes the sum of the squares of the error at each point.

We can show that our solution gives the minimum, or least, squared error by differentiating the formula for $\bar{E}$ with respect to $\mathbf{u}$ and setting the result to 0. The derivative of $\bar{E}$ with respect to $\mathbf{u}$ is

$$\frac{d\bar{E}}{d\mathbf{u}} = 2 \sum_{i=1}^{n} \mathbf{x}_i^{\mathrm{T}} (\mathbf{x}_i \cdot \mathbf{u} - y_i) = 0,$$

which upon rearranging and using the definition of a matrix vector product leads to

$$\mathbf{X}^{\mathrm{T}}\mathbf{X}\mathbf{u} = \mathbf{X}^{\mathrm{T}}\mathbf{y}.$$

That is, the value of $\mathbf{u}$ is found by multiplying the system by the transpose of $\mathbf{X}$. This is the equation we solve to estimate the coefficients in the model.

Least-squares regression has the property that if we sum up the errors, we get zero. It works particularly well when the errors in the data (that is the deviation from the model) are independent random variables that do not depend on the value of the independent or dependent variables.

BOX 11.1 NUMERICAL PRINCIPLE

Least-squares regression is the most common type of model/curve fitting method: it will give errors with an average of zero, and it minimizes the sum of the squares of the error.

11.2 MULTIPLE LINEAR REGRESSION

We will now generalize the problem of fitting a linear function to data to allow for there to be multiple independent variables. Consider I observations of a dependent variable y and independent variables $\mathbf{x} = (1, x_1, x_2, \ldots, x_J)^{\mathrm{T}}$. We desire to fit a linear model of the form

$$y(\mathbf{x}) = a_0 + a_1 x_1 + \cdots + a_J x_J + \epsilon.$$

The equations that govern the relationship between y and $\mathbf{x}$ are given by

$$\mathbf{X}\mathbf{u} = \mathbf{y},$$

where $\mathbf{X}$ is an $I \times (J + 1)$ matrix of the form

$$\mathbf{X} = \begin{pmatrix} 1 & x_{11} & x_{12} & \ldots & x_{1J} \\ 1 & x_{21} & x_{22} & \ldots & x_{2J} \\ \vdots & \vdots & \vdots & & \vdots \\ 1 & x_{i1} & x_{i2} & \ldots & x_{iJ} \\ \vdots & \vdots & \vdots & & \vdots \\ 1 & x_{I1} & x_{I2} & \ldots & x_{IJ} \end{pmatrix}, \tag{11.1}$$

where the notation x_{ij} is the ith observation for the jth independent variable. The vector $\mathbf{u}$ holds the coefficients of the model

$$\mathbf{u} = (a_0, a_1, \ldots, a_J)^{\mathrm{T}}, \tag{11.2}$$

TABLE 11.1   Statistics from 2013 Texas A&M Football Team

Game	Opponent	Yards Gained	Points Scored	Off. T.O.[a]	Def. T.O.[a]
1	Rice	486	52	1	2
2	Sam Houston State	714	65	1	2
3	Alabama	628	42	2	1
4	SMU	581	42	1	3
5	Arkansas	523	45	0	2
6	Ole Miss	587	41	2	1
7	Auburn	602	41	2	1
8	Vanderbilt	558	56	5	3
9	UTEP	564	57	1	4
10	Mississippi State	537	51	3	1
11	LSU	299	10	2	0
12	Missouri	379	21	1	0
13	Duke	541	52	0	2

[a] *T.O. is an abbreviation for turnovers.*

and the vector $\mathbf{y}$ contains the observations of the dependent variable

$$\mathbf{y} = (y_1, \ldots, y_I)^{\mathrm{T}}. \tag{11.3}$$

As before, we will multiply both sides of the equation by $\mathbf{X}^{\mathrm{T}}$ to form the normal equations, and solve for $\mathbf{u}$. We will now apply this more general technique to a real set of data.

11.2.1 Example From Outside of Engineering

Consider the following data in Table 11.1. This is data from the 2013 Texas A&M Aggies' football season. The numeric columns are

- Yards Gained on Offense
- Points Scored on Offense
- Turnovers by the Aggie Offense
- Turnovers received by the Aggie Defense.

For this example we will use the points scored as the dependent variable with a single independent variable: yards gained. The data in Table 11.1 has been stored in a csv file that is read in the next code block.

```
In [5]: import csv
        with open('FootballScores.csv', newline='') as csvfile:
            reader = csv.DictReader(csvfile)
            opponent = []
            Yards = np.array([])
            Points = np.array([])
            OffTurn = np.array([])
            DefTurn = np.array([])
```

```
        for row in reader:
            #print(row)
            opponent.append(row['Opponent'])
            Yards = np.append(Yards,float(row['Yards Gained']))
            Points = np.append(Points,float(row['Points Scored']))
            OffTurn = np.append(OffTurn,float(row['Off Turnovers']))
            DefTurn = np.append(DefTurn,float(row['Def Turnovers']))
    print(opponent)
    print(Yards)
    print(Points)
    print(OffTurn)
    print(DefTurn)

['Rice ', 'Sam Houston State ', 'Alabama ', 'SMU ', 'Arkansas ',
'Ole Miss ', 'Auburn ', 'Vanderbilt ', 'UTEP ', 'Mississippi State ',
'LSU ', 'Missouri ', 'Duke ']
[ 486.  714.  628.  581.  523.  587.  602.  558.  564.  537.  299.  379.
  541.]
[ 52.  65.  42.  42.  45.  41.  41.  56.  57.  51.  10.  21.  52.]
[ 1.  1.  2.  1.  0.  2.  2.  5.  1.  3.  2.  1.  0.]
[ 2.  2.  1.  3.  2.  1.  1.  3.  4.  1.  0.  0.  2.]
```

Now we want to perform our least squares procedure where the independent variable is yards gained and the dependent variable is points scored. First, we build the matrix as before. This is a good opportunity to use the vstack function in NumPy to stack two vectors on top of each other and then take the transpose to get a rectangular matrix with 13 rows and 2 columns:

```
In [6]: A = np.vstack([np.ones(Yards.size), Yards]).transpose()
        print("The A matrix is\n",A)

The A matrix is
 [[   1.   486.]
  [   1.   714.]
  [   1.   628.]
  [   1.   581.]
  [   1.   523.]
  [   1.   587.]
  [   1.   602.]
  [   1.   558.]
  [   1.   564.]
  [   1.   537.]
  [   1.   299.]
  [   1.   379.]
  [   1.   541.]]
```

NumPy also has a built-in least squares function linalg.lstsq. This function is used by passing it the data matrix and the righthand side. We use this function to determine the linear model for points scored as a function of yards gained:

```
In [7]: solution = np.linalg.lstsq(A,Points)[0]
        print("The function is Points =",solution[0],"+",solution[1],"* Yards")

The function is Points = -16.2577502382 + 0.112351872138 * Yards
```

BOX 11.2 NUMPY PRINCIPLE

In the linear algebra section of NumPy there is a function that can solve the least squares problem called lstsq. It is called via the syntax

```
x = np.linalg.lstsq(A,b)[0]
```

where x is the vector containing the solution described in Eq. (11.2), A is the data matrix described above in Eq. (11.1), and b is the vector of dependent variables given in Eq. (11.3). The [0] selects the first of the many variables the function lstsq returns. This function returns several extra pieces of information that we do not need.

It is important to interpret what the coefficients in your model mean. In this case the model says that for every 10 yards gained, we expect to score 1.1 points. It also says that if zero yards are gained, then we expect −16 points to be scored. This seemingly anomalous result arises because we are trying to fit a simple model to this data. The original data and model are plotted next to help illustrate what the model is telling us:

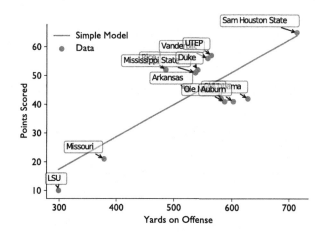

This is what real data typically look like: a lot messier than a simple line with a little noise. The model shows that several games with very high or very low yards set the trend for the line. Also we notice that half of the data are on one side of the line and the other half are on the other. This splitting of the data is typical of least-squares regression.

To adjust our model, we can remove two of the lower conference schools (Sam Houston State and UTEP); the scores in these games are expected to be different than the other games based on the strength of those teams. This removes two data points to the top right. Removing data points like this can help a model if we think that the data removed is not representative of the overall data set.

```
In [8]:   Asmall = A[(0,2,3,4,5,6,7,9,10,11,12),:]
          PointsSmall = Points[[0,2,3,4,5,6,7,9,10,11,12]]
          solution = np.linalg.lstsq(Asmall,PointsSmall)[0]
```

```
The function is Points = -13.9002357065 + 0.105908511234 * Yards
```

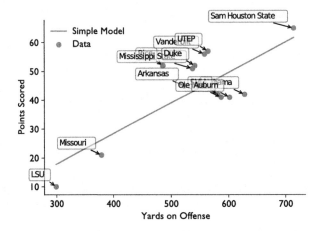

Notice that the slope of our line did not change much by removing some data (0.112 points per yard versus 0.106). This is a useful observation. It means that our model is robust to removing some games. Had the result changed drastically we should be worried that the model was driven by just a few data points.

11.2.2 Adding More Variables

To attempt to improve this model we could add more independent variables. For example, the number of turnovers produced by the defense (i.e., interceptions or fumbles that the opponent gives up) or given up by the offense (throwing an interception or losing a fumble by the offense) could also be important. The next figure colors the dots with the number of turnovers taken by the defense, and the size of the dot is the number of turnovers given up by the offense. Looking at this chart seems to indicate that the lower scoring games had fewer turnovers produced by the defense.

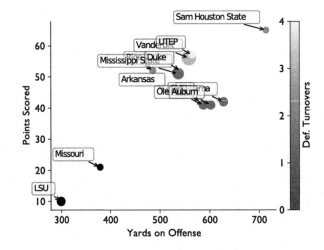

We will add these two variables to our model and compute the fit via least squares.

```
In [9]:   A = np.vstack([np.ones(Yards.size), Yards, OffTurn, DefTurn]).transpose()
          print("The A matrix is\n",A)
          solution = np.linalg.lstsq(A,Points)[0]
          print("The function is Points =",solution[0],
              "+",solution[1],"* Yards","+",solution[2],
              "* Off Turnovers","+",solution[3],"* Def Turnovers")
```

```
The A matrix is
 [[   1.   486.    1.    2.]
  [   1.   714.    1.    2.]
  [   1.   628.    2.    1.]
  [   1.   581.    1.    3.]
  [   1.   523.    0.    2.]
  [   1.   587.    2.    1.]
  [   1.   602.    2.    1.]
  [   1.   558.    5.    3.]
  [   1.   564.    1.    4.]
  [   1.   537.    3.    1.]
  [   1.   299.    2.    0.]
  [   1.   379.    1.    0.]
  [   1.   541.    0.    2.]]

The function is Points = -10.2946466989 + 0.0826356338026 * Yards +
    0.376961922593 * Off Turnovers + 5.57033662396 * Def Turnovers
```

Once again, we will try to interpret the coefficients in the model. Given the sign of the coefficient for offensive turnovers, the model indicates that turning over the ball on offense is actually good for scoring points. Even a basic understanding of football indicates that this is not likely to be a good strategy. When you get a model with an obviously wrong (or indefensible coefficient), it is best to remove that coefficient and refit the model. The reason we can get coefficients with strange values is that the model does not tell us what *causes* the dependent variable to change, it is simply telling us what is *correlated* with the dependent variable. To get at causation we would have to do a controlled experiment, not look at historical, observational data such as this data set.

Before we remove the offending independent variable from the mode, we will look at the values of the predictions versus the actual values. This is an important diagnostic for multivariate models because it is harder to visualize the trends like we did for a single variable model. We make a scatter plot where the x axis is the actual points scored and the y axis is the predicted points scored. If the model is perfect, then all the points will fall on the diagonal line $y = x$.

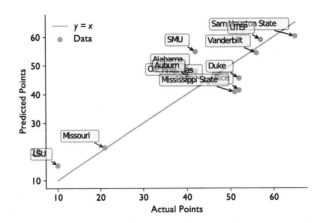

This figure indicates that our model does seem to predict the number of points scored based on the independent variables. To get a more quantitative measure of the model accuracy we will compute the average amount the model is off by. This quantity is the average absolute value of the error, which is simply called the mean absolute error. For this model the mean absolute error is

```
In [10]: yhat = np.dot(A,solution)
         np.mean(np.fabs(Points-yhat))
```

```
Out[10]: 5.3201734672523262
```

BOX 11.3 NUMERICAL PRINCIPLE

The mean absolute error can be a better measure of the model fit than R^2. This is especially true is the model is going to be used for predictions. In such a case you are often more concerned with how far off a prediction is going to be.

We have to compute the average absolute error because of error cancellation—if we computed the average error, the cancellation of positive and negative errors would give us an average error very close to zero. Indeed this is what we see:

```
In [11]: np.mean(Points-yhat)
```

```
Out[11]: -1.1409676622301608e-13
```

As indicated above, we should remove offensive turnovers from the model because the coefficient for this variable did not make sense. We will remove this variable, refit the model, and plot the actual versus predicted values.

```
In [12]: A = np.vstack([np.ones(Yards.size), Yards, DefTurn]).transpose()
         solution = np.linalg.lstsq(A,Points)[0]
         print("The function is Points =",solution[0],
               "+",solution[1],"* Yards","+",solution[2],"* Def Turnovers")
```

```
The function is Points = -9.79027984448 + 0.0829036035041 * Yards +
                         5.54687804786 * Def Turnovers
```

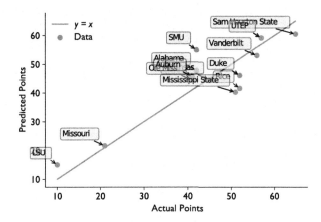

The coefficients of this model indicate that an additional defensive turnover is worth about 5.5 points on offense and that each 10 yards gained is worth 0.8 points. The actual versus predicted values plot looks very similar to the previous model. Additionally, the mean absolute error is about the same.

```
In [12]: np.mean(np.fabs(Points-yhat))

Out[12]: 5.3393751264980898
```

The R^2 for this model is

```
In [13]: 1-np.sum((Points-yhat)**2)/np.sum((Points.mean() - Points)**2)

Out[13]: 0.79071317769713123
```

This is a respectable value for R^2 for a real data set: almost 80% of the variability in points scored can be explained by our model that contains only the number yards gained and the number of defensive turnovers. The mean absolute error also tells us that given the number of yards gained by the Aggies and the number of takeaways on defense, we can estimate how many points the team scored to within about ±5 points.

It is a good thing to remove unnecessary variables in the model for several reasons. The first is that putting too many variables in the model can lead to overfitting: if we add enough variables eventually the model will have zero error (when we have the same number of variables as observations, the fit will be "perfect"). However, this model will perform poorly in making predictions.

The other reason is that strange values will call the model into question. Nobody will believe anything else you say if you make a patently false claim (like turn the ball over to get more points). It is important when performing curve fitting to think critically about what the model is indicating. Especially, if you want to use the model to explain what is occurring in the data set.

11.3 "NONLINEAR" MODELS

Least squares regression can be applied to models that are nonlinear in the independent variables, if we can map the nonlinearity to new independent variables. In this section we will demonstrate this procedure.

Suppose we wish to fit a model that has the form

$$f(x) = a_0 + a_1 x_1 + a_2 x_1^2.$$

We can fit this model using least squares by defining a new variable $x_2 \equiv x_1^2$. This will make the model look like a multivariate regression model:

$$f(x) = a_0 + a_1 x_1 + a_2 x_2.$$

Therefore, to fit this quadratic model we set up the matrix with this additional variable x_2 in the appropriate column.

Consider the following data generated from the function $y = 1 - 5x + 2x^2$ with noise added:

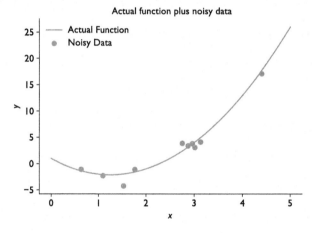

To fit the quadratic model need to build the matrix with x^2 in the appropriate column before calling the least squares function:

```
In [14]: #now build the matrix
         A = np.ones((N,3))
         for i in range(N):
             A[i,1] = x[i]
             A[i,2] = x[i]**2
         solution = np.linalg.lstsq(A,y)[0]
         print("The function is y =",solution[0],"+",
               solution[1],"* x","+",solution[2],"* x^2")

The function is y = 0.50907172038 + -4.69368434583 * x + 1.92891037106 * x^2
```

The fitted function is close to the actual function used to produce the data. The coefficients are slightly off. Graphically, the fitted and original function are close.

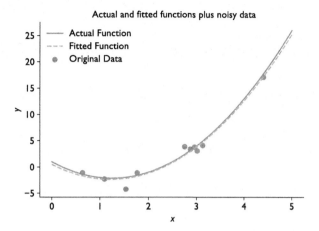

If we wanted to fit more complicated polynomial models, we can do this by placing the correct powers of the independent variables in the data matrix $\mathbf{X}$. This makes the least squares curve fitting process very flexible.

We can also fit non-polynomial models such as

$$f(x) = a + bx + c \sin x.$$

In this case we have a column in the matrix that correspond to $\sin x$. We will demonstrate this with data produced from the function $y = 1 - 0.5x + 3 \sin x$ with added noise:

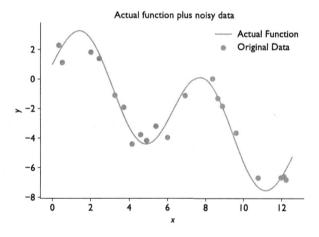

To fit this model we place the sine of x in the appropriate column of the data matrix and solve using least squares, as shown next:

```
In [15]: #now build the matrix
         A = np.ones((N,3))
         for i in range(N):
             A[i,1] = x[i]
```

```
        A[i,2] = np.sin(x[i])
    solution = np.linalg.lstsq(A,y)[0]
    print("The fitted function is y =",solution[0],
        "+",solution[1],"* x","+",solution[2],"* sin(x)")

The fitted function is y = 0.792955748456 + -0.482057534691 * x
                    + 2.7346948684 * sin(x)
```

The fitted function does deviate from the original function, primarily due to the noise in the data making some of the points used to fit the model being far away from the actual function. Despite this, the overall trend appears to be correct. This is apparent in a graphical comparison between the original function and the fitted function.

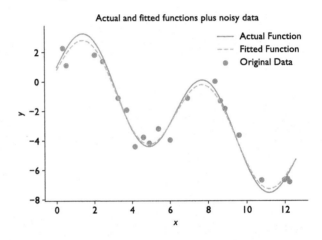

In this example with the sine function, the least squares regression model does a good job of finding the underlying model coefficients. Despite there being some data points far away from the actual function due to noise, the model fit is very close to the actual model used to generate the data.

BOX 11.4 NUMERICAL PRINCIPLE

Many different models can be fit with least-squares regression, even if at first glance the model does not look linear. If you can define new independent variables that are func-tional transformations of other independent variables, the resulting model can be fit with least-squares.

11.4 EXPONENTIAL MODELS: THE LOGARITHMIC TRANSFORM

Beyond nonlinear transformations to independent variables, we can use logarithms to fit models that are exponentials or power laws of independent variables. A particularly relevant example would be fitting a model to determine the half-life of a radioactive sample:

$$A(t) = A_0 e^{-\lambda t}$$

$$\lambda = \frac{\ln 2}{T_{1/2}},$$

where A is the activity. To fit this model we can take the natural logarithm of both sides and find

$$\ln A(t) = \ln A_0 - \lambda t.$$

Therefore, if we fit a model where the dependent variable (i.e., the left-hand side) is the natural logarithm of $f(x)$ then we can infer both the half-life and the initial activity.

As before we will generate data from a known exponential and add noise to the data points. In this case we generate decay data from a sample of 10^{12} atoms of arsenic-76, which has a half-life 1.09379 ± 0.00045 days.

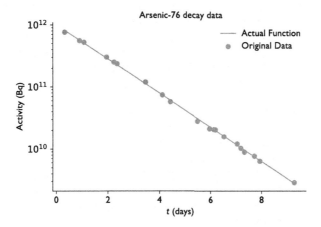

To fit the exponential model, we need to make the righthand side in the least-squares equations equal to the natural logarithm. The data matrix will contain the number of days since the sample was obtained.

```
In [16]: #now build the matrix
         A = np.ones((N,2))
         for i in range(N):
             A[i,1] = t[i]
         print("The A matrix is\n",A)
         solution = np.linalg.lstsq(A,np.log(activity))[0]
         print("The inital activity is A_0 =",np.exp(solution[0]),
               "and the half-life is",-np.log(2)/solution[1],"days.")

The inital activity is A_0 = 1.00216154364e+12 and the half-life is
1.09039181369 days.
```

The fitted model gives a reasonable approximation of the half-life given the noise in the data.

11.4.1 Power Law Models

It is also possible fit power-law models using similar manipulations. The function

$$f(x) = ax^b,$$

can be transformed to a linear, additive model by writing a function

$$\ln f(x) = \ln a + b \ln(x),$$

that is we take the natural logarithm of x and $f(x)$ to get a linear function. Such power laws appear in all kinds of natural data. One, perhaps unexpected, place a power law appears is in the number of words used with a given frequency in language. In English it has been conjectured that the 100 most common words make up 50% of all writing. Another way to look at this, is that there are a small number of words that are used very frequently (e.g., the, a, and, etc.), and many words that are used very infrequently (e.g. consanguine or antiderivative). Therefore, if we look at any work of literature we expect there to be thousands of words used one or two times, and a few words used thousands of times. To demonstrate this we can look at the word frequency distribution for that venerable work of literature *Moby Dick*. The next figure is a histogram of word frequency in *Moby Dick*. For example, there are approximately 10^4 words that are only used once in the book out of the 17,227 unique words in the book.

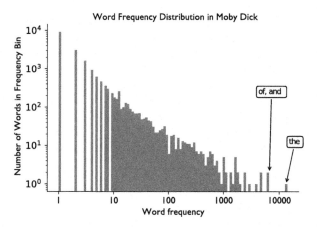

The word "the" was used over 10,000 times.
For this data, we want to fit a model as

$$\text{Number of words with a given frequency} = a(\text{Word Frequency})^b.$$

This will require us to make the righthand side of the least square equations equal to the logarithm of the dependent variable, and place the logarithm of the independent variable in the data matrix. The resulting model for *Moby Dick* is

$$\text{Number of words with a given frequency} = 7.52(\text{Word Frequency})^{-0.94}.$$

We can then compare this model to the actual data, and see that a simple power law is a reasonable fit to the data:

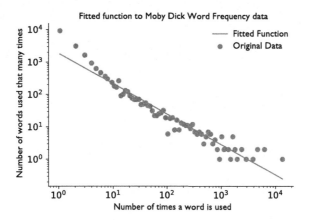

Except for words used less than 3 times, the model predicts the number of words with a given frequency well. This natural behavior of language (and other systems such as income) are examples of power laws.

An important lesson that we can take from this data is that power law distributions, such as those for the word frequency in *Moby Dick* cannot be approximated by a normal or Gaussian distribution. For example, if we generate samples from a Gaussian distribution with a mean equal to the mean number of times a word is used in *Moby Dick*, and a variance equal to the observed variance in word frequencies the data looks nothing like a power law. The figure below shows 211,763 samples chosen from a Gaussian distribution with mean 11.137 (the average number of times a word in used in *Moby Dick*), and variance of 22,001 (the observed variance of word frequencies in *Moby Dick*).

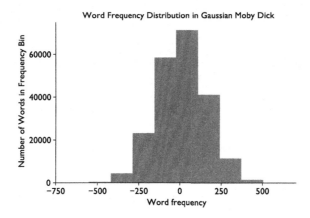

This, you can agree, does not look anything like the distribution of word frequency in Moby Dick. Even if we ignore the negative frequencies, the distribution says there should be very few words used more frequently than 500 times. In actuality, there are about 100 words used more than 500 times. Gaussian Moby Dick is a very different book. "Call me Ishmael" may not have made the cut.

CODA

There are many times in engineering and quantitative analysis in general where you need to reconcile data with a model (as is the case in the half-life example above) or when you want to develop a model to explain data (similar to the football score example). The most basic, but still very useful, tool for these situations is least-squares regression. These regression models can be fit easily using the capabilities in NumPy.

FURTHER READING

The application of curve fitting to observed data is an important part of the study of statistics. As such, there are many works on applying regression techniques. For a deep dive into the field of machine learning (of which linear regression is an example) see the comprehensive monograph by Hastie, Tibshirani, and Friedman [13]. For the application of regression techniques to data collected from a variety of sources (including social science and public safety), Gelman and Hill's book [14] is recommended.

N.N. Taleb is a crusader against using Gaussian ideas in a world of power law distributions. His various books on this subject are entertaining and informative reads [15–17].

PROBLEMS

Programming Projects

1. Power Law Fit

Fit a power law to the data below from the United States Internal Revenue Service for the number (in thousands) of tax returns filed in 2014 with particular values of incomes (in thousands). The incomes come from the midpoints of the brackets that the IRS uses to report data. Make the independent variable be the income, and the number of returns be the dependent variable. Compute R^2 and the mean absolute error for the model.

Income (Thousands $)	Returns (Thousands)
2.5	10263
7.5	11790
12.5	12290
17.5	11331
22.5	10062
27.5	8819
35	14600
45	11473
57.5	19395
87.5	12826

Income (Thousands $)	Returns (Thousands)
150	17501
350	4979
750	835
1250	180
1750	77
3500	109
7500	27
10000	17

2. Inflating R^2

Produce 10 uniform random samples in the range [0,1], call this the independent variable x_1. Produce another 10 uniform random samples in range [0,1], call this the independent variable x_2. Finally, produce 10 uniform random samples in the range [2,3] and call this the dependent variable y. Fit 5 linear models

$$y = a_0 + a_1 x_1,$$
$$y = b_0 + b_1 x_2,$$
$$y = c_0 + c_1 x_1 + c_2 x_2,$$
$$y = d_0 + d_1 x_1 + d_2 x_2 + d_3 x_1 x_2,$$
$$y = e_0 + e_1 x_1 + e_2 x_2 + e_3 x_1 x_2 + e_4 x_1^2 + e_5 x_2^2.$$

For each of these models compute R^2 and the mean absolute error. Discuss the results.

3. k_∞ and Diffusion Length for a Water Balloon

Consider a water balloon that you fill with a homogeneous mixture of heavy water and fissile material. You do not know the exact ratio of the water molecules to fissile atoms. However, you can fill the balloon with different amounts of the solution (thereby changing the radius of the sphere). You are able to measure the multiplication factor, k_{eff}, at several different balloon radii by pulsing the sphere with a neutron source and observing the subcritical multiplication. You also astutely remember that the 1-group diffusion theory relation between the multiplication factor, the infinite medium multiplication factor (k_∞), and the diffusion length, L is

$$k_{eff} = \frac{k_\infty}{1 + L^2 B_g^2},$$

where B_g^2 is the geometric buckling for the system. In this case we have

$$B_g^2 = \frac{\pi^2}{(R+d)^2},$$

where d is the extrapolation length. If we assume that $d \ll R$, we can do a linear Taylor expansion around $d = 0$ to write

$$B_g^2 = \frac{\pi^2}{R^2} - \frac{2\pi^2}{R^3} d.$$

Given the measurement data below, infer the values of k_∞, L, and d. Is the assumption on d correct? What radius will make the reactor critical? Report R^2 for your model.

Hint: make your regression model have the dependent variable be $1/k_{eff}$.

Experiment	R	k_{eff}
1	10.00	0.16
2	12.50	0.23
3	15.00	0.31
4	20.00	0.46
5	25.00	0.60
6	35.00	0.80
7	36.00	0.82
8	40.00	0.87
9	45.00	0.93
10	50.00	0.98

4. Model Without a Constant

It does not always make sense to have a constant in a linear model. What the constant indicates is the value of the dependent variable when all the independent variables are zero. One example of a model where this is the case is the temperature coefficient of reactivity for a nuclear system. This coefficient, denoted by α, given by

$$\Delta\rho = \alpha\Delta T,$$

where T is the temperature of the system and ΔT is the difference between the current temperature and the point where the reactivity ρ is 0. The reactivity is given by

$$\rho = \frac{k-1}{k}.$$

Your task is to find α using the data below by fitting a model without a constant. To do this you have to modify the data matrix in the least squares procedure. Report R^2 for your model.

k_{eff}	T (K)
1.000000	250
0.999901	300
0.998032	350
0.996076	400
0.994055	450
0.992329	500

We're Talking Root Down, I Put My Boot Down
And If You Want To Battle Me, You're Putting Loot Down

"Root Down" by the Beastie Boys

CHAPTER POINTS

- Finding the root of a nonlinear function requires an iterative method.

- The simplest and most robust method is bisection.

- Other methods can improve on bisection, but may be slower on some problems.

Up to this point we have only solved linear equations. In this chapter we solve nonlinear equations. Commonly, this process is called root finding because the problem is generally stated as determining a root of a function, i.e., find x so that

$$f(x) = 0.$$

The complication with nonlinear functions is that they may have many roots or not have any roots at all. For a single-variable linear function, $f(x) = a + bx$, we know that there is a single root, provided that $b \neq 0$.

A feature of nonlinear root finding is that it almost always requires an iterative method for finding the root: an initial guess is required and that guess is refined until the answer is *good enough*. In this chapter we will treat iterative methods for finding roots that start with an initial interval that we know the root is in, and then shrink that interval until we know the root is in a very small interval. These methods are called closed root finding methods because the root is known to be inside a closed interval.

12.1 BISECTION

The first closed method we use is the bisection algorithm. As with any closed method, to use it we need to first bracket the root. That is we need to find two points a and b where

$$f(a)f(b) < 0,$$

that is two points where $f(x)$ has a different sign. Therefore, we know by the intermediate value theorem that $f(x) = 0$ is in $[a, b]$, provided $f(x)$ is continuous in $[a, b]$.

Once we have two points a and b, we then pick the midpoint of the interval and call it c:

$$c = \frac{a+b}{2}.$$

Then we can determine which side of the interval root is on by comparing the sign of $f(c)$ to the sign of the function at the endpoints. Once we determine which side of the midpoint the root was on, we can change our interval to be that half. Algorithmically,

$$\text{if} \quad f(a)f(c) < 0 \qquad \text{then we set } b = c$$

because the root is between a and c. Otherwise,

$$\text{if} \quad f(b)f(c) < 0 \qquad \text{then we set } a = c$$

because the root is between c and b. Of course, if $f(a)f(c) = 0$, then c is the root, the probability of this happening is vanishingly small, however. If the function has multiple roots, it is possible that both $f(a)f(c) < 0$ and $f(b)f(c) < 0$. In this case we could take either half of the interval and still find a root. The result of this procedure is that the new interval is half the size of the previous interval: we can think of the result is that we have improved our estimate of the root by a factor of 2.

We can repeat this process of computing the sign of the function at the midpoint and shrinking the interval by a factor of 2 until the range is small enough that we can say we are done. Define the width of an interval after iteration n as $\Delta x_n = b - a$, with the initial interval width written as Δx_0. Using this definition and the fact that each iteration cuts the interval in half, we know that after n iterations the width of the interval is

$$\Delta x_n = 2^{-n} \Delta x_0.$$

If we want to know the root within a tolerance ϵ, then we can solve for n in the equation

$$\epsilon = 2^{-n} \Delta x_0,$$

which implies

$$n = \frac{\log(\Delta x_0/\epsilon)}{\log 2}.$$

BOX 12.1 NUMERICAL PRINCIPLE

To find the root of a nonlinear function almost always requires an iterative method. Bisection is the simplest method, but it is also very robust and almost always guaranteed to converge to the root.

Below is an implementation of the bisection algorithm in Python. As you can see, it is a simple algorithm.

```
In [1]: def bisection(f,a,b,epsilon=1.0e-6):
            """Find the root of the function f via bisection
            where the root lies within [a,b]
            Args:
                f: function to find root of
                a: left-side of interval
                b: right-side of interval
                epsilon: tolerance

            Returns:
                estimate of root
            """
            assert (b>a)
            fa = f(a)
            fb = f(b)
            assert (fa*fb < 0)
            delta = b - a
            print("We expect",
                  int(np.ceil(np.log(delta/epsilon)/np.log(2))),"iterations")
            iterations = 0
            while (delta > epsilon):
                c = (a+b)*0.5
                fc = f(c)
                if (fa*fc < 0):
                    b = c
                    fb = fc
                elif (fb*fc < 0):
                    a = c
                    fa = fc
                else:
                    return c
                delta = b-a
                iterations += 1
            print("It took",iterations,"iterations")
            return c #return midpoint of interval
```

Notice that we save the value of the function evaluations so that we only evaluate the function at a given point once. This will be important if it takes a long time to evaluate the function. For example, if the function evaluation involves the solution of a system of equations, as done in one of the exercises for this chapter, one does not want to be evaluating the function more times than necessary.

Below we test the bisection function with a simple cubic, to find a single root:

```
In [2]: def nonlinear_function(x):
            #compute a nonlinear function for demonstration
            return 3*x**3 + 2*x**2 - 5*x-20
        root = bisection(nonlinear_function,1,2)
        print("The root estimate is",root,"\nf(",root,
            ") =",nonlinear_function(root))
```

```
We expect 20 iterations
It took 20 iterations
The root estimate is 1.9473047256469727
f( 1.9473047256469727 ) = -1.883242441280686e-05
```

Bisection is a useful algorithm because it is really easy to implement, and we know how long it should take to converge. Also, we know that it will converge. We can give it a really big range and as long as the root is in the range, it will find it. In this case we increase the initial interval from a width of 3 to 7. Bisection has no trouble finding the root.

```
In [3]: root = bisection(nonlinear_function,-5,2)
        print("The root estimate is",root,"\nf(",root,
            ") =",nonlinear_function(root))
```

```
We expect 23 iterations
It took 23 iterations
The root estimate is 1.9473060369491577
f( 1.9473060369491577 ) = 2.9577188165319512e-05
```

It did take more iterations, but with this larger interval, bisection still arrived at the answer.

12.1.1 Critical Radius of a Sphere

An example problem that we will use to compare root finding methods is that of finding the critical radius of a bare sphere reactor. From elementary nuclear reactor theory [7], one can show that for a critical, spherical reactor

$$\left(\frac{\pi}{R+2D}\right)^2 = \frac{\nu\Sigma_f - \Sigma_a}{D},$$

where D [cm] is the diffusion coefficient, ν is the number of neutrons born per fission, Σ_f [cm^{-1}] is the fission macroscopic cross-section, Σ_a [cm^{-1}] is the macroscopic cross-section for absorption, and R [cm] is the radius of the reactor. Therefore, given D, $\nu\Sigma_f$, and Σ_a we can compute the critical radius. To use bisection we need to define the function we want to be zero:

$$f(R) = \left(\frac{\pi}{R+2D}\right)^2 - \frac{\nu\Sigma_f - \Sigma_a}{D}.$$

We will solve the problem with $D = 9.21$ cm, $\nu\Sigma_f = 0.1570$ cm^{-1}, and $\Sigma_a = 0.1532$ cm^{-1}.

For the initial interval we will pick $a = 0$ and $b = 250$ cm, because $f(R)$ has different signs at these points. In the code below we define a simple function and pass it to our bisection algorithm. Notice that this function has all the arguments except for R have default arguments. This is because the bisection algorithm we defined earlier operates on a function of only a single variable.

The bisection algorithm finds the critical radius in 28 iterations.

```
In [4]: #first define the function
        def Crit_Radius(R, D=9.21, nuSigf = 0.1570, Siga = 0.1532):
            return (np.pi/(R + 2*D))**2 - (nuSigf - Siga)/D
        a = 0
        b = 250
        Radius = bisection(Crit_Radius,a,b)
        print("The critical radius estimate is",Radius,
              "\nf(",Radius,") =",Crit_Radius(Radius))
```

```
We expect 28 iterations
It took 28 iterations
The critical radius estimate is 136.2435193732381
f( 136.2435193732381 ) = 2.175801910603986e-12
```

This will be the baseline for comparison as we examine new closed root finding methods.

12.2 FALSE POSITION (REGULA FALSI)

Another closed root finding method is the false position method (also called regula falsi, if one prefers Latin names). This method draws a line between the two endpoints of the interval $[a, b]$ and uses the position where that line intersects the x axis as the value of c (the guess for the new endpoint). Below is a graphical example:

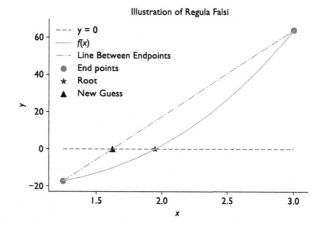

The false position methods works in almost the exact same way as bisection except that c is not directly in the middle of the interval, rather it is where the interpolating line intersects the x axis. The reason this may be a good idea, is that if the function is linear (or near linear) in the interval, this value of c will be the root.

To derive false position, we first define the slope between the two endpoints:

$$m \equiv \frac{f(b) - f(a)}{b - a}.$$

For a line the slope is the same everywhere so we know that

$$\frac{f(b) - f(a)}{b - a} = \frac{f(c) - f(a)}{c - a}.$$

We also want $f(c)$ to be zero, so this simplifies to

$$\frac{f(b) - f(a)}{b - a} = \frac{-f(a)}{c - a}.$$

Solving for c gives

$$c = a - \frac{f(a)}{m}.$$

We have to use a different convergence criterion than we did for bisection as the interval size is not the best measure in false position because after each iteration our guess is c. In this case we will use $|f(c)| < \epsilon$ for our convergence criteria.

A Python implementation of this algorithm is below.

```
In [5]:  def false_position(f,a,b,epsilon=1.0e-6):
             """Find the root of the function f via false position
             where the root lies within [a,b]
             Args:
                 f: function to find root of
                 a: left-side of interval
```

```
        b: right-side of interval
        epsilon: tolerance

    Returns:
        estimate of root
    """
    assert (b>a)
    fa = f(a)
    fb = f(b)
    assert (fa*fb< 0)
    delta = b - a
    iterations = 0
    residual = 1.0
    while (np.fabs(residual) > epsilon):
        m = (fb-fa)/(b-a)
        c = a - fa/m
        fc = f(c)
        if (fa*fc < 0):
            b = c
            fb = fc
        elif (fb*fc < 0):
            a = c
            fa = fc
        else:
            print("It took",iterations,"iterations")
            return c
        residual = fc
        iterations += 1
    print("It took",iterations,"iterations")
    return c #return c
```

We will apply false position to the cubic function we defined above:

```
In [6]:  root = false_position(nonlinear_function,1,2)
         print("The root estimate is",root,"\nf(",root,
             ") =",nonlinear_function(root))


It took 5 iterations
The root estimate is 1.9473052141477751
f( 1.9473052141477751 ) = -7.983476244532994e-07
```

This is faster than bisection on this short interval because bisection took 20 iterations. This factor of 4 improvement is impressive. If we give it a bigger interval, it is still faster than bisection by a factor of 3.

```
In [7]:  root = false_position(nonlinear_function,-5,2)
         print("The root estimate is",root,"\nf(",root,") =",
             nonlinear_function(root))


It took 9 iterations
The root estimate is 1.947305257431454
f( 1.947305257431454 ) = 7.995646100766862e-07
```

Finally, we apply it on the critical slab case (bisection took 28 iterations for this problem and initial interval).

```
In [8]:   a = 0
          b = 250
          Radius = false_position(Crit_Radius,a,b)
          print("The critical radius estimate is",Radius,"\nf(",Radius,
                ") =",Crit_Radius(Radius))
```

```
It took 260 iterations
The critical radius estimate is 136.42803805393132
f( 136.42803805393132 ) = -9.827177585040653e-07
```

What happened? False position was much faster on the cubic function, but is now 9 times slower than bisection. It is important to note that it still converged (that's good), but it converged slowly (that's bad). To see why, we can look at the first iteration by plotting the function and the false position update:

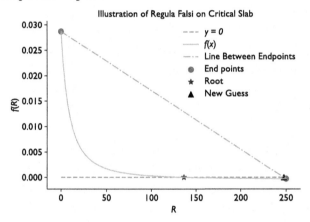

Notice how the function is very steep near 0 and much less steep to the right of the root. This makes the value of c very close to b. Since the right endpoint moves so slowly, it takes a long time to converge. If we gave it a better initial range, it should work better. With the minimum value of the initial interval shifted to $R = 100$ we get:

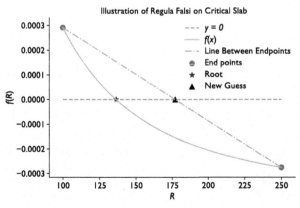

```
In [9]:   a = 100
          b = 250
          Radius = false_position(Crit_Radius,a,b)
          print("The critical radius estimate is",Radius,"\nf(",Radius,
               ") =",Crit_Radius(Radius))
```

```
It took 6 iterations
The critical radius estimate is 136.42658989359794
f( 136.42658989359794 ) = -9.750187400692396e-07
```

This demonstrates that when we tried to get better than bisection we sometimes are slower. Bisection is the tank of nonlinear solvers: it is slow, but it will get there. Regula falsi can be faster than the tank, but it can also be slowed down in the mud. There is a lot of mud in the realm of solving nonlinear equations.

BOX 12.2 NUMERICAL PRINCIPLE

Faster methods are not always better in terms of robustness. False position can be much faster than bisection, but it also can be much slower than bisection in certain cases.

12.3 RIDDER'S METHOD

An obvious way to quantify the speed of a rootfinding method is the number of iterations. This, however, is not a perfect comparison between methods because the number of function evaluations per iteration may differ between methods. For both bisection and false position, the number of additional function evaluations needed is one, at $x = c$. This will not always be the case. The next method requires an extra function evaluation in each iteration.

BOX 12.3 NUMERICAL PRINCIPLE

An appropriate measure of a root finding methods speed is the number of function evaluations. This is the case because the func-tion in many applications may be difficult to compute and might require the solution of a linear system of equations.

In Ridder's method we interpolate between a and b in a way that is different than linear interpolation. In particular, given points a and b which bracket the root, we compute the values $f(a)$, $f(b)$, $f(c)$ where c is the midpoint of a and b.

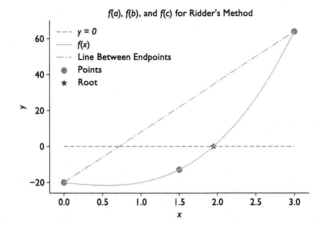

Then we use these function values to estimate the function:

$$g(x) = f(x)e^{(x-a)Q}.$$

The function $g(x)$ will only be zero when $f(x)$ is zero. Furthermore, it touches the original function at a, and not at b. To determine the value of Q in $g(x)$ we require that $g(a)$, $g(b)$, and $g(c)$ all lie on a line (or they are collinear). To do this we first look at each of these values:

$$g(a) = f(a), \qquad g(b) = f(b)e^{2hQ}, \qquad g(c) = f(c)e^{hQ},$$

where $h = c - a$. If these three points are to lie on a straight line then,

$$g(c) = \frac{g(a) + g(b)}{2},$$

or

$$f(c)e^{hQ} = \frac{f(a) + f(b)e^{2hQ}}{2}.$$

Solving this equation for e^{hQ} we get

$$e^{hQ} = \frac{f(c) \pm \sqrt{f(c)^2 - f(a)f(b)}}{f(b)}.$$

The root we choose is the "+" root if $f(a) - f(b) > 0$ and the "−" root otherwise. Once we have our $g(x)$ function, we then linearly interpolate from $(a, g(a))$ and $(c, g(c))$ to find where $g(d) = 0$ and use this as the next guess at the root:

$$d = c \pm \frac{(c-a)f(c)}{\sqrt{f(c)^2 - f(a)f(b)}}.$$

The following figures shows this procedure graphically.

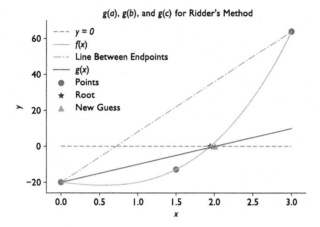

$g(a)$, $g(b)$, and $g(c)$ for Ridder's Method

As you can see the new guess is really close. This appears to be much better than false position. This is one of the benefits of performing interpolation on the exponentially varying function $g(x)$ rather than using the original function $f(x)$.

The general algorithm for Ridder's method is below.

```
In [10]: def ridder(f,a,b,epsilon=1.0e-6):
             """Find the root of the function f via Ridder's Method
             where the root lies within [a,b]
             Args:
                 f: function to find root of
                 a: left-side of interval
                 b: right-side of interval
                 epsilon: tolerance

             Returns:
                 estimate of root
             """
             assert (b>a)
             fa = f(a)
             fb = f(b)
             assert (fa*fb < 0)
             delta = b - a
             iterations = 0
             residual = 1.0
             while (np.fabs(residual) > epsilon):
                 c = 0.5*(b+a)
                 d = 0.0
                 fc = f(c)
                 if (fa - fb > 0):
                     d = c + (c-a)*fc/np.sqrt(fc**2-fa*fb)
                 else:
                     d = c - (c-a)*fc/np.sqrt(fc**2-fa*fb)
                 fd = f(d)
                 #now see which part of interval root is in
                 if (fa*fd < 0):
                     b = d
```

```
        fb = fd
    elif (fb*fd < 0):
        a = d
        fa = fd
    residual = fd
    iterations += 1
print("It took",iterations,"iterations")
return d #return c
```

On the example using a cubic function from before Ridder performs the best of the methods we have seen (bisection took 23 iterations and false position took 9):

```
In [20]: root = ridder(nonlinear_function,-5,2)
         print("The root estimate is",root,"\nf(",root,
               ") =",nonlinear_function(root))
```

```
It took 7 iterations
The root estimate is 1.94730524023
f( 1.94730524023 ) = 1.64623202181e-07
```

Now we try our criticality problem (the one that false position required 260 iterations and bisection took 28). This is where Ridder's method shines: it can handle the large change in our function's behavior.

```
In [21]: a = 0
         b = 250
         Radius = ridder(Crit_Radius,a,b)
         print("The critical radius estimate is",Radius,"\nf(",Radius,
               ") =",Crit_Radius(Radius))
```

```
It took 3 iterations
The critical radius estimate is 136.242306777
f( 136.242306777 ) = 6.47192578831e-09
```

That was much faster than either of the previous methods.

The total cost of Ridder's method per iteration is one function evaluation more per iteration than false position or bisection: we need to evaluate the function at the guess from the previous iteration, $f(d)$, and at the midpoint of the new interval. Therefore, to be more efficient than false position or bisection, the number of iterations needs to be 2 times fewer than for either of those methods. In the example above Ridder's method met this hurdle, except in the case where Ridder had 7 iterations and false position had 9 (to have the same number of function evaluations Ridder's method needed 4.5 iterations). This slight miss is mitigated by the fact that Ridder's method seems much more robust than false position.

CODA

We have reviewed root finding methods that work by bracketing the root and then zooming in on the root either simply, as in bisection, or through interpolation, as in false position or Ridder's method. One observation we can make at this point is that the methods all converged to the root, even though they could be slow.

In the next chapter we will cover open root finding methods that require only a single initial guess and do not bracket a root. These methods can converge quickly, though they generally require information about the slope of the function.

FURTHER READING

There are additional closed root finding methods that we have not discussed. A popular one is Brent's method the details of this method can be found in Atkinson's text [18], among others.

PROBLEMS

Short Exercises

12.1. Find a root of $\cos x$ using the three methods discussed in this section and an initial interval of $[0, 10]$. Compare the solutions and the number of iterations to find the root for each method.

12.2. You are given a radioactive sample with an initial specific activity of 10^4 Bq/kg, and you are told the half-life is 19 days. Compute the time it will take to get the specific activity of Brazil nuts (444 Bq/kg) using the three methods specified above.

Programming Projects

1. 2-D Heat Equation Optimization

Previously, in Algorithm 9.5 we gave code to solve the heat equation:

$$-k\nabla^2 T = q, \qquad \text{for } x \in [0, L_x] \quad y \in [0, L_y].$$

With the boundary condition

$$T(x, y) = 0 \qquad \text{for } x, y \text{ on the boundary.}$$

Your have been tasked to determine what value of k will make the maximum temperature equal to 3 when $L_x = L_y = 1$ and the source, q, is given by

$$q = \begin{cases} 1 & 0.25 \le x \le 0.75 \qquad 0.25 \le y \le 0.75 \\ 0 & \text{otherwise} \end{cases}.$$

Your particular tasks are as follows:

- Define a function called `max_temperature` that finds the maximum value of $T(x)$ in the domain. This function will take as it's only argument k. Inside the function solve the heat equation with $\Delta x = \Delta y = 0.025$. The function `np.max` will be helpful here.
- Find the value of k for which the max temperature equals 3 using bisection, false position, and Ridder's method. Use an initial interval of $k \in [0.001, 0.01]$. Remember that the root-finding methods find when a function is equal to 0. You will have to define a function that is equal to 0 when the maximum temperature is equal to 3. How many iterations does each method take?
- The Python package, `time`, has a function `time.clock()` that returns the system time. Using this function time how long it takes for each method to find the value of k that makes the maximum temperature equal to 3. Which method is the fastest?

This problem will demonstrate why it is important to be parsimonious with the number of function evaluations.

2. Peak Xenon Time

In Programming Project 1 from Chapter 7 you determined the equilibrium concentrations of ^{135}Xe, ^{135}I and ^{135}Te. After shutdown of the reactor, the ^{135}Xe concentration, $X(t)$ can be written as

$$X(t) = X_0 e^{-\lambda_X t} + \frac{\lambda_I}{\lambda_I - \lambda_X} I_0 \left(e^{-\lambda_X t} - e^{-\lambda_I t}\right),$$

where t is the time since shutdown, X_0 and I_0 are the equilibrium concentrations of ^{135}Xe and ^{135}I during reactor operation, and λ_X and λ_I are the decay constants for ^{135}Xe and ^{135}I.

Using a root finding method of your choice, compute how long after shutdown the maximum concentration of ^{135}Xe is reached, as well as the value of the concentration at those times for power densities of 5, 50, and 100 W/cm^3.

13

Open Root Finding Methods

Two roads diverged in a yellow wood,
And sorry I could not travel both
And be one traveler, long I stood
And looked down one as far as I could
To where it bent in the undergrowth;

–"The Road Not Taken" by Robert Frost

CHAPTER POINTS

- Open root finding methods require an initial guess of the solution.

- These can converge faster than closed methods, but are not guaranteed to find a root.

- Newton's method uses the derivative of the function to generate a new estimate of the root.

- Newton's method can be generalized to multidimensional problems. Each iteration then requires the solution of a linear system of equations.

The previous chapter discussed closed root finding methods that bracket the root and then zoom in on the root by tightening the interval like a boa constrictor on a rat. Today we'll discuss open methods that only need an initial guess, but not a range to begin. These are called open root finding methods.

13.1 NEWTON'S METHOD

Newton's method is a rapidly convergent method that is a good choice provided that one has an estimate of the root. Newton's method is also known as the Newton–Raphson method because Isaac Newton is famous enough, and Raphson published the method before Newton did. However, the historical record indicates that Newton had used this method well before Raphson published it. Interestingly enough, Raphson also coined the term pantheism.

Newton's method is fairly robust. What the method does is compute the tangent at the guess x_i (via the derivative $f'(x_i)$), and uses where the tangent crosses zero to get the next guess. In the following figures, this procedure is illustrated. First, we show the procedure for going from the initial guess to the first calculated update:

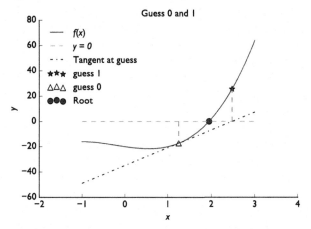

Next, we compute the slope of the function at this point and find where it crosses the axis:

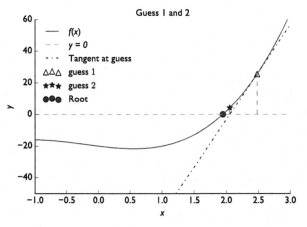

As you can see, we get closer to the root each iteration. This is due to the fact that over a short range, the tangent line is a reasonable approximation to the original function—just as the linear Taylor series can be a good approximation for small perturbations around the center of the expansion. Of course, using the slope to approximate the function behavior near a root is not always a good approximation, as we will see later.

The basic idea is to compute where the tangent line to the current root estimate crosses the x axis. We do this by equating the derivative of the function at the current estimate, x_i, to the slope of a line between the point $f(x_i)$ and the estimate of the function evaluated at an unknown point $f(x_{i+1})$:

$$f'(x_i) = \frac{f(x_{i+1}) - f(x_i)}{x_{i+1} - x_i}.$$

Then, we set $f(x_{i+1}) = 0$ to determine where the tangent crosses zero. This yields

$$x_{i+1} = x_i - \frac{f(x_i)}{f'(x_i)}.$$

Therefore, each iteration requires the evaluation of the slope at x_i and the value of the function at that same point. An implementation of Newton's method in Python is given below.

```
In [1]: def newton(f,fprime,x0,epsilon=1.0e-6, LOUD=False):
            """Find the root of the function f via Newton-Raphson method
            Args:
                f: function to find root of
                fprime: derivative of f
                x0: initial guess
                epsilon: tolerance

            Returns:
                estimate of root
            """
            x = x0
            if (LOUD):
                print("x0 =",x0)
            iterations = 0
            fx = f(x)
            while (np.fabs(fx) > epsilon):
                fprimex = fprime(x)
                if (LOUD):
                    print("x_",iterations+1,"=",x,"-",fx,
                          "/",fprimex,"=",x - fx/fprimex)
                x = x- fx/fprimex
                iterations += 1
                fx = f(x)
            print("It took",iterations,"iterations")
            return x #return estimate of root
```

We will test the method on the cubic nonlinear function from the previous chapter. We will need to define a derivative function to use because Newton's method requires it. As a reference, bisection took 23 iterations and Ridder's method took 8. It is slightly difficult to compare an open method to the closed methods because we cannot run the method under the same conditions and the open method only requires a single initial guess, not an interval that brackets the root. Furthermore, in Newton's method we have to evaluate the function and its derivative in each iteration. It is useful, however, to compare the overall convergence behavior between methods.

```
In [2]: def Dnonlinear_function(x):
            #compute a nonlinear function for demonstration
            return 9*x**2 + 4*x - 5
        root = newton(nonlinear_function,Dnonlinear_function,-1.5,LOUD=True)
        print("The root estimate is",root,"\nf(",root,") =",
            nonlinear_function(root))


x0 = -1.5
x_ 1 = -1.5 - -18.125 / 9.25 = 0.45945945945945943
x_ 2 = 0.45945945945945943 - -21.584111503760884 / -1.2622352081811545
= -16.64045295295295
x_ 3 = -16.64045295295295 - -13206.446010659996 / 2420.5802585031524
= -11.184552053047796
x_ 4 = -11.184552053047796 - -3911.2567600472344 / 1076.1096334338297
= -7.54992539557002
x_ 5 = -7.54992539557002 - -1159.3159776918205 / 477.81265972577813
= -5.123627234367753
x_ 6 = -5.123627234367753 - -345.3883141179978 / 210.7694953933235
= -3.484925611661592
x_ 7 = -3.484925611661592 - -105.25615528465704 / 90.36265622268793
= -2.320106429882939
x_ 8 = -2.320106429882939 - -35.100340028952424 / 34.165618894325654
= -1.2927478991471761
x_ 9 = -1.2927478991471761 - -16.675175982276617 / 4.869782580156233
= 2.131465750600949
x_ 10 = 2.131465750600949 - 7.479645608829035 / 44.41417921626759
= 1.963059044886359
x_ 11 = 1.963059044886359 - 0.5864442011938849 / 37.53464350293673
= 1.9474349667957278
x_ 12 = 1.9474349667957278 - 0.004789634744607696 / 36.92226641627101
= 1.947305244673835
It took 12 iterations
The root estimate is 1.947305244673835
f( 1.947305244673835 ) = 3.2858902798693634e-07
```

In this example we had a bad initial guess so the method went the wrong way at first, but it eventually honed in on the solution. This highlights a feature of open methods: the root estimate can get worse, and even diverge. This is in comparison with closed methods where the root is confined to an interval. On the other hand, open methods only require an initial guess instead of knowledge of an interval where the root lies. In the first iteration, the method did move closer to the root, but the new estimate was close to a local minimum:

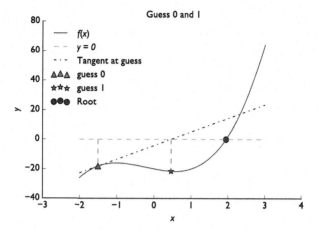

The slope at this point then moves the method long distance in the wrong direction:

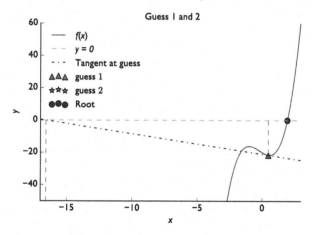

This estimate is much worse than that after the first iteration. Nevertheless, after this point the method, does move in the correct direction.

The other test we performed on the closed root finding methods was the critical sphere problem:

$$f(R) = \left(\frac{\pi}{R+2D}\right)^2 - \frac{\nu\Sigma_f - \Sigma_a}{D}.$$

For Newton's method, we need to compute the derivative:

$$f'(R) = \frac{-2\pi^2}{(R+2D)^3}.$$

Recall that regula falsi (false position) had a very hard time with this problem because the function is flat for a large range of the radius and sharply changing past a transition point. Newton's method does not suffer from these problems.

First, we define the function and a function to compute its derivative:

```
In [3]: def Crit_Radius(R, D=9.21, nuSigf = 0.1570, Siga = 0.1532):
            return (np.pi/(R + 2*D))**2 - (nuSigf - Siga)/D
        def DCrit_Radius(R, D=9.21, nuSigf = 0.1570, Siga = 0.1532):
            return (-2.0*np.pi**2/(R + 2*D)**3)
```

For this problem bisection took 28 iterations, false position took 260, and Ridder's method took 3 iterations. Here, we will run Newton's method with an initial guess of 120 (about the midpoint of the initial range used in the previous chapter).

```
In [4]: Radius = newton(Crit_Radius,DCrit_Radius,120,LOUD=True)
        print("The critical radius estimate is",Radius,
          "\nf(",Radius,") =",Crit_Radius(Radius))

x0 = 120
x_ 1 = 120 - 0.00010251745512686794 / -7.442746142981495e-06
= 133.77414373101277
x_ 2 = 133.77414373101277 - 1.3497467451998235e-05 / -5.599328100744185e-06
= 136.1846949987987
It took 2 iterations
The critical radius estimate is 136.1846949987987
f( 136.1846949987987 ) = 3.140322315689872e-07
```

This is even faster than Ridder's method, though the comparison is not exactly fair. We did, however, need to know the derivative of the function and evaluate this at each iteration.

BOX 13.1 NUMERICAL PRINCIPLE

Newton's method converges rapidly and has a degree of robustness. It also only requires an initial guess, not an interval bounding the root. You do need to know the derivative of the target function to use Newton's method without modification.

13.2 INEXACT NEWTON

What if we do not know the derivative of the function? There may be cases where the derivative of the function is unknown or not easily calculable. In these cases we can use a method that is known as inexact Newton. This method estimates the derivative using a finite difference:

$$f'(x_i) \approx \frac{f(x_i + \delta) - f(x_i)}{\delta},$$

which will converge to the derivative as $\delta \to 0$. Indeed, the limit of this approximation as $\delta \to 0$ is the definition of a derivative. To implement this we need to only make a small change to the code to estimate the derivative instead of calling a derivative function. The downside is that we need an extra function evaluation to estimate the derivative. This extra cost is partially offset by eliminating the need to evaluate a function that gives the derivative. In the code below, we modify the Newton's method implementation above to perform inexact Newton iterations.

```
In [5]: def inexact_newton(f,x0,delta = 1.0e-7, epsilon=1.0e-6, LOUD=False):
            """Find the root of the function f via Newton-Raphson method
            Args:
                f: function to find root of
                x0: initial guess
                delta: finite difference parameter
                epsilon: tolerance

            Returns:
                estimate of root
            """
            x = x0
            if (LOUD):
                print("x0 =",x0)
            iterations = 0
            fx = f(x)
            while (np.fabs(fx) > epsilon):
                fxdelta = f(x+delta)
                slope = (fxdelta - fx)/delta
                if (LOUD):
                    print("x_",iterations+1,"=",x,"-",fx,"/",slope,"=",
                          x - fx/slope)
                x = x - fx/slope
                fx = f(x)
                iterations += 1
            print("It took",iterations,"iterations")
            return x #return estimate of root
```

To compute the derivative we need to define the value of δ. In general, a reasonable value for this parameter is 10^{-7}, though it should be adjusted if the value of x in the function evaluation is very large or small.

On the critical radius problem, inexact Newton performs the same as the original Newton method:

```
In [5]: Radius = inexact_newton(Crit_Radius,120,LOUD=True)
        print("The critical radius estimate is",Radius,
              "\nf(",Radius,") =",Crit_Radius(Radius))

x0 = 120
x_ 1 = 120 - 0.00010251745512686794 / -7.442742169100347e-06
= 133.77415108540026
x_ 2 = 133.77415108540026 - 1.3497426272372716e-05 / -5.599325256927523e-06
= 136.18469622307978
It took 2 iterations
The critical radius estimate is 136.18469622307978
f( 136.18469622307978 ) = 3.1402569209483836e-07
```

Notice that we get the same answer and it took the same number of iterations.

BOX 13.2 NUMERICAL PRINCIPLE

Inexact Newton gives a way around the necessity of knowing the function's deriva- tive. The cost is an extra function evaluation at each iteration.

13.3 SECANT METHOD

The secant method is a variation on the theme of Newton's method. It takes its name from the fact that it constructs a straight line that intersects the curve at two points: such a line is called a secant. In this case we use the previous two guesses to construct the slope:

$$f'(x_i) \approx \frac{f(x_i) - f(x_{i-1})}{x_i - x_{i-1}}.$$

The benefit of this is that it does not require an additional function evaluation, nor do we have to evaluate a derivative function. This will be a big savings if it takes a long time to do a function evaluation. One issue is that we need two points to get started. Therefore, we can use inexact Newton for the first step and then use secant from then on. In a graphical demonstration, we first take a step of inexact Newton:

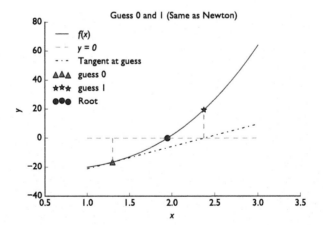

Then we draw the estimate of the derivative using x_0 and x_1, and find where this crosses the x axis:

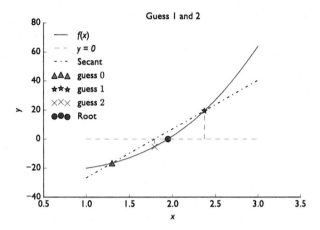

Notice that this estimate of the root behaves similarly to that from Newton's method, with only one function evaluation per iteration. A Python implementation of the secant method is given next:

```
In [6]:  def secant(f,x0,delta = 1.0e-7, epsilon=1.0e-6, LOUD=False):
             """Find the root of the function f via Newton-Raphson method
             Args:
                 f: function to find root of
                 x0: initial guess
                 delta: finite difference parameter
                 epsilon: tolerance

             Returns:
                 estimate of root
             """
             x = x0
             if (LOUD):
                 print("x0 =",x0)
             #first time use inexact Newton
             x_old = x
             fold = f(x_old)
             fx = fold
             slope = (f(x_old+delta) - fold)/delta
             x = x - fold/slope
             if (LOUD):
                 print("Inexact Newton\nx_",1,"=",x,"-",fx,"/",slope,"=",
                     x - fx/slope,"\nStarting Secant")
             fx = f(x)
             iterations = 1
             while (np.fabs(fx) > epsilon):
                 slope = (fx - fold)/(x - x_old)
                 fold = fx
                 x_old = x
                 if (LOUD):
                     print("x_",iterations+1,"=",x,"-",fx,
                         "/",slope,"=",x - fx/slope)
                 x = x - fx/slope
```

```
        fx = f(x)
        iterations += 1
    print("It took",iterations,"iterations")
    return x #return estimate of root
```

Running the secant method on the critical radius problem, we observe that it takes one more iteration the Newton's method:

```
In [7]:  Radius = secant(Crit_Radius,120,LOUD=True)
         print("The critical radius estimate is",Radius,"\nf(",Radius,
         ") =",Crit_Radius(Radius))


x0 = 120
Inexact Newton
x_ 1 = 133.77415108540026 - 0.00010251745512686794 / -7.442742169100347e-06
= 147.54830217080053
Starting Secant
x_ 2 = 133.77415108540026 - 1.3497426272372716e-05 / -6.462832322846442e-06
= 135.86262028870274
x_ 3 = 135.86262028870274 - 2.0397790242225296e-06 / -5.486146135184707e-06
= 136.23442573718336
It took 3 iterations
The critical radius estimate is 136.23442573718336
f( 136.23442573718336 ) = 4.8524539480966286e-08
```

Despite the fact that it took one more iteration, secant only required four function evaluations to compute the root: one per iteration, plus one extra for the slope estimation in the first step.

BOX 13.3 NUMERICAL PRINCIPLE

The secant method is a middle ground between inexact Newton and Newton's method. It uses existing function evaluations to approximate the derivative of the function. It does converge slightly slower than Newton's method, however.

13.4 SLOW CONVERGENCE

Newton's method, including its inexact variant, and the secant method can converge slowly in the presence of the following:

1. Multiple roots or closely spaced roots,
2. Complex roots,
3. Bad initial guess.

BOX 13.4 NUMERICAL PRINCIPLE

General nonlinear functions can behave in many different ways, and some of these behaviors make root-finding difficult. In practi-cal applications, finding a good guess at the root is often necessary to find a root effi-ciently.

The function

$$f(x) = x^7$$

has multiple roots at 0: it converges slowly with Newton's method.

```
In [8]:   mult_root = lambda x: 1.0*x**7
          Dmult_root = lambda x: 7.0*x**6
          root = newton(mult_root,Dmult_root,1.0,LOUD=True)
          print("The root estimate is",root,"\nf(",root,") =",mult_root(root))

x0 = 1.0
x_ 1 = 1.0 - 1.0 / 7.0 = 0.8571428571428572
x_ 2 = 0.8571428571428572 - 0.33991667708911394 / 2.7759861962277634
= 0.7346938775510204
x_ 3 = 0.7346938775510204 - 0.11554334736330486 / 1.1008713373781547
= 0.6297376093294461
x_ 4 = 0.6297376093294461 - 0.03927511069548781 / 0.43657194805493627
= 0.5397750937109538
...
x_ 12 = 0.18347855622969242 - 6.9999864354836515e-06 / 0.00026706066395597614
= 0.15726733391116493
x_ 13 = 0.15726733391116493 - 2.3794121288184727e-06 / 0.00010590810238531585
= 0.13480057192385567
It took 13 iterations
The root estimate is 0.13480057192385567
f( 0.13480057192385567 ) = 8.088018642535101e-07
```

If we decrease the number of roots to 1,

$$f(x) = \sin(x),$$

we see that a similar problem converges faster.

```
In [9]:   mult_root = lambda x: np.sin(x)
          Dmult_root = lambda x: np.cos(x)
          root = newton(mult_root,Dmult_root,1.0,LOUD=True)
          print("The root estimate is",root,"\nf(",root,") =",mult_root(root))

x0 = 1.0
x_ 1 = 1.0 - 0.841470984808 / 0.540302305868
= -0.557407724655
x_ 2 = -0.557407724655 - -0.52898809709 / 0.848629243626
= 0.0659364519248
x_ 3 = 0.0659364519248 - 0.0658886845842 / 0.997826979613
```

```
= -9.57219193251e-05
x_ 4 = -9.57219193251e-05 - -9.57219191789e-05 / 0.999999995419
= 2.92356620141e-13
It took 4 iterations
The root estimate is 2.92356620141e-13
f( 2.92356620141e-13 ) = 2.92356620141e-13
```

For the case of complex roots we will consider a function that has complex roots near the actual root. One such function is

$$f(x) = x(x-1)(x-3) + 3.$$

The derivative of this function is

$$f'(x) = 3x^3 - 8x + 3.$$

The root is at $x = -0.546818$.

```
In [10]: x = np.linspace(-1,4,200)
         comp_root = lambda x: x*(x-1)*(x-3) + 3
         d_comp_root = lambda x: 3*x**2 - 8*x + 3
         root = newton(comp_root,d_comp_root,2.0,LOUD=True)
         print("The root estimate is",root,"\nf(",root,") =",mult_root(root))


x0 = 2.0
x_ 1 = 2.0 - 1.0 / -1.0 = 3.0
x_ 2 = 3.0 - 3.0 / 6.0 = 2.5
x_ 3 = 2.5 - 1.125 / 1.75 = 1.8571428571428572
x_ 4 = 1.8571428571428572 - 1.1807580174927113 / -1.5102040816326543
= 2.6389961389961383
. . .

x_ 42 = -0.6654802789331873 - -1.062614102742372 / 9.652434236412477
= -0.5553925977621718
x_ 43 = -0.5553925977621718 - -0.07133846535161004 / 8.368523595044415
= -0.5468679799438203
x_ 44 = -0.5468679799438203 - -0.00041113661500030271 / 8.272137602014066
= -0.5468182785685793
It took 44 iterations
The root estimate is -0.5468182785685793
f( -0.5468182785685793 ) = -0.519972092294
```

This converged slowly because the complex roots at $x = 2.2734 \pm 0.5638i$ make the slope of the function change so that tangents do not necessarily point to a true root.

We can see this graphically by looking at each iteration.

The first iteration moves in the wrong direction:

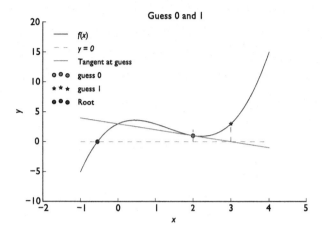

Iterations two and three move in the correct direction:

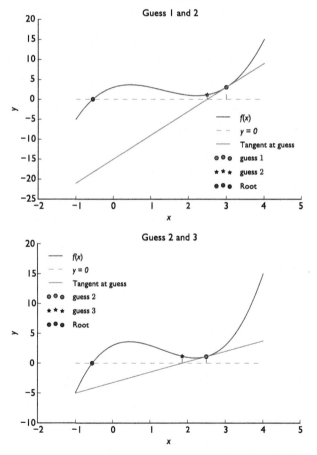

Step four then moves in wrong direction: it has been two steps forward, one step back:

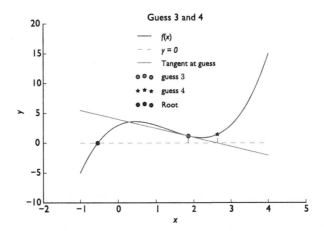

The presence of the complex root causes the solution to oscillate around the local minimum of the function. Eventually, the method will converge on the root, but it takes many iterations to do so. The upside, however, is that it does eventually converge.

13.5 NEWTON'S METHOD FOR SYSTEMS OF EQUATIONS

Finding the root of a single function is interesting, but in general a simple problem. It is very common that a problem you encounter will require the root of a multidimensional function. This is the situation we analyze now.

Say that we have a function of n variables $\mathbf{x} = (x_1, \ldots x_n)$:

$$\mathbf{F}(\mathbf{x}) = \begin{pmatrix} f_1(x_1, \ldots, x_n) \\ \vdots \\ f_n(x_1, \ldots, x_n) \end{pmatrix},$$

the root is $\mathbf{F}(\mathbf{x}) = 0$. For this scenario we no longer have a tangent line at point $\mathbf{x}$, rather we have a Jacobian matrix that contains the derivative of each component of $\mathbf{F}$ with respect to each component of $\mathbf{x}$:

$$\mathbf{J}(\mathbf{x}) = \begin{pmatrix} \frac{\partial f_1}{\partial x_1}(x_1, \ldots, x_n) & \cdots & \frac{\partial f_1}{\partial x_n}(x_1, \ldots, x_n) \\ \vdots & & \vdots \\ \frac{\partial f_n}{\partial x_1}(x_1, \ldots, x_n) & \cdots & \frac{\partial f_n}{\partial x_n}(x_1, \ldots, x_n) \end{pmatrix}.$$

We reformulate Newton's method for a single equation

$$x_{i+1} = x_i - \frac{f(x_i)}{f'(x_i)},$$

as

$$f'(x_i)(x_{i+1} - x_i) = -f(x_i).$$

The multidimensional analog of this equation is

$$\mathbf{J}(\mathbf{x}_i)(\mathbf{x}_{i+1} - \mathbf{x}_i) = -\mathbf{F}(\mathbf{x}_i).$$

Note this is a linear system of equations to solve to get a vector of changes for each x: $\boldsymbol{\delta} = \mathbf{x}_{i+1} - \mathbf{x}_i$. Therefore, in each step we solve the system

$$\mathbf{J}(\mathbf{x}_i)\boldsymbol{\delta} = -\mathbf{F}(\mathbf{x}_i),$$

and set

$$\mathbf{x}_{i+1} = \mathbf{x}_i + \boldsymbol{\delta}.$$

In multidimensional rootfinding we can observe the importance of having a small number of iterations: we need to solve a linear system of equations at each iteration. If this system is large, the time to find the root could be prohibitively long.

BOX 13.5 NUMERICAL PRINCIPLE

Finding the root of a multi-dimensional function is often computationally expensive. There is the additionally complication of determining the Jacobian matrix. The efficient methods for finding roots of functions in many dimensions are beyond the scope of this work. Suffice it to say that these methods are based on approximating the Jacobian of the system with finite differences and employing efficient linear solvers.

13.5.1 Rectangular Parallelepiped (Shoebox) Reactor Example

As an example of a multidimensional root finding problem, we will consider the problem of designing a parallelepiped reactor. For this type of reactor the critical equation given by one-group diffusion theory, i.e., when geometric and materials buckling are equal, is

$$\left(\frac{\pi}{a+2D}\right)^2 + \left(\frac{\pi}{b+2D}\right)^2 + \left(\frac{\pi}{c+2D}\right)^2 = \frac{\nu \Sigma_f - \Sigma_a}{D},$$

where a, b, and c are length of the sides. We will solve this with $D = 9.21$ cm, $\nu \Sigma_f = 0.1570$ cm^{-1}, and $\Sigma_a = 0.1532$ cm^{-1}.

To make this a system, we stipulate that the surface area should be 1.2×10^6 cm^2, and that $a = b$. This makes

$$\mathbf{F}(\mathbf{x}) = \begin{pmatrix} \left(\frac{\pi}{a+2D}\right)^2 + \left(\frac{\pi}{b+2D}\right)^2 + \left(\frac{\pi}{c+2D}\right)^2 - \frac{\nu \Sigma_f - \Sigma_a}{D} \\ 2(ab + bc + ac) - 1.2 \times 10^6 \\ a - b \end{pmatrix},$$

with $\mathbf{x} = (a, b, c)$.

The Jacobian is

$$\mathbf{J(x)} = \begin{pmatrix} \frac{2\pi}{(a+18.42)^2} & \frac{2\pi}{(b+18.42)^2} & \frac{2\pi}{(c+18.42)^2} \\ 2(b+c) & 2(a+c) & 2(a+b) \\ 1 & -1 & 0 \end{pmatrix}.$$

Start with an initial guess of $a = b = 7000$ and $c = 100$ all in cm:

$$\mathbf{F(x_0)} = \begin{pmatrix} 0.000292 \\ 9.96 \times 10^7 \\ 0 \end{pmatrix}.$$

Solving the system

$$\mathbf{J(x_0)}\boldsymbol{\delta}_1 = -\mathbf{F(x_0)},$$

via Gauss elimination gives

$$\boldsymbol{\delta}_1 = \begin{pmatrix} -6357.33535866 \\ -6357.33535866 \\ 45.4741297 \end{pmatrix},$$

which gives

$$\mathbf{x_1} = \begin{pmatrix} 642.66464134 \\ 642.66464134 \\ 145.4741297 \end{pmatrix}.$$

Rather than continuing by hand, we will write a Python function to solve the problem. We will define a function inside of our Newton function to compute the finite difference Jacobian on the fly.

```
In [11]: #first import our Gauss Elim function
         from GaussElim import *
         def newton_system(f,x0,delta = 1.0e-7, epsilon=1.0e-6, LOUD=False):
             """Find the root of the function f via inexact Newton-Raphson method
             Args:
                 f: function to find root of
                 x0: initial guess
                 delta: finite difference parameter
                 epsilon: tolerance

             Returns:
                 estimate of root
             """
             def Jacobian(f,x,delta = 1.0e-7):
                 N = x0.size
                 J = np.zeros((N,N))
                 idelta = 1.0/delta
                 x_perturbed = x.copy() #copy x to add delta
                 fx = f(x) #only need to evaluate this once
                 for i in range(N):
                     x_perturbed[i] += delta
```

```
            col = (f(x_perturbed) - fx) * idelta
            x_perturbed[i] = x[i]
            J[:,i] = col
        return J

    x = x0
    if (LOUD):
        print("x0 =",x0)
    iterations = 0
    fx = f(x)
    while (np.linalg.norm(fx) > epsilon):
        J = Jacobian(f,x,delta)

        RHS = -fx;
        delta_x = GaussElimPivotSolve(J,RHS)
        x = x + delta_x
        fx = f(x)
        if (LOUD):
            print("Iteration",iterations+1,": x =",x," norm(f(x)) =",
                  np.linalg.norm(fx))
        iterations += 1
    print("It took",iterations,"iterations")
    return x #return estimate of root
```

To use this function we have to define the function we want to minimize:

```
In [12]: def Reactor(x, D=9.21, nuSigf = 0.1570, Siga = 0.1532):
             """This function is defined in the equation above
             """
             answer = np.zeros((3))
             answer[0] = (np.pi/(x[0] + 2*D))**2 +
                         (np.pi/(x[1] + 2*D))**2 +
                         (np.pi/(x[2] + 2*D))**2 - (nuSigf - Siga)/D
             answer[1] = 2*(x[0]*x[1] + x[1]*x[2] + x[0]*x[2])-1.2e6
             answer[2] = x[0] - x[1]
             return answer
```

We can now set up our initial guess and then solve:

```
In [13]: x0 = np.array([7000.0,7000.0,100.0])
         x = newton_system(Reactor,x0,LOUD=True, epsilon=1.0e-8, delta = 1.0e-10)
         #check
         print("The surface area is",2.0*(x[0]*x[1] + x[1]*x[2] + x[0]*x[2]))
         D=9.21; nuSigf = 0.1570; Siga = 0.1532;
         print("The geometric buckling is",(np.pi/(x[0] + 2*D))**2 +
               (np.pi/(x[1] + 2*D))**2 + (np.pi/(x[2] + 2*D))**2)
         print("The materials buckling is",(nuSigf - Siga)/D)

x0 = [ 7000.  7000.   100.]
Iteration 1 : x = [ 3457.80758198  3457.80758198    124.53337225]
norm(f(x)) = 24435316.3032
...
Iteration 9 : x = [ 642.66464134  642.66464134  145.4741297 ]
norm(f(x)) = 1.08420217249e-19
```

```
It took 9 iterations
The surface area is 1200000.0
The geometric buckling is 0.000412595005429
The materials buckling is 0.0004125950054288814
```

The solution is that the *a* and *b* are about 642.66 cm and the height of the reactor is 145.47 cm. There are multiple solutions to the problem. We can change the initial condition so that the method finds a reactor that is taller than this one by guessing a thin and tall parallelepiped.

```
In [14]: x0 = np.array([100.0,100.0,10000.0])
         x = newton_system(Reactor,x0,LOUD=True, epsilon=1.0e-8, delta = 1.0e-10)
         #check
         print("The surface area is",2.0*(x[0]*x[1] + x[1]*x[2] + x[0]*x[2]))
         D=9.21; nuSigf = 0.1570; Siga = 0.1532;
         print("The geometric buckling is",(np.pi/(x[0] + 2*D))**2 +
               (np.pi/(x[1] + 2*D))**2 + (np.pi/(x[2] + 2*D))**2)
         print("The materials buckling is",(nuSigf - Siga)/D)

x0 = [  100.    100.  10000.]
Iteration 1 : x = [  141.87143819    141.87143819 -1265.89519199]
norm(f(x)) = 1878122.47601
...
Iteration 8 : x = [  201.6439505    201.6439505   1386.94891624]
norm(f(x)) = 1.86264514923e-09
It took 8 iterations
The surface area is 1200000.0
The geometric buckling is 0.000412595005429
The materials buckling is 0.0004125950054288814
```

This reactor is almost 14 meters tall and has a geometric cross-section that is about 2 by 2 meters. The fact that there are multiple solutions to the problem is one of the features of nonlinear root finding to be aware of. One of these solutions may be better for a particular application, and picking a different initial guess will influence which root the method converges to. Furthermore, in the presence of multiple roots, an initial guess in the middle of the two roots may slow the convergence:

```
In [81]: x0 = np.array([421.0,421.0,750.0])
         x = newton_system(Reactor,x0,LOUD=True, epsilon=1.0e-8, delta = 1.0e-10)
         #check
         print("The surface area is",2.0*(x[0]*x[1] + x[1]*x[2] + x[0]*x[2]))
         D=9.21; nuSigf = 0.1570; Siga = 0.1532;
         print("The geometric buckling is",(np.pi/(x[0] + 2*D))**2 +
               (np.pi/(x[1] + 2*D))**2 + (np.pi/(x[2] + 2*D))**2)
         print("The materials buckling is",(nuSigf - Siga)/D)

x0 = [ 421.  421.  750.]
Iteration 1 : x = [ -403.19253367  -403.19253367  2792.77379015]
norm(f(x)) = 5378973.72321
...
Iteration 16 : x = [  201.6439505    201.6439505   1386.94891624]
norm(f(x)) = 4.65661287308e-10
```

```
It took 16 iterations
The surface area is 1200000.0
The geometric buckling is 0.000412595005429
The materials buckling is 0.00041259500054288814
```

This guess basically doubled the number of iterations.

CODA

In this, and the previous lecture, we have discussed a range of rootfinding methods. Open methods are good when you only have a single initial guess. These methods usually require either a knowledge of the functions's derivative or require an approximation to the derivative. Closed methods are more robust because the root estimate will always be in the initial bounds. This robustness comes at the cost of determining bounds for the root.

In practice, the appropriate method will be problem dependent. With the root finding tools we have covered, you have a rich toolset to find roots for just about any problem you will encounter.

PROBLEMS

Short Exercises

13.1. Apply Newton's method to the function $f(x) = (1 - x)^2 + 100(1 - x^2)^2 = 0$, using the initial guess of 2.5.

13.2. You are given a radioactive sample with an initial specific activity of 10^4 Bq/kg, and you are told the half-life is 19 days. Compute the time it will take to get the specific activity of Brazil nuts (444 Bq/kg) using Newton's method, inexact Newton, and secant.

Programming Projects

1. Roots of Bessel Function

Consider the Bessel function of the first kind defined by

$$J_\alpha(x) = \sum_{m=0}^{\infty} \frac{(-1)^m}{m! \, \Gamma(m + \alpha + 1)} \left(\frac{x}{2}\right)^{2m+\alpha}.$$

13.1. Write a Python code that prompts the user asks if they want to use bisection or Newton's method. Then the user enters an initial guess or initial range depending on the method selected.

13.2. Using the input find a root to J_0. Each iteration print out to the user the value of $J_0(x)$ for the current guess and the change in the guess.

13.3. For testing, there is a root at $x \approx 2.4048$. Also, math.gamma(x), will give you $\Gamma(x)$.

2. Nonlinear Heat Conduction

We will consider heat conduction in a cylindrical nuclear fuel pellet. In order to simplify the model, we will

- suppose a steady state operation,
- neglect the axial heat conduction,
- suppose that the heat is uniformly generated radially,
- neglect the presence of the clad.

Under these assumption, the heat conduction equation is:

$$\frac{1}{r} \frac{\partial}{\partial r} \left(r k(T) \frac{\partial T}{\partial r} \right) = -q''' \quad \text{for } 0 \le r \le R, \tag{13.1}$$

$$T(R) = T_R,$$

$$\left. \frac{\partial T}{\partial r} \right|_{r=0} = 0,$$

where $T = T(r)$ is the temperature inside the pellet, $k = k(T)$ is the temperature dependent conductivity, in W/(m·C), and q''' is the heat source (W/m^3). The temperature T_R is the temperature at the surface of the pellet. Solving for the temperature distribution within the pellet can be transformed into the following statement:

$$\int_{T_R}^{T(r)} k(T) \, dT = q''' \frac{R^2 - r^2}{4}. \tag{13.2}$$

Suppose you are given a formula for $k(T)$. You can then compute the conductivity integral (i.e., the antiderivative)

$$I(u) = \int k(u) \, du + C.$$

Finally, the problem boils down to solving the following nonlinear equation of one variable:

$$I(T(r)) = q''' \frac{R^2 - r^2}{4} + I(T_R).$$

If you solve the above equation at various radii r for $T(r)$, you will then get the temperature profile (i.e., the temperature at these different positions).

The data below (see Table 13.1) provides you with the conductivity formula $k(T)$ (which is easy to integrate), the pellet radius R, and the boundary condition T_R. The rest of the data will be useful to determine the average power density q''' (power per unit volume) for the entire core. The core is a typical Westinghouse PWR reactor, containing a given number of fuel assemblies. Each fuel assembly is loaded with a given number of fuel rods. Be careful with your unit conversions.

TABLE 13.1 Problem Definition

Conductivity, W/(m·C)	$k = 1.05 + 2150/(T + 200)$
Total power generated in the core	4200 MWth
Number of fuel assemblies (FA)	205
Number of fuel pins per FA	264
Core height	14 ft.
Pellet radius R	0.41 cm
Temperature at R	400°C

Your assignment

- Derive Eq. (13.2) using the heat conduction equation and it's boundary conditions.
- Write a clean and clear Python code to solve the above problem using the following methods: (1) bisection and (2) Newton's.
  - Use 11 grids points for the temperature profile (i.e., $r_i = \frac{i-1}{10} R$ for $i = 1 \dots 11$).
  - Compare graphically your results with the case where the conductivity is assumed to be the following constant:

$$k = k(500°C) = 4.12 \, W/(m \cdot C).$$

Note that when the conductivity is constant, you have an analytical solution. Provide the analytical solution $T^A(r)$.

14

Finite Difference Derivative Approximations

Différance is the systematic play of differences, of the traces of differences, of the spacing by means of which elements are related to each other.

               –Positions *by Jacques Derrida*

CHAPTER POINTS

- The definition of a derivative requires taking an infinitesimally small perturbation of a function.

- We cannot deal with infinitesimals on a compute so finite difference formulae are commonly used. These can be derived from the Taylor series polynomial of the function.

- The convergence of an approximate derivative can be inferred theoretically from the Taylor series and empirically from the error on a log-log plot.

- Using complex numbers allows the derivative to be approximated to a high degree of precision by obviating the need for taking differences of functions.

In introductory calculus, students are taught that the derivative of a function is defined via the limit

$$f'(x) = \frac{df}{dx} = \lim_{h \to 0} \frac{f(x+h) - f(x)}{h}.$$

On a computer we want to compute derivatives, but we cannot evaluate the derivative using an infinitesimally small value for h due to finite precision arithmetic. In this chapter we will derive formulas called finite difference derivatives because h will have a finite value rather than an infinitesimally small one. In one formula we will approximate the derivative in the same manners as the above equation using

$$f'(x) \approx \frac{f(x+h) - f(x)}{h}.$$

The error we incur in doing so we also be discussed.

14.1 TAYLOR SERIES AND BIG-O NOTATION

We can approximate a function $f(x)$ near a point x using the Taylor series

$$f(x+h) = f(x) + hf'(x) + \frac{h^2}{2}f''(x) + \frac{h^3}{6}f'''(x) + \cdots + \frac{h^n}{n!}f^{(n)}(x) + \cdots.$$

The Taylor series is an infinite sum. One way that this is commonly written is using a particular notation instead of the "…":

$$f(x+h) = f(x) + hf'(x) + \frac{h^2}{2}f''(x) + \frac{h^3}{6}f'''(x) + O(h^4).$$

What the $O(h^4)$ means that the rest of the terms will be smaller than some constant times h^4 as $h \to 0$. We can see this through a simple example. Here we look at the approximation of

$$\cos h = \cos 0 - h \sin 0 - \frac{h^2}{2}\cos 0 + \frac{h^3}{6}\sin 0 + O(h^4).$$

Below we plot the both the function and its approximation.

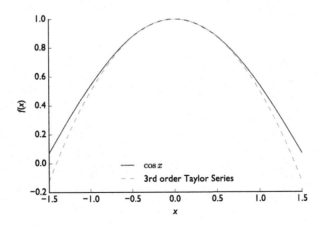

The absolute difference between $\cos h$ and the third-order Taylor series as a function of h on a log-log scale is shown next. It appears to be a line with a slope that is approximately 4.

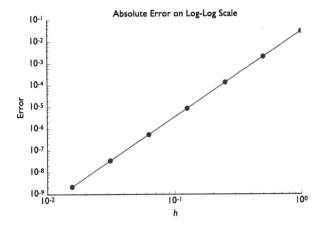

You might ask yourself, why is the slope 4? This can be answered by looking at the equation for the error:

$$f(x+h) - \left(f(x) + hf'(x) + \frac{h^2}{2} f''(x) + \frac{h^3}{6} f'''(x) \right) = Ch^4 + O(h^5),$$

in this equation we have used the fact that $O(h^4)$ means some constant, which here we call C, times h^4. The remaining terms we have written as $O(h^5)$.

When we take the absolute value and the logarithm of each side we get

$$\log_{10} \left| f(x+h) - \left(f(x) + hf'(x) + \frac{h^2}{2} f''(x) + \frac{h^3}{6} f'''(x) \right) \right| \approx \log_{10} |C| + 4 \log_{10} h,$$

which is a line with slope 4. The constant tells us what level the curve starts out at when $h = 1$. This formula is approximate because we have left out the h^5 and higher terms, which we assume to be small since h is small in our demonstration above.

BOX 14.1 NUMERICAL PRINCIPLE

When plotting the error of an approximation as a function of a parameter h on a log-log scale, the slope of the line is an approximation to how the error scales as h to some power. That is, if the slope in plot is m, then we can estimate that the error is

$$\text{error} = Ch^m = O(h^m).$$

The we have already seen the concept of Big-O notation when we discussed the scaling Gaussian Elimination, saying that it scaled as $O(n^3)$ where n was the number of equations in the linear system. One difference is that in the algorithm scaling discuss we were concerned about the scaling as $n \to \infty$, whereas here we are interested in $h \to 0$.

14.2 FORWARD AND BACKWARD DIFFERENCE FORMULAS

From the Taylor series at $x + h$,

$$f(x + h) = f(x) + hf'(x) + \frac{h^2}{2} f''(x) + \frac{h^3}{6} f'''(x) + O(h^4),$$

we notice that there is an $f'(x)$ term in the equation. If we "solve" for this derivative by subtracting $f(x)$ from both sides and then dividing by h we get

$$\frac{f(x + h) - f(x)}{h} = f'(x) + \frac{h}{2} f''(x) + \frac{h^2}{6} f'''(x) + O(h^3),$$

or in shorter form

$$\frac{f(x + h) - f(x)}{h} = f'(x) + O(h).$$

Therefore the approximation

$$f'(x) \approx \frac{f(x + h) - f(x)}{h},$$

is an order h approximation because the error is proportional to h as h goes to 0. This is called a forward difference formula because the function is evaluated h forward of x to approximate the derivative.

We did this using $f(x + h)$ in our formula, but we could have also used $f(x - h)$ which has the Taylor series

$$f(x - h) = f(x) - hf'(x) + \frac{h^2}{2} f''(x) - \frac{h^3}{6} f'''(x) + O(h^4),$$

to get the formula

$$\frac{f(x) - f(x - h)}{h} = f'(x) + O(h).$$

Therefore the approximation

$$f'(x) \approx \frac{f(x) - f(x - h)}{h},$$

is also an order h approximation because the error is proportional to h as h goes to 0. This formula is a backward difference formula because the function is evaluated h behind x.

BOX 14.2 NUMERICAL PRINCIPLE

The forward and backward finite difference formulas are first-order in h approximations to the derivative given by

Forward Difference

$$f'(x) = \frac{f(x+h) - f(x)}{h} + O(h),$$

and

Backward Difference

$$f'(x) = \frac{f(x) - f(x-h)}{h} + O(h).$$

14.3 HIGHER-ORDER APPROXIMATIONS

Both of these formulas for the derivative are first-order in h. These formulas are fine, but as we will see when we solve differential equations, first-order solutions typically have too much error for our purposes. We desire a way of getting higher-order approximations to the derivative.

Here we will derive a second-order approximation using both

$$f(x+h) = f(x) + hf'(x) + \frac{h^2}{2}f''(x) + \frac{h^3}{6}f'''(x) + O(h^4),$$

and

$$f(x-h) = f(x) - hf'(x) + \frac{h^2}{2}f''(x) - \frac{h^3}{6}f'''(x) + O(h^4).$$

Notice that if we subtract the $f(x-h)$ equation from the equation for $f(x+h)$ and then divide by $2h$ we get

$$\frac{f(x+h) - f(x-h)}{2h} = f'(x) + \frac{h^2}{6}f'''(x) + O(h^4),$$

or in shorter form

$$\frac{f(x+h) - f(x-h)}{2h} = f'(x) + O(h^2).$$

Therefore the approximation

$$f'(x) \approx \frac{f(x+h) - f(x-h)}{2h},$$

is an order h^2 approximation because the error is proportional to h^2 as h goes to 0. This formula is called a central-difference formula because the function is evaluated around a center of x a value of h on either side. One thing to note is that the error terms in this approximation only have even powers of h because of the way the odd powers cancel when combining the two.

With a second-order approximation, if we cut h in half, the error goes down by a factor of 4, compared to a factor of 2 with a first-order method.

BOX 14.3 NUMERICAL PRINCIPLE

The central finite difference formula is a second-order in h approximation to the derivative given by

Central Difference

$$f'(x) = \frac{f(x+h) - f(x-h)}{2h} + O(h^2).$$

We could go through the process of obtaining even higher-order derivatives (third-order, fourth-order, etc.), but in practice this is generally not useful because the formulas become cumbersome and there are typically better ways of accomplishing higher-order accuracy. We will see one of these ways shortly.

14.4 COMPARISON OF THE APPROXIMATIONS

Consider the function

$$f(x) = \arctan(x)\cosh(x)$$

and look at approximations to $f'(x)$ at $x = 1$. The actual answer is $f'(1) = 1.694541176517952557683135$. In the following graph we show the error in the derivative estimate as a function of h for the three methods we have seen so far on a log-log scale.

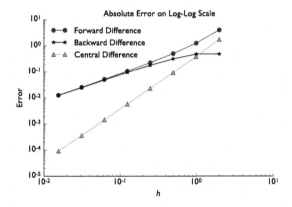

The central difference result has a slope of about 2, between the forward and backward differences have a slope around 1 (1.01 and 0.989 between the last two points of each line for the forward and backward difference, respectively).

In this example we see that the errors decay as expected for the two first-order methods and the one second-order method. As we expect, as h gets smaller the second-order method wins out. Nevertheless, this does not mean that at a particular value of h the second-order method will have smaller error. The graph above shows that at $h = 2$, the backward difference approximation has the smallest error. This is due to the fact that order of the formula just says how the error changes with h and says nothing about the constant in front of the leading-order term. Eventually, the second-order method will win, but we cannot say anything about a particular point.

BOX 14.4 NUMERICAL PRINCIPLE

Just because a method has higher order accuracy than another does not mean that it will give a better approximation for a particular value of h. What higher order accuracy means is that the error will decrease faster as $h \to 0$.

14.5 SECOND DERIVATIVES

We may also want to compute the value of $f''(x)$. To do this we start with the Taylor series for $f(x+h)$ and $f(x-h)$:

$$f(x+h) = f(x) + hf'(x) + \frac{h^2}{2}f''(x) + \frac{h^3}{6}f'''(x) + O(h^4),$$

and

$$f(x-h) = f(x) - hf'(x) + \frac{h^2}{2}f''(x) - \frac{h^3}{6}f'''(x) + O(h^4).$$

If we add these two equations together, notice that the h and h^3 terms cancel:

$$f(x+h) + f(x-h) = 2f(x) + h^2 f''(x) + O(h^4).$$

Now rearranging this formula to solve for the second-derivative we get

$$\frac{f(x+h) - 2f(x) + f(x-h)}{h^2} = f''(x) + O(h^2).$$

That is, we can get a second-order in h approximation to the second derivative by evaluating the function at $f(x)$ and $f(x \pm h)$.

BOX 14.5 NUMERICAL PRINCIPLE

A common formula for estimating a second-derivative with second-order accuracy is

$$f''(x) = \frac{f(x+h) - 2f(x) + f(x-h)}{h^2} + O(h^2).$$

Another derivative we may want to approximate in nuclear engineering is

$$\frac{d}{dx} D(x) \frac{d\phi}{dx}.$$

This term appears in the diffusion equation for neutrons or other particles. To approximate this, we first approximate $D(x + h/2)\phi'(x + h/2)$ using a central difference with $h/2$ as

$$D\left(x + \frac{h}{2}\right)\phi'\left(x + \frac{h}{2}\right) = D\left(x + \frac{h}{2}\right)\frac{\phi(x+h) - \phi(x)}{h} + O(h^2).$$

Doing the same with $D(x - h/2)\phi'(x - h/2)$ gives

$$D\left(x - \frac{h}{2}\right)\phi'\left(x - \frac{h}{2}\right) = D\left(x - \frac{h}{2}\right)\frac{\phi(x) - \phi(x-h)}{h} + O(h^2).$$

The final step involves writing

$$\frac{d}{dx} D(x)\frac{d\phi}{dx} = \frac{1}{h}\left(D\left(x + \frac{h}{2}\right)\phi'\left(x + \frac{h}{2}\right) - D\left(x - \frac{h}{2}\right)\phi'\left(x - \frac{h}{2}\right)\right) + O(h^2),$$

which is a central difference formula as well. Putting all of our results together gives

$$\frac{d}{dx} D(x)\frac{d\phi}{dx} = \frac{1}{h}\left(D\left(x + \frac{h}{2}\right)\frac{\phi(x+h) - \phi(x)}{h} - D\left(x - \frac{h}{2}\right)\frac{\phi(x) - \phi(x-h)}{h}\right).$$

With a constant value of $D(x)$ this formula becomes

$$\frac{d}{dx} D(x)\frac{d\phi}{dx} = \frac{D}{h^2}\left(\phi(x+h) - 2\phi(x) + \phi(x-h)\right).$$

One outstanding question is what order is this approximation. It is fairly obvious that it is second-order when D is constant. It is tedious, but straightforward, to show that the error is second-order in h even when D is changing.

What about higher derivatives? As we did in the diffusion operator, we can just apply the same formula over and over until we get a derivative of any degree we like. We will not go further into the higher-degree derivative formulas here because the formulae are usually for specialized problems and can be generated easily. One thing to note, as we saw when we went from first to second derivatives (two points to three points), the number of points you need to evaluate the function at grows with the derivative degree.

14.6 RICHARDSON EXTRAPOLATION

If we want high-order approximations to derivatives (or many other quantities), we can use Richardson extrapolation. This idea goes back to Lewis Fry Richardson, one of the first people to solve problems using computers. Though in his case in the early 1900s the computers were adolescent boys doing calculations on slide rules. This also is where the notion of an expensive algorithm might come from because Richardson paid the "computeers" by the operation.

In any case, Richardson extrapolation combines two approximations to get a more accurate answer. To see how this works, we can look at a central difference approximation to the first derivative using h and $h/2$:

$$\frac{f(x+h) - f(x-h)}{2h} = f'(x) + \frac{h^2}{6} f'''(x) + O(h^4),$$

and

$$\frac{f(x+h/2) - f(x-h/2)}{h} = f'(x) + \frac{h^2}{24} f'''(x) + O(h^4).$$

For simplicity we define

$$\hat{f}'_h \equiv \frac{f(x+h) - f(x-h)}{2h},$$

and

$$\hat{f}'_{h/2} \equiv \frac{f(x+h/2) - f(x-h/2)}{h}.$$

Notice that if we take the combination

$$\frac{4\hat{f}'_{h/2} - \hat{f}'_h}{3} = f'(x) + O(h^4).$$

This is a fourth-order approximation to the derivative as the error term scales as h^4 as h is decreased. In this case we obtained two extra orders of accuracy by combining two second-order approximations because the central difference approximation only has even powers of h in its error term.

The same type of extrapolation can be done with the forward or backward difference scheme. For the forward difference method we have

$$\frac{f(x+h) - f(x)}{h} = f'(x) + \frac{h}{2} f''(x) + \frac{h^2}{6} f'''(x) + O(h^3),$$

and

$$\frac{f(x+h/2) - f(x)}{h/2} = f'(x) + \frac{h}{4} f''(x) + \frac{h^2}{24} f'''(x) + O(h^3).$$

Now we can write

$$2\hat{f}'_{h/2} - \hat{f}'_h = f'(x) + O(h^2),$$

where the $\hat{f}$'s are now the forward difference estimates. Notice there that we only improved the order of accuracy by one order this time. This is the most common case with Richardson extrapolation.

We can generalize what we did to get a general Richardson extrapolation formula. Call R_{k+1} the Richardson extrapolated quantity of order $k+1$, and $\hat{g}_h$ a quantity estimated using h and $\hat{g}_{h/n}$ a quantity estimated with the same method using h/n. If the original method is order k accurate, then the Richardson extrapolated estimate is

$$R_{k+1} = \frac{n^k \hat{g}_{h/n} - \hat{g}_h}{n^k - 1} = g + O(h^{k+1}).$$

In the example above using central differences, $n = 2$ and $k = 2$, that is why the 4 appeared in the numerator and a 3 appeared in the denominator. In that example we had

$$R_3 = \frac{4 \hat{f}'_{h/2} - \hat{f}'_h}{3}.$$

We call this R_3 even though we obtained fourth-order accuracy in this case because there are no odd-order powers of h in the error term.

BOX 14.6 NUMERICAL PRINCIPLE

Richardson extrapolation applies a numerical approximation with several values of h and uses the knowledge for how the error scales in h to cancel leading-order error terms.

To boot, you can apply Richardson extrapolation repeatedly to get more and more accurate approximations.

We can continue that to even higher order by applying Richardson extrapolation multiple times. As a test we will use our example function from above. The first thing we will do is define a Richardson extrapolation function:

```
In [1]: def RichardsonExtrapolation(fh, fhn, n, k):
            """Compute the Richardson extrapolation based on
            two approximations of order k
            where the finite difference parameter h is used in fh and h/n in fhn.
            Inputs:
            fh:  Approximation using h
            fhn: Approximation using h/n
            n:   divisor of h
            k:   original order of approximation

            Returns:
            Richardson estimate of order k+1"""

            numerator = n**k * fhn - fh
            denominator = n**k - 1
            return numerator/denominator
```

Using this function we can approximate the derivate using Richardson extrapolation. In the following figure, the slope between the last two points in the Richardson extrapolation estimate is 3.998:

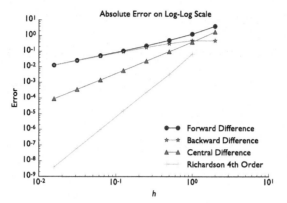

Notice that there is one fewer point in the Richardson line relative to the other lines because it takes two estimates to apply Richardson extrapolation.

We can apply Richardson extrapolation again to get a sixth-order approximation (note that we skip 5 just as we skipped 3). That is we apply the Richardson extrapolation function to the estimate we computed using Richardson extrapolation on the central-difference estimate. The results from this double-extrapolation yield a slope of about 6, as expected:

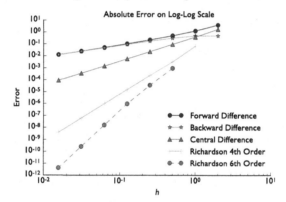

We can apply Richardson extrapolation again to get an eighth-order approximation:

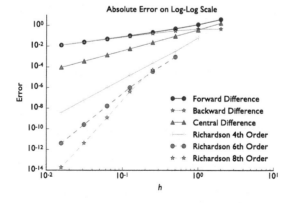

In this case the slope between the last two points is 7.807. This is because floating point precision is starting to affect the estimate.

The overall results are pretty compelling: the 8th-order approximation is about 10 orders of magnitude better than the original central difference approximation at the finest level of h. The only trade-off is that the original central difference needs two points to approximate the derivative and the eighth-order extrapolated value needs several central difference approximations to get the accurate answer it does.

Richardson extrapolation is a powerful tool to have in your numerical toolkit. All you need to know is an estimate of the order of accuracy for your base method and the level by which you refined h. Knowing this and computing several approximations, you can combine them with Richardson extrapolation to get a better answer.

14.7 COMPLEX STEP APPROXIMATIONS

For functions that are real-valued for a real argument *and* can be evaluated for a complex number, we can use complex arithmetic to get an estimate of the derivative without taking a difference. Consider the Taylor series approximation of a function at $f(x + ih)$:

$$f(x + ih) = f(x) + ihf'(x) - \frac{h^2}{2} f''(x) - \frac{ih^3}{6} f'''(x) + O(h^4). \tag{14.1}$$

The imaginary part of this series is

$$\text{Im}\{f(x + ih)\} = hf'(x) - \frac{h^3}{6} f'''(x) + O(h^5),$$

which leads to

$$f'(x) = \frac{\text{Im}\{f(x + ih)\}}{h} + O(h^2).$$

This is the complex step approximation and it can estimate the derivative by evaluating the function at a complex value and dividing by h. This will be a second-order in h approximation to the derivative. At first, this may seem like it is no better than the central difference formula, and worse than our high-order Richardson extrapolation estimates.

Indeed, if we apply the complex approximation to the derivative on the function from before and the same values of h, we see that it does not perform noticeably different.

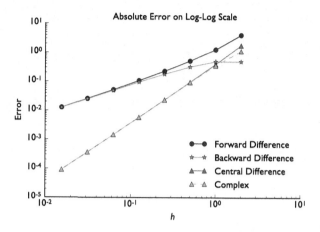

Nevertheless, if we let h get even smaller, the central difference approximation reaches a minimum error value before rising. This is due to the fact that the error in finite precision arithmetic starts to dominate the difference between the function values in the central difference formula. In the complex step method there are no differences, and the error in the approximation can go much lower:

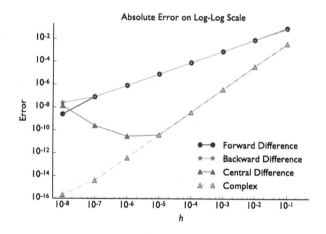

Though we have not shown it here, the Richardson extrapolation estimates based on the central difference formula do not reach a lower error than the central difference approximations.

Finally, we will show that it is possible to get a fourth-order complex step approximation. To do this we will combine Eq. (14.1) with the equation for a step of size $h/2$:

$$f\left(x + \frac{ih}{2}\right) = f(x) + \frac{ih}{2}f'(x) - \frac{h^2}{8}f''(x) - \frac{ih^3}{24}f'''(x) + O(h^4). \tag{14.2}$$

Inspecting these two equations, we can get an approximation for $f'(x)$ without the $f'''(x)$ term by adding $-1/8$ times Eq. (14.1) to Eq. (14.2) and multiplying the sum by $8/3$ to get

$$f'(x) = \frac{8}{3h}\text{Im}\left\{ f\left(x + \frac{ih}{2}\right) - \frac{1}{8}f(x + ih)\right\} + O(h^4).$$

This approximation requires two evaluations of the function, but as we can see it reaches the minimum error with a relatively large value of h:

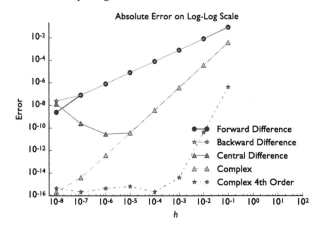

The complex step method is a useful approximation to the derivative when one has a function that can be evaluated with a complex argument. As we have seen, it can be much more accurate than real finite differences, including Richardson extrapolations of those approximations.

CODA

The flip side of numerical differentiation is numerical integration. This will be covered in the next chapter. Whereas, in numerical differentiation the answers got better as the distance between points became smaller, in numerical integration we will break a domain into a finite number of subdivisions and the error will go to zero as the width of those subdivisions goes to zero. The principle of order of accuracy and Richardson extrapolation will be useful in our discussion there as well.

FURTHER READING

The complex step method dates back to the 1960s, but it has only found widespread application in the past few decades [19]. There are also other methods of approximating derivatives called differential quadrature methods [20] that are the derivative versions of the integral quadrature methods that we will encounter in the next chapter.

PROBLEMS

Short Exercises

Compute using $h = 2^{-1}, 2^{-2}, \ldots 2^{-5}$ and the forward, backward, and centered difference, as well as the two complex step approximations the following derivatives.

14.1. $f(x) = \sqrt{x}$ at $x = 0.5$. The answer is $f'(0.5) = 2^{-1/2} \approx 0.70710678118$.

14.2. $f(x) = \arctan(x^2 - 0.9x + 2)$ at $x = 0.5$. The answer is $f'(0.5) = \frac{5}{212}$.

14.3. $f(x) = J_0(x)$, at $x = 1$, where $J_0(x)$ is a Bessel function of the first kind given by

$$J_\alpha(x) = \sum_{m=0}^{\infty} \frac{(-1)^m}{m! \, \Gamma(m + \alpha + 1)} \left(\frac{x}{2}\right)^{2m + \alpha}.$$

The answer is $f'(1) \approx -0.4400505857449335$. Repeat the calculation using the second-order complex step approximation to get a Richardson extrapolated fourth-order estimate. Compare this to the fourth-order complex step approximation.

Programming Projects

1. Comparison of Methods

Consider the function

$$f(x) = e^{-\frac{x^2}{\sigma^2}}.$$

We will use finite differences to estimate derivatives of this function when $\sigma = 0.1$.

- Using forward, backward, and centered differences, and the two complex step approximations evaluate the error in the approximate derivative of the function at 1000 points between $x = -1$ and $x = 1$ (np.linspace will be useful) using the following values of h:

$$h = 2^0, 2^{-1}, 2^{-2}, \ldots, 2^{-7}.$$

For each set of approximations compute the average absolute error over the one thousand points

$$\text{Average Absolute Error} = \frac{1}{N} \sum_{i=1}^{N} |f'(x_i) - f'_{\text{approx}}(x_i)|,$$

where $f'_{\text{approx}}(x_i)$ is the value of an approximate derivative at x_i and N is the number of points the function derivative is evaluated at. You will need to find the exact value of the derivative to complete this estimate.

Plot the value of the average absolute error from each approximation on the same figure on a log-log scale. Discuss what you see. Is the highest-order method always the most accurate? Compute the order of accuracy you observe by computing the slope on the log-log plot.

Next, compute the maximum absolute error for each value of h as

$$\text{Maximum Absolute Error} = \max_i |f'(x_i) - f'_{\text{approx}}(x_i)|.$$

Plot the value of the maximum absolute error from each approximation on the same figure on a log-log scale. Discuss what you see. Is the highest-order method always the most accurate?

- Repeat the previous part using the second-order version of the second-derivative approximation discussed above. You will only have one formula in this case.
- Now derive a formula for the fourth derivative and predict its order of accuracy. Then repeat the calculation and graphing of the average and maximum absolute errors and verify the order of accuracy.

15

Numerical Integration With Newton–Cotes Formulas

Once the area is appointed,
They enclose the broad-squared plaza
With a mass of mighty tree-trunks
Hewn to stakes of firm endurance...

Alonso de Ercilla y Zúñiga, The Araucaniad, *as translated by Charles Maxwell Lancaster and*
Paul Thomas Manchester

CHAPTER POINTS

- Newton–Cotes rules perform polynomial interpolation over the integrand and integrate the resulting interpolant.

- The midpoint, trapezoid, and Simpson's rules are based on constant, linear, and quadratic interpolation, respectively.

- The range of integration is typically broken up into several pieces and a rule is applied to each piece.

- We can combine Richardson extrapolation with Newton–Cotes rules to get highly accurate integral approximations.

In this and the next chapter we are going to discuss ways to compute the integral of a general function numerically. In particular we are interested in ways that we can approximate an integral by a sum with a finite number of terms:

$$\int_a^b f(x)\,dx \approx \sum_{\ell=1}^{L} w(x_\ell) f(x_\ell).$$

Such an approximate is called quadrature, but numerical integration is the more modern term. The term quadrature arose from a process in ancient Greek geometry of constructing a square (a quadrilateral) with the same area as a given shape.

Writing an integral as a finite sum is analogous to the definition of an integral as a Riemann sum, when the number of intervals goes to infinity. Therefore, just as in finite difference derivatives, we use finite mathematics to approximation the infinitesimals of calculus.

15.1 NEWTON–COTES FORMULAS

The Newton–Cotes formulas are ways to approximate an integral by fitting a polynomial through a given number of points and then doing the integral of that polynomial exactly. (Clearly, Newton is the larger numerical luminary in the name of this method. One might suspect that Cotes rode on Isaac Newton's coat tails here.) The polynomial can be integrated exactly because integration formulas for polynomials are straightforward. We will not delve into the general theory of Newton–Cotes formulas, rather we will give three important examples.

15.1.1 The Midpoint Rule

In the midpoint rule we approximate the integral by the value of the function in the middle of the range of integration times the length of the region. This simple formula is

$$I_{\text{midpoint}} = h\, f\left(\frac{a+b}{2}\right) \approx \int_a^b f(x)\,dx,$$

where $h = b - a$.

To demonstrate this rule we look at a simple function integrated over an interval with the midpoint rule:

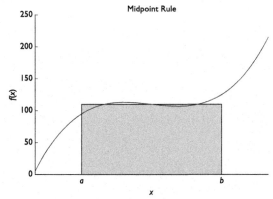

From this demonstration, we see that the resulting approximation is not terrible, but there are clearly parts of the function where the rectangle does not match the function well. We can do better than a rectangle that approximates the function as flat. Namely we can approximate the integrand as linear; we do this next.

BOX 15.1 NUMERICAL PRINCIPLE

The midpoint rule approximates the integrand as a rectangle that touches the function at the midpoint of the interval of integration:

$$\int_a^b f(x)\,dx = hf(c),$$

where $h = b - a$ and $c = (a + b)/2$.

15.1.2 The Trapezoid Rule

In this method we fit a line between a and b and then do the integration. The formula for this is

$$I_{\text{trap}} \equiv \frac{h}{2}(f(a) + f(b)) \approx \int_a^b f(x)\,dx,$$

where $h = b - a$. Here is a graphical example.

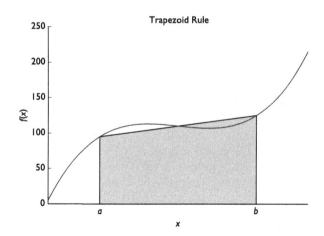

BOX 15.2 NUMERICAL PRINCIPLE

The trapezoid rule approximates the integrand as the area of a trapezoid with bases that touch the function at the endpoints of the interval of integration:

$$\int_a^b f(x)\,dx = \frac{h}{2}\left(f(a)+f(b)\right),$$

where $h = b - a$.

Additionally, in this demonstration we can see where the rule gets its name. The approximation to the integral is the area of a trapezoid. Indeed, the approximation formula is the same as the area of a trapezoid found in geometry textbooks. We can also see in the figure that the approximation is not exact because the trapezoid does not exactly follow the function, but if a and b are close enough together it should give a good approximation because any well-behaved function can be approximated linearly over a narrow enough domain.

That leads to a variation to the trapezoid rule (and any other rule for that matter). We can break up the domain $[a, b]$ into many smaller domains and integrate each of these. Here is an example where we break $[a, b]$ into 4 pieces:

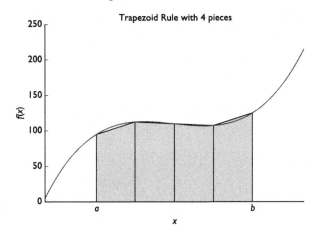

BOX 15.3 NUMERICAL PRINCIPLE

Commonly, the interval of integration is broken up into many pieces and a quadrature rule is applied to each of the pieces. This allows the user to control the size of the interval over which the approximation is applied. This application of a quadrature rule over smaller ranges of the interval is called a composite quadrature rule.

As you can see the approximation is starting to look better. We can write a trapezoid rule function that will take in a function, a, b, and the number of pieces and perform this integration. Also, because the right side of each piece is the left side of the next piece, if we are clever we can only evaluate the function $N + 1$ times where N is the number of pieces. The following function implements the trapezoid rule.

```
In [1]: def trapezoid(f, a, b, pieces):
            """Find the integral of the function f between a and b
            using pieces trapezoids
            Args:
                f: function to integrate
                a: lower bound of integral
                b: upper bound of integral
                pieces: number of pieces to chop [a,b] into

            Returns:
                estimate of integral
            """
            integral = 0
            h = b - a
            #initialize the left function evaluation
            fa = f(a)
            for i in range(pieces):
                #evaluate the function at the left end of the piece
                fb = f(a+(i+1)*h/pieces)
                integral += 0.5*h/pieces*(fa + fb)
                #now make the left function evaluation the right for the next step
                fa = fb
            return integral
```

We can test this method on a function that we know the integral of

$$\int_0^\pi \sin x \, dx = 2.$$

In addition to the estimates, the approximations to the integral are plotted.

```
In [2]: integral_estimate = trapezoid(np.sin,0,np.pi,pieces=6,graph=True)
        print("Estimate is",integral_estimate,"Actual value is 2")

        integral_estimate = trapezoid(np.sin,0,np.pi,pieces=20,graph=True)
        print("Estimate is",integral_estimate,"Actual value is 2")
```

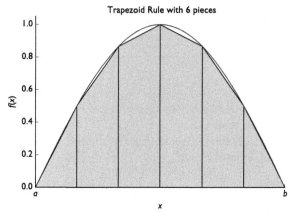

```
Estimate is 1.95409723331 Actual value is 2
```

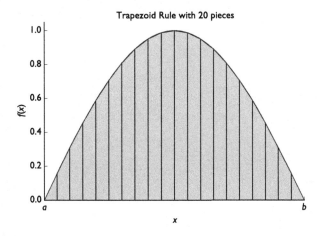

```
Estimate is 1.99588597271 Actual value is 2
```

We can run this multiple times and see how the error changes. Similar to what we did for finite difference derivatives, we can plot the error versus number of pieces on a log-log scale. In this case, h is the width of each of the pieces: as the number of pieces grows, the value of h decreases.

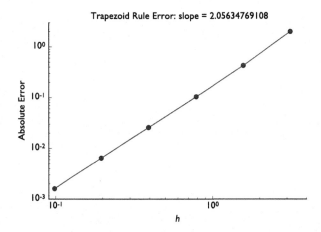

The error in the trapezoid rule that we observe is second-order in h, because the slope of the error on the log-log scale is 2.

We can see that this is the expected error in the estimate by looking at the linear approximation to a function around $x = a$:

$$f(x) = f(a) + (x-a)f'(a) + \frac{(x-a)^2}{2}f''(a) + O((x-a)^3).$$

We can approximate the derivative using a forward difference:

$$f'(a) \approx \frac{f(b) - f(a)}{h} + O(h),$$

where $h = b - a$. Now the integral of $f(x)$ from a to b becomes

$$\int_a^b f(x)\,dx = hf(a) + \int_a^b (x-a)f'(a)\,dx + \int_a^b \frac{(x-a)^2}{2}f''(a)\,dx + O(h^4).$$

The integral

$$\int_a^b (x-a)f'(a) = \frac{(b-a)^2}{2}f'(a) = \frac{h^2}{2}\left(\frac{f(b)-f(a)}{h} + O(h)\right) = \frac{h}{2}(f(b) - f(a)) + O(h^3).$$

Additionally,

$$\int_a^b \frac{(x-a)^2}{2}f''(a)\,dx = -\frac{h^3}{6}f''(a) = O(h^3).$$

When we plug this into the original integral we get

$$\int_a^b f(x)\,dx = \frac{h}{2}(f(a) + f(b)) + O(h^3).$$

This says that error in one piece of the trapezoid rule is third-order accurate, which means the error can be written as $Ch^3 + O(h^4)$. However, when we break the interval into N pieces, each of size $h = (b-a)/N$, the error terms add and each piece has its own constant so that

$$\sum_{i=1}^N C_i h^3 \leq Nh^3 C_{\max} = (b-a)C_{\max}h^2,$$

where $C_{\max}$ is the maximum value of $|C_i|$. Therefore, the error in the sum of trapezoid rules decreases as h^2, which we observed above. This analysis can be extended to show that the error terms in the trapezoid rule only have even powers of h:

$$\text{Error} = C_2 h^2 + C_4 h^4 + \dots$$

We will use this later when we combine the trapezoid rule with Richardson extrapolation.

15.2 SIMPSON'S RULE

Simpson's rule is like the trapezoid rule, except instead of fitting a line we fit a parabola between three points, a, b, and $(a + b)/2$. The formula for this is

$$I_{\text{Simpson}} \equiv \frac{h}{6}\left(f(a) + 4f\left(a + \frac{h}{2}\right) + f(b)\right) \approx \int_a^b f(x)\,dx,$$

where $h = b - a$. (This is sometimes called Simpsons 1/3 rule, because there is another Simpson rule that is based on quartic interpolation.) First, let's examine how this rule behaves on the integral of $\sin x$ with one piece:

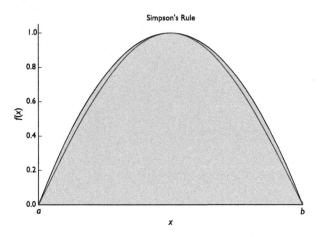

It looks like the function can be well approximated by a parabola.

BOX 15.4 NUMERICAL PRINCIPLE

Simpson's rule approximates an integral by performing quadratic interpolation between the endpoints and midpoint of the interval of integration. The formula for this method is

$$\int_a^b f(x)\,dx = \frac{h}{6}\left(f(a) + 4f(c) + f(b)\right),$$

where $h = b - a$ and $c = (a + b)/2$.

Here is a function to perform Simpson's rule just like we did for the trapezoid rule.

```
In [4]: def simpsons(f, a, b, pieces):
            """Find the integral of the function f between a and b
            using Simpson's rule
            Args:
                f: function to integrate
                a: lower bound of integral
```

```
    b: upper bound of integral
    pieces: number of pieces to chop [a,b] into

Returns:
    estimate of integral
"""
integral = 0
h = b - a
one_sixth = 1.0/6.0
#initialize the left function evaluation
fa = f(a)
for i in range(pieces):
    #evaluate the function at the left end of the piece
    fb = f(a+(i+1)*h/pieces)
    fmid = f(0.5*(a+(i+1)*h/pieces+ a+i*h/pieces))
    integral += one_sixth*h/pieces*(fa + 4*fmid + fb)
    #now make the left function evaluation the right for the next step
    fa = fb

return integral
```

We then use this function to estimate the integral of the sine function using two and twenty pieces:

```
In [5]: integral_estimate = simpsons(np.sin,0,np.pi,pieces=2,graph=True)
        print("Estimate is",integral_estimate,"Actual value is 2")

        integral_estimate = simpsons(np.sin,0,np.pi,pieces=20,graph=True)
        print("Estimate is",integral_estimate,"Actual value is 2")
```

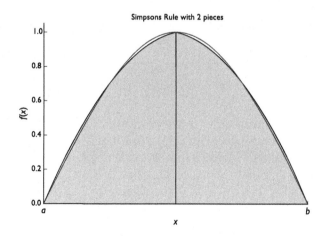

```
Estimate is 2.00455975498 Actual value is 2
```

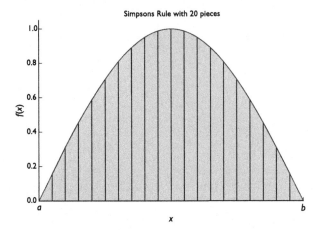

```
Estimate is 2.00000042309 Actual value is 2
```

Just like the trapezoid rule, we can look at the error in Simpson's rule.

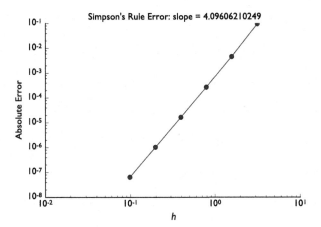

Simpson's rule is fourth-order in the piece size. This means that every time I double the number of pieces the error goes down by a factor of $2^4 = 16$.

Before moving on, we will use Simpson's rule to calculate π:

$$\int_0^1 4\sqrt{1 - x^2}\,dx = \pi.$$

```
In [6]: integrand = lambda x: 4*np.sqrt(1-x**2)
        simpsons(integrand,0,1,pieces = 8,graph=True) #actual value is 3.14159
```

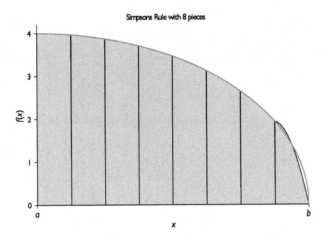

Simpsons Rule with 8 pieces

```
Out[6]: 3.1343976689845969
```

It looks like most of the error comes at $x = 1$. The reason for this is that function is changing rapidly near $x = 1$ because there is a singularity in the derivative:

$$\frac{d}{dx} 4\sqrt{1 - x^2} = -\frac{4x}{\sqrt{1 - x^2}}.$$

Note that the denominator goes to 0 at $x = 1$. We will revisit this integral later.

15.3 ROMBERG INTEGRATION

When we use trapezoid integration, we know that the error is second-order in the piece size. Using this information we can apply Richardson extrapolation. We can combine the approximation with one piece with that using two pieces to get a better approximation (one that is higher-order). Then, we can combine this approximation with the estimate using four pieces, to get an even better answer. To do this we need to use the fact that the trapezoid rule only has error terms that are even powers of h. To demonstrate this, we will compute the integral

$$\int_1^2 \frac{\ln x}{1 + x} \, dx = 0.1472206769592413 \ldots \tag{15.1}$$

The result from a one-piece trapezoidal integration is

```
In [7]: integrand = lambda x: np.log(x)/(1.0+x)
        integral_estimate1 = trapezoid(integrand,1,2,pieces=1,graph=True)
        print("Estimate is",integral_estimate1,
            "Actual value is 0.1472206769592413, Error is",
            np.fabs(0.1472206769592413-integral_estimate1))
```

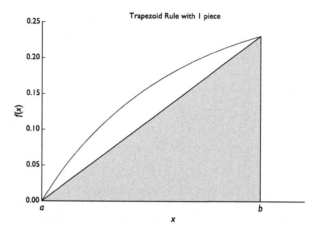

```
Estimate is 0.115524530093 Actual value is 0.1472206769592413,
Error is 0.0316961468659
```

Then we use two-pieces

```
In [8]: integral_estimate2 = trapezoid(integrand,1,2,pieces=2,graph=True)
        print("Estimate is",integral_estimate2,
              "Actual value is 0.1472206769592413, Error is",
              np.fabs(0.1472206769592413-integral_estimate2))
```

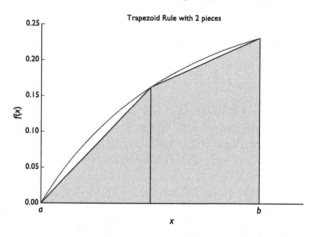

```
Estimate is 0.138855286668 Actual value is 0.1472206769592413,
Error is 0.00836539029095
```

We combine these estimates using Richardson Extrapolation. First, we need to define a new function for Richardson extrapolation. Our new implementation will use floating point numbers with higher precision than standard floating point numbers.

```
In [9]: import decimal
        #set precision to be 100 digits
```

```
decimal.getcontext().prec = 100
def RichExtrap(fh, fhn, n, k):
    """Compute the Richardson extrapolation based on
    two approximations of order k
    where the finite difference parameter h is used in fh and h/n in fhn.
    Inputs:
    fh:  Approximation using h
    fhn: Approximation using h/n
    n:   divisor of h
    k:   original order of approximation

    Returns:
    Richardson estimate of order k+1"""
    n = decimal.Decimal(n)
    k = decimal.Decimal(k)
    numerator = decimal.Decimal(n**k * decimal.Decimal(fhn)
                                - decimal.Decimal(fh))
    denominator = decimal.Decimal(n**k - decimal.Decimal(1.0))
    return float(numerator/denominator)
```

To make Richardson work well with high-order approximations we use arbitrary precision arithmetic using the decimal library.

BOX 15.5 PYTHON PRINCIPLE

The library decimal allows one to use higher precision floating point numbers than the standard floating point numbers in Python. It is necessary to set the desired precision with the command

```
decimal.getcontext().prec = Precision
```

where Precision is the integer number of digits of accuracy desired. Also, numbers that you want to be represented using this precision will need to be surrounded by the construct

```
decimal.Decimal(N)
```

where N is a number.

We will apply this function to the approximations with the trapezoid rule above.

```
In [10]: Richardson2 = RichExtrap(integral_estimate1,integral_estimate2,n=2,k=2)
         print("Estimate is",Richardson2,
               "Actual value is 0.1472206769592413, Error is",
               np.fabs(0.1472206769592413-Richardson2))
```

```
Estimate is 0.1466322055266186 Actual value is 0.1472206769592413,
Error is 0.000588471432623
```

By applying Richardson extrapolation, we improved the estimate by an order of magnitude. Now if we use 4 points, we get

```
In [15]: integral_estimate4 = trapezoid(integrand,1,2,pieces=4,graph=True)
         print("Estimate is",integral_estimate4,
               "Actual value is 0.1472206769592413, Error is",
               np.fabs(0.1472206769592413-integral_estimate4))
```

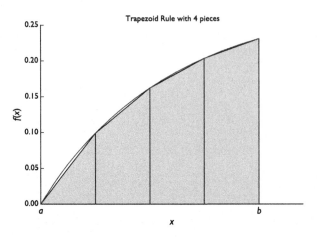

```
Estimate is 0.145095533798 Actual value is 0.1472206769592413,
Error is 0.00212514316171
```

There are two Richardson extrapolations we can do at this point, one between the 4 and 2 piece estimates, and then one combines the two Richardson extrapolations:

```
In [11]: Richardson4 = RichExtrap(integral_estimate2,
                                          integral_estimate4, n=2,k=2)
         print("Estimate is",
         Richardson4,
             "Actual value is 0.1472206769592413, Error is",
             np.fabs(0.1472206769592413-Richardson4))
```

```
Estimate is 0.14717561617394495 Actual value is 0.1472206769592413,
Error is 4.50607852963e-05
```

```
In [12]: Richardson42 = RichExtrap(Richardson2,Richardson4,
                                          n=2,k=4)
         #note this is fourth order
         print("Estimate is",Richardson42,
             "Actual value is 0.1472206769592413, Error is",
             np.fabs(0.1472206769592413-Richardson42))
```

```
Estimate is 0.14721184355043337 Actual value is 0.1472206769592413,
Error is 8.83340880792e-06
```

Notice that the error from combining the two extrapolations is 3 orders of magnitude smaller than the error using 4 pieces. We could continue on by hand, but it is pretty easy to write a function for this. This procedure is called Romberg integration, and that is what we will name our function. The function will return a table of approximations where the first column is the original trapezoid rule approximations and the subsequent columns are a Richardson extrapolation of the column that came before.

```
In [13]: def Romberg(f, a, b, MaxLevels = 10, epsilon = 1.0e-6, PrintMatrix = False):
             """Compute the Romberg integral of f from a to b
```

```
Inputs:
f:  integrand function
a: left edge of integral
b: right edge of integral
MaxLevels: Number of levels to take the integration to

Returns:
Romberg integral estimate"""

estimate = np.zeros((MaxLevels,MaxLevels))

estimate[0,0] = trapezoid(f,a,b,pieces=1)
i = 1
converged = 0
while not(converged):
    estimate[i,0] = trapezoid(f,a,b,pieces=2**i)
    for extrap in range(i):
        estimate[i,1+extrap] = RichExtrap(estimate[i-1,extrap],
                                          estimate[i,extrap],
                                          2,2**(extrap+1))

    converged = np.fabs(estimate[i,i] - estimate[i-1,i-1]) < epsilon
    if (i == MaxLevels-1): converged = 1
    i += 1
if (PrintMatrix):
    print(estimate[0:i,0:i])
return estimate[i-1, i-1]
```

This function is defined to compute the integral estimate using a series of intervals and can print out the intermediate estimates and the extrapolated values. We will test this on the same integral as before.

```
In [14]: #this should give us what we got before
         integral_estimate = Romberg(integrand,1,2,MaxLevels=3, PrintMatrix=True)
         print("Estimate is",integral_estimate,
             "Actual value is 0.1472206769592413, Error is",
                 np.fabs(0.1472206769592413-integral_estimate))

[[ 0.11552453  0.          0.        ]
 [ 0.13885529  0.14663221  0.        ]
 [ 0.14509553  0.14717562  0.14721184]]
Estimate is 0.14721184355 Actual value is 0.1472206769592413,
Error is 8.83340880792e-06

In [15]: #Now let it converge, don't set Max Levels so low
         integral_estimate = Romberg(integrand,1,2,MaxLevels = 10, PrintMatrix=True,
                                     epsilon = 1.0e-10)
         print("Estimate is",integral_estimate,
             "Actual value is 0.1472206769592413, Error is",
             np.fabs(0.1472206769592413-integral_estimate))

[[ 0.11552453  0.          0.          0.          0.          0.          0.]
 [ 0.13885529  0.14663221  0.          0.          0.          0.          0.]
```

```
[ 0.14509553  0.14717562  0.14721184  0.          0.          0.          0.]
[ 0.14668713  0.14721767  0.14722047  0.14722051  0.          0.          0.]
[ 0.14708715  0.14722049  0.14722067  0.14722067  0.14722067  0.          0.]
[ 0.14718729  0.14722066  0.14722068  0.14722068  0.14722068  0.14722068  0.]
[ 0.14721233  0.14722068  0.14722068  0.14722068  0.14722068  0.14722068  0.14722068]]
Estimate is 0.147220676959 Actual value is 0.1472206769592413,
Error is 7.19035941898e-13
```

BOX 15.6 NUMERICAL PRINCIPLE

Romberg integration combines Richardson extrapolation with a known quadrature rule. It can produce integral estimates of high accuracy with a few function evaluations.

As one final example we will estimate π:

```
In [16]: integrand = lambda x: 4*np.sqrt(1-x**2)
         integral_estimate = Romberg(integrand,0,1,MaxLevels = 8,
                                PrintMatrix=False, epsilon = 1.0e-10)
         print("Estimate is",integral_estimate,
               "Actual value is",np.pi,", Error is",
               np.fabs(np.pi-integral_estimate))
```

```
Estimate is 3.14131611425 Actual value is 3.141592653589793,
Error is 0.000276539339068
```

One thing to note is that our implementation of Romberg integration is not the most efficient. Technically, we are evaluating the function more times than we need to because when we call the trapezoid rule function with more pieces we are evaluating the function again at places we already did (for example, $f(a)$ and $f(b)$ are evaluated each time). However, making the most efficient algorithm would not make the most useful teaching example. For our purposes it suffices to know that this can be done in a smarter way if each function evaluation takes a long time.

There also is no reason we could not use the Romberg idea using Simpson's rule. Here is a function for that.

```
In [17]: def RombergSimps(f, a, b, MaxLevels = 10, epsilon = 1.0e-6,
                  PrintMatrix = False):
             """Compute the Romberg integral of f from a to b
             Inputs:
             f:  integrand function
             a: left edge of integral
             b: right edge of integral
             MaxLevels: Number of levels to take the integration to

             Returns:
             Romberg integral estimate"""

             estimate = np.zeros((MaxLevels,MaxLevels))
```

```
    estimate[0,0] = simpsons(f,a,b,pieces=1)
    i = 1
    converged = 0
    while not(converged):
        estimate[i,0] = simpsons(f,a,b,pieces=2**i)
        for extrap in range(i):
            estimate[i,1+extrap] = RichExtrap(estimate[i-1,extrap],
                                              estimate[i,extrap],
                                              n=2,k=2+2.0**(extrap+1))

        converged = np.fabs(estimate[i,i] - estimate[i-1,i-1]) < epsilon
        if (i == MaxLevels-1): converged = 1
        i += 1
    if (PrintMatrix):
        print(estimate[0:i,0:i])
    return estimate[i-1, i-1]
```

Using this function we can get a good estimate of the integral of the rational function from Eq. (15.1):

```
In [18]: #this should be better than what we got before,
         #Error was 8.83340880792e-06
         integrand = lambda x: np.log(x)/(1.0+x)
         integral_estimate = RombergSimps(integrand,1,2,MaxLevels=3,
                                          PrintMatrix=True)
         print("Estimate is",integral_estimate,
             "Actual value is 0.1472206769592413, Error is",
             np.fabs(0.1472206769592413-integral_estimate))


[[ 0.14663221  0.          0.        ]
 [ 0.14717562  0.14721184  0.        ]
 [ 0.14721767  0.14722047  0.14722061]]
Estimate is 0.147220608522 Actual value is 0.1472206769592413,
Error is 6.84372362669e-08
```

Applying this to estimate π using the default number of levels, we get

```
In [19]: integrand = lambda x: 4*np.sqrt(1-x**2)
         #trapezoid error was 0.00221405375506
         integral_estimate = RombergSimps(integrand,0,1,MaxLevels = 8,
                                          PrintMatrix=False, epsilon = 1.0e-10)
         print("Estimate is",integral_estimate,
             "Actual value is",np.pi,", Error is",
             np.fabs(np.pi-integral_estimate))


Estimate is 3.14149721605 Actual value is 3.141592653589793,
Error is 9.5437540061e-05
```

To get 10 digits of accuracy we need 20 levels or $2^{20} = 1\,048\,576$ intervals:

```
In [20]: integrand = lambda x: 4*np.sqrt(1-x**2)
         integral_estimate = RombergSimps(integrand,0,1,MaxLevels = 20,
                                   PrintMatrix=False, epsilon = 1.0e-14)
         print("Estimate is",integral_estimate,
              "Actual value is",np.pi,", Error is",
              np.fabs(np.pi-integral_estimate))
Estimate is 3.14159265323 Actual value is 3.141592653589793,
Error is 3.63962637806e-10
```

CODA

Here we have learned the basics of numerical integration using Newton–Cotes formulas. More importantly, we have shown how to combine these rules with Richardson extrapolation to get accurate estimates. In the next chapter we will discuss other types of quadrature rules and how to estimate multi-dimensional integrals.

FURTHER READING

The decimal package has a variety of further applications and can be a powerful tool. It is covered in detail in the official Python documentation at docs.python.org.

PROBLEMS

Short Exercises

Using the trapezoid rule and Simpson's rule estimate the following integrals with the following number of intervals: $2, 4, 8, 16, \ldots 512$. Compare your answers with Romberg integration where the maximum number of levels set to 9.

15.1. $\int_0^{\pi/2} e^{\sin x}\, dx \approx 3.104379017855555098181$.

15.2. $\int_0^{2.405} J_0(x)dx \approx 1.470300035485$, where $J_0(x)$ is a Bessel function of the first kind given by

$$J_\alpha(x) = \sum_{m=0}^{\infty} \frac{(-1)^m}{m!\,\Gamma(m+\alpha+1)} \left(\frac{x}{2}\right)^{2m+\alpha}.$$

Programming Projects

1. Inverse Fourier Transform

Consider the neutron diffusion equation in slab geometry an infinite, homogeneous medium given by

$$-D\frac{d^2}{dx^2}\phi(x) + \Sigma_a\phi(x) = \delta(x),$$

where $\delta(x)$ is the Dirac delta function. This source is equivalent to a planar source inside the slab at $x = 0$. One way to solve this problem is to use a Fourier transform. The Fourier transform of a function can be defined by

$$\mathcal{F}\{f(x)\} = \hat{f}(k) = \frac{1}{\sqrt{2\pi}} \int_{-\infty}^{\infty} dx \, f(x)(\cos kx - i \sin kx).$$

The Fourier transform of the diffusion equation above is

$$(Dk^2 + \Sigma_a)\hat{\phi}(k) = \frac{1}{\sqrt{2\pi}}.$$

We can solve this equation for $\hat{\phi}(k)$, and then apply the inverse Fourier transform:

$$\mathcal{F}^{-1}\{\hat{f}(k)\} = f(x) = \frac{1}{\sqrt{2\pi}} \int_{-\infty}^{\infty} dk \, \hat{f}(k)(\cos kx + i \sin kx).$$

This leads to the solution being defined by

$$\phi(x) = \int_{-\infty}^{\infty} \frac{\cos kx \, dk}{2\pi(Dk^2 + \Sigma_a)} + i \int_{-\infty}^{\infty} \frac{\sin kx \, dk}{2\pi(Dk^2 + \Sigma_a)}.$$

The imaginary integral is zero because $\phi(x)$ is real. You can see that this is so because the integrand of the imaginary part is odd and the integral is symmetric about 0.

Your task is to compute the value of $\phi(x)$ at various points using $D = \Sigma_a = 1$. Because you cannot integrate to infinity you will be computing integrals of the form

$$\int_{-L}^{L} f(x) \, dx,$$

for large values of L.

15.1. Compute value of $\phi(x)$ at 256 points in $x \in [-3, 3]$ using Simpson's and the trapezoidal rule with several different numbers of intervals (pieces) in the integration *and* using different endpoints in the integration, L. Plot these estimates of $\phi(x)$.

15.2. Plot the error between your estimate of $\phi(1)$ and the true solution of $\frac{1}{2}e^{-1}$. Make one graph each for trapezoid and Simpson's rule where the x-axis is h and the y-axis is the absolute error. On each plot show a curve for the error decay for $L = 10, 1000, 10^5, 10^8$.

15.3. Give your best estimate, using numerical integration, for the absorption rate density of neutrons, $\Sigma_a\phi(x)$, at $x = 2$.

CHAPTER

16

Gauss Quadrature and Multi-dimensional Integrals

They're two, they're four, they're six, they're eight
Shunting trucks and hauling freight

–Thomas and Friends "Roll Call"

CHAPTER POINTS

- Gauss quadrature is designed to integrate functions with a small number of points.

- The idea behind Gauss quadrature is to exactly integrate the highest degree polynomial possible given a number of function evaluations.

- Multidimensional integrals can be found by applying 1-D integrals.

16.1 GAUSS QUADRATURE RULES

In the last chapter we saw that we can approximate integrals by fitting the integrand with an interpolating polynomial and then integrating the polynomial exactly. In this chapter we take a different approach that guarantees the maximum accuracy for a given number of points. These types of methods are called Gauss quadrature rules. We will discuss the rule for finite intervals.

Gauss quadrature rules re-write an integral as the sum of the function evaluated at a given number of points multiplied by a weight function:

$$\int_a^b f(x)\,dx \approx \sum_{\ell=1}^L w_\ell f(x_\ell),$$

where the weights w_ℓ and quadrature points x_ℓ are chosen to give the integral certain properties. There are several types of Gauss-quadrature rules, but the type we will cover in detail is known as Gauss–Legendre quadrature. In particular this quadrature rule is for integrals of the form

$$\int_{-1}^1 f(x)\,dx \approx \sum_{\ell=1}^L w_\ell f(x_\ell).$$

The integral does not need to be limited just to the range, $[-1, 1]$, however. If we want to integrate $f(x)$ from $[a, b]$, we define a variable

$$x = \frac{a+b}{2} + \frac{b-a}{2}z, \qquad dx = \frac{b-a}{2}dz,$$

to make the transformation

$$\int_a^b f(x)\,dx = \frac{b-a}{2} \int_{-1}^1 f\left(\frac{b-a}{2}z + \frac{a+b}{2}\right) dz.$$

We still have not shown how to pick the weights, w_ℓ, and abscissas, x_ℓ. These are chosen so that the rule is as accurate as possible with L points. It turns out that we can pick the weights and abscissas such that the Gauss–Legendre quadrature formula is exact for polynomials of degree $2L - 1$ or less. This should not be a complete surprise because the integral of a $2L - 1$ degree polynomial is a degree $2L$ polynomial. Such a polynomial has $2L + 1$ coefficients, only $2L$ of these depend on the original polynomial because the constant term is determined by the integration bounds. Therefore, the integral has $2L$ degrees of freedom, the exact number of degrees of freedom we have with our L weights and abscissas.

The weights and abscissas are given for L up to 8 in Table 16.1. Notice that the odd L sets all have $x = 0$ in the set. Also, the sum of the w_ℓ adds up to 2 because the range of integration has a length of 2.

16.1.1 Where Did These Points Come From?

One way to derive the Gauss–Legendre quadrature rules is by looking at the integral of generic monomials of degree 0 up to $2L - 1$ and setting each equal to the L point Gauss–

TABLE 16.1 The Abscissae and Weights for Gauss–Legendre Quadrature up to $L = 8$

L	x_ℓ	w_ℓ
1	0	2
2	±0.5773502691896257645091488	1
3	0	0.888888889
	±0.7745966692414833770358531	0.555555556
4	±0.3399810435848562648026658	0.652145155
	±0.8611363115940525752239465	0.347854845
5	0	0.568888889
	±0.5384693101056830910363144	0.47862867
	±0.9061798459386639927976269	0.236926885
6	±0.2386191860831969086305017	0.467913935
	±0.6612093864662645136613996	0.360761573
	±0.9324695142031520278123016	0.171324492
7	0	0.417959184
	±0.4058451513773971669066064	0.381830051
	±0.7415311855993944398638648	0.279705391
	±0.9491079123427585245261897	0.129484966
8	±0.1834346424956498049394761	0.362683783
	±0.5255324099163289858177390	0.313706646
	±0.7966664774136267395915539	0.222381034
	±0.9602898564975362316835609	0.101228536

Legendre quadrature rule:

$$\int_{-1}^{1} dx\, a_0 x^0 = a_0 \sum_{\ell=1}^{L} w_\ell x_\ell^0,$$

$$\int_{-1}^{1} dx\, a_1 x^1 = a_1 \sum_{\ell=1}^{L} w_\ell x_\ell^1,$$

and continuing until

$$\int_{-1}^{1} dx\, a_{2L-1} x^{2L-1} = a_{2L-1} \sum_{\ell=1}^{L} w_\ell x_\ell^{2L-1}.$$

Notice that the a_i constants cancel out of each equation so they do not matter. This system is $2L$ equations with L weights, w_ℓ, and L abscissas, x_ℓ. We could solve these equations to get the weights and abscissas, though this is not how it is done in practice generally—this is accomplished by using the theory of orthogonal polynomials.

BOX 16.1 NUMERICAL PRINCIPLE

Gauss quadrature rules pick the values of the points and weights in the quadrature rule so that the highest degree polynomial can be exactly integrated. An L point quadrature rule has $2L$ degrees of freedom (L weights and L abscissas), and can therefore integrate a polynomial of degree $2L - 1$ exactly.

16.1.2 Code for Gauss–Legendre Quadrature

We will now write a function that will compute the integral of a function from $[-1, 1]$ using these quadrature rules. For values of L beyond 2, we will use a NumPy function that generates the points and weights. The NumPy documentation asserts that the rules for $L > 100$ have not been tested and may be inaccurate.

```
In [1]: def GLQuad(f, L=8,dataReturn = False):
            """Compute the Gauss-Legendre Quadrature estimate
            of the integral of f(x) from -1 to 1
            Inputs:
            f:   name of function to integrate
            L:   Order of integration rule (8 or less)

            Returns:
            G-L Quadrature estimate"""
            assert(L>=1)
            if (L==1):
                weights = np.ones(1)*2
                xs = np.array([0])
            elif (L==2):
                weights = np.ones(2)
                xs = np.array([-np.sqrt(1.0/3.0),np.sqrt(1.0/3.0)])
            else: #use numpy's function
                xs, weights = np.polynomial.legendre.leggauss(L)

            quad_estimate = np.sum(weights*f(xs))
            if (dataReturn):
                return quad_estimate, weights, xs
            else:
                return quad_estimate
```

The weights and abscissas are shown in the following figure where the size of a point is proportional to the weight:

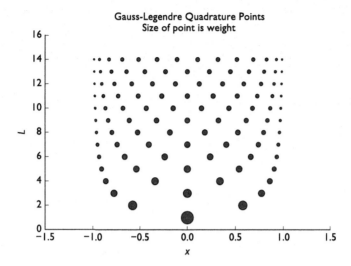

As a simple demonstration of the Gauss–Legendre quadrature, let's show that it integrates polynomials of degree $2L - 1$ exactly. Consider the integral

$$\frac{L}{2^{2L-1}} \int_{-1}^{1} (x+1)^{2L-1} \, dx = 1.$$

In the following code we show that we integrate this function exactly (to floating point precision) using Gauss–Legendre quadrature:

```
In [2]: L = np.arange(1,12)
        for l in L:
            f = lambda x: (x+1)**(2*l-1)*l/(2**(2*l - 1))
            integral = 1.0
            GLintegral = GLQuad(f,l)
            print("L =", l,"\t Estimate is",GLintegral,
                    "Exact value is",integral,
                    "\nAbs. Relative Error is", np.abs(GLintegral-integral)/integral)

L = 1      Estimate is 1.0 Exact value is 1
Abs. Relative Error is 0.0
L = 2      Estimate is 1.0 Exact value is 1
Abs. Relative Error is 1.11022302463e-16
L = 3      Estimate is 1.0 Exact value is 1
Abs. Relative Error is 2.22044604925e-16
L = 4      Estimate is 1.0 Exact value is 1
Abs. Relative Error is 7.77156117238e-16
L = 5      Estimate is 1.0 Exact value is 1
Abs. Relative Error is 4.4408920985e-16
L = 6      Estimate is 1.0 Exact value is 1
Abs. Relative Error is 1.99840144433e-15
L = 7      Estimate is 1.0 Exact value is 1
Abs. Relative Error is 3.99680288865e-15
L = 8      Estimate is 1.0 Exact value is 1
```

```
Abs. Relative Error is 2.22044604925e-15
L = 9     Estimate is 1.0 Exact value is 1
Abs. Relative Error is 1.33226762955e-15
L = 10    Estimate is 1.0 Exact value is 1
Abs. Relative Error is 4.4408920985e-16
L = 11    Estimate is 1.0 Exact value is 1
Abs. Relative Error is 4.21884749358e-15
```

As mentioned earlier, we are generally interested in integrals not just over the domain $x \in [-1, 1]$. We'll now make a function that does Gauss–Legendre quadrature over a general range using the formula

$$\int_a^b f(x)\,dx = \frac{b-a}{2} \int_{-1}^1 f\left(\frac{b-a}{2}z + \frac{a+b}{2}\right) dz.$$

```
In [3]: def generalGL(f,a,b,L):
            """Compute the Gauss-Legendre Quadrature estimate
            of the integral of f(x) from a to b
            Inputs:
            f:   name of function to integrate
            a:   lower bound of integral
            b:   upper bound of integral
            L:   Order of integration rule
            Returns:
            G-L Quadrature estimate"""
            assert(L>=1)
            #define a re-scaled f
            f_rescaled = lambda  z: f(0.5*(b-a)*z + 0.5*(a+b))
            integral = GLQuad(f_rescaled,L)
            return integral*(b-a)*0.5
```

We can show that this version integrates polynomials of degree $2L - 1$ exactly. Consider the integral

$$\frac{2L}{9^L - 4^L} \int_{-3}^2 (x+1)^{2L-1}\,dx = 1.$$

```
In [4]: L = np.arange(1,12)
        for l in L:
            f = lambda x: (x+1)**(2*l-1)*(2*l/(9**l-4**l))
            integral = 1.0
            GLintegral = generalGL(f,-3,2,l)
            print("L =", l,"\t Estimate is",GLintegral,
                "Exact value is",integral,
                "\nAbs. Relative Error is", np.abs(GLintegral-integral)/integral)
```

```
L = 1     Estimate is 1.0 Exact value is 1
Abs. Relative Error is 0.0
L = 2     Estimate is 1.0 Exact value is 1
Abs. Relative Error is 1.11022302463e-16
L = 3     Estimate is 1.0 Exact value is 1
Abs. Relative Error is 4.4408920985e-16
```

```
L = 4     Estimate is 1.0 Exact value is 1
Abs. Relative Error is 6.66133814775e-16
L = 5     Estimate is 1.0 Exact value is 1
Abs. Relative Error is 1.7763568394e-15
L = 6     Estimate is 1.0 Exact value is 1
Abs. Relative Error is 2.44249065418e-15
L = 7     Estimate is 1.0 Exact value is 1
Abs. Relative Error is 6.66133814775e-15
L = 8     Estimate is 1.0 Exact value is 1
Abs. Relative Error is 3.33066907388e-15
L = 9     Estimate is 1.0 Exact value is 1
Abs. Relative Error is 3.10862446895e-15
L = 10    Estimate is 1.0 Exact value is 1
Abs. Relative Error is 4.4408920985e-16
L = 11    Estimate is 1.0 Exact value is 1
Abs. Relative Error is 6.43929354283e-15
```

So far we have only used Gauss–Legendre quadrature on polynomials, below we test it on two functions that are not polynomials:

$$\int_0^\pi \sin(x)\,dx = 2,$$

and

$$\int_0^1 4\sqrt{1 - x^2}\,dx = \pi.$$

For the integral of $\sin(x)$ we get exponential convergence:

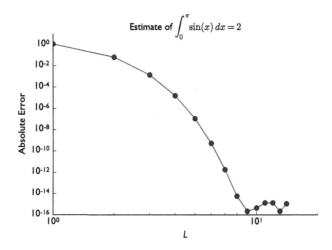

Notice that we get to the smallest error possible by evaluating the integral at only 8 points. This is much better than we saw with trapezoid or Simpson's rules. The approximation converges exponentially to the exact answer. What this means is that it error decreases faster than a polynomial of the form L^{-n}. Exponential convergence can be seen in the plot of the error because the error goes to zero faster than linearly.

This exponential convergence will only be obtained on smooth solutions without singularities in the function or its derivatives. In the last chapter we discussed the integral

$$\int_0^1 4\sqrt{1-x^2}\,dx = \pi.$$

This integral has a singularity in its derivative at $x = 1$. Gauss–Legendre quadrature will not have exponential convergence on this function.

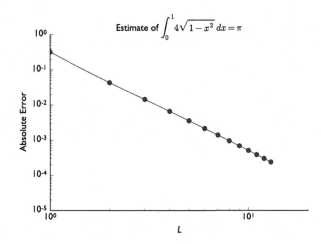

Slope of line from L = 8 to 11 is -2.81212265648

Not quite as impressive, but still pretty good considering that we only evaluated the function at 12 points. The difficulty in the obtaining a highly accurate solution is the same as with Newton–Cotes: the function cannot be described well by a polynomial near $x = 1$.

There is a big difference between the exponential convergence we saw in the integral of the sine function and the polynomial convergence in this problem. Even at a high rate of convergence (order 2.8), the error converges slowly in that we are still only accurate to 3 digits at $L = 13$.

16.2 MULTI-DIMENSIONAL INTEGRALS

Up to this point we have only dealt with integrals in one dimension. In all likelihood, the integrals that you will want to evaluate numerically will be multi-dimensional. We spent the effort we did on learning 1-D integrals because these methods are crucial for multi-D quadrature.

If we want to do a multi-dimensional integral, we can do this rather simply by defining multi-dimensional integrals in terms of 1-D integrals. Consider the 2-D integral:

$$\int_a^b dx \int_c^d dy\, f(x, y).$$

If we define an auxiliary function,

$$g(x) = \int_c^d dy\, f(x, y),$$

then the 2-D integral can be expressed as a 1-D integral of $g(x)$:

$$\int_a^b dx \int_c^d dy\, f(x, y) = \int_a^b dx\, g(x).$$

We could do the same thing with a 3-D integral. In this case we define

$$\int_a^b dx \int_c^d dy \int_e^f dz\, f(x, y, z) = \int_a^b dx\, h(x),$$

where

$$g(x, y) = \int_e^f dz\, f(x, y, z),$$

and

$$h(x) = \int_c^d dy\, g(x, y).$$

What this means in practice is that we need to define intermediate functions that involve integrals over one dimension. We define a general 2-D integral where the 1-D integrals use Gauss–Legendre quadrature.

```
In [5]: def GLQuad2D(f,L):
            """Compute the Gauss-Legendre Quadrature estimate
            of the integral of f(x,y) from x = -1 to 1, y = -1 to 1
            Inputs:
            f:    name of function to integrate
            L:    Order of integration rule (8 or less)
            Returns:
            G-L Quadrature estimate"""
            assert(L>=1)
            #get weights and abscissas for GL quad
            temp, weights, xl = GLQuad(lambda x: x,L,dataReturn=True)
            estimate = 0
            for l in range(L):
                f_onevar = lambda y: f(x,y)
                #set x so that when f_onevar is called it knows what x is
                x = xl[l]
```

```
        g = lambda x: GLQuad(f_onevar,L)
        estimate += weights[l] * g(xl[l])
    return estimate
```

This approach to computing 2-D integrals requires L^2 function evaluations because at each of the L quadrature points in y it evaluates the integrand at L points.

The next step is to generalize this to any rectangular domain in a similar fashion as we did in 1-D.

```
In [6]: def generalGL2D(f,a,b,c,d,L):
            """Compute the Gauss-Legendre Quadrature estimate
            of the integral of f(x,y) from x = a to b
            and y = a to b
            Inputs:
            f:    name of function to integrate
            a:    lower bound of integral in x
            b:    upper bound of integral in x
            c:    lower bound of integral in y
            d:    upper bound of integral in y
            L:    Order of integration rule
            Returns:
            G-L Quadrature estimate"""
            assert(L>=1)
            #define a re-scaled f
            f_rescaled = lambda z,zz: f(0.5*(b-a)*z +
                                    0.5*(a+b),0.5*(d-c)*zz + 0.5*(c+d))
            integral = GLQuad2D(f_rescaled,L)
            return integral*(b-a)*0.5*(d-c)*0.5
```

To test this, we use the following integral

$$\frac{2L}{-3(-1)^{2L} + 2^{4L+3} - 5} \int_{-2}^{3} dx \int_{0}^{3} dy \left[(x+1)^{2L-1} + (y+1)^{2L-1} \right] = 1.$$

We expect that this will be integrated exactly by a two-dimensional Gauss–Legendre quadrature rule with L^2 points.

```
In [7]: L = np.arange(1,12)
        for l in L:
            f = lambda x,y: ((x+1)**(2*l-1) + (y+1)**(2*l-1))*(2*l)/(-3*(-1)**(2*l)
                            + 2**(4*l+3)-5)
            integral = 1
            GLintegral = generalGL2D(f,-2,3,0,3,l)
            print("L =", l,"\t Estimate is",GLintegral,
                  "Exact value is",integral,
                  "\nAbs. Relative Error is", np.abs(GLintegral-integral)/integral)

L = 1    Estimate is 1.0 Exact value is 1
Abs. Relative Error is 0.0
L = 2    Estimate is 1.0 Exact value is 1
Abs. Relative Error is 0.0
L = 3    Estimate is 1.0 Exact value is 1
```

```
Abs. Relative Error is 4.4408920985e-16
L = 4     Estimate is 1.0 Exact value is 1
Abs. Relative Error is 4.4408920985e-16
L = 5     Estimate is 1.0 Exact value is 1
Abs. Relative Error is 1.33226762955e-15
L = 6     Estimate is 1.0 Exact value is 1
Abs. Relative Error is 1.88737914186e-15
L = 7     Estimate is 1.0 Exact value is 1
Abs. Relative Error is 4.66293670343e-15
L = 8     Estimate is 1.0 Exact value is 1
Abs. Relative Error is 1.33226762955e-15
L = 9     Estimate is 1.0 Exact value is 1
Abs. Relative Error is 1.33226762955e-15
L = 10    Estimate is 1.0 Exact value is 1
Abs. Relative Error is 8.881784197e-16
L = 11    Estimate is 1.0 Exact value is 1
Abs. Relative Error is 3.77475828373e-15
```

The method has successfully extended the properties from 1-D to 2-D.

As a further test of this consider a 2-D Cartesian reactor that is a reactor that is finite in x and y and infinite in z (this is nearly the case in power reactors as the reactor is very tall relative to the radius). The scalar flux in this reactor is given by

$$\phi(x, y) = 2 \times 10^9 \left[\cos\left(\frac{\pi x}{100}\right) \cos\left(\frac{\pi y}{100}\right) \right],$$

for $x \in [-50, 50]$ cm and $y \in [-50, 50]$ cm. If $\Sigma_f = 0.1532$ cm^{-1}, what is the total fission rate per unit height in this reactor?

The answer is

$$\int_{-50}^{50} dx \int_{-50}^{50} dy\, \Sigma_f \phi(x, y) = 1.241792427 \times 10^{12}.$$

```
In [8]: Sigma_f = 0.1532
        phi = lambda x,y: 2.0e9 *np.cos(np.pi*x*0.01) * np.cos(np.pi*y*0.01)
        FissionRate = generalGL2D(phi,-50,50,-50,50,6)*Sigma_f
        print("Estimated fission rate per unit height is",
              FissionRate,"fissions/cm")
```

```
Estimated fission rate per unit height is 1.24179242607e+12 fissions/cm
```

We get 9 digits of accuracy when we evaluate the integrand at only 4^2 points. This is a very efficient way to evaluate an integral.

CODA

We have discussed two basic quadrature techniques: Newton–Cotes and Gauss quadrature, and shown how to generalize them to multi-dimensional integrals. Between numerical integration and differentiation we have the basic tools to solve calculus problems. When we combine this with the linear algebra and nonlinear solver skills that we learned previously, we will be able to solve systems of ordinary differential equations and partial differential equations.

We start down this path in the next chapter when we discuss the solution of initial value problems for ordinary differential equations.

PROBLEMS

Short Exercises

Using Gauss–Legendre quadrature estimate the following integrals with $L = 2, 4, 6, 8$, and 30.

16.1. $\int_0^{\pi/2} e^{\sin x}\, dx \approx 3.104379017855555098181$

16.2. $\int_0^{2.405} J_0(x)\, dx \approx 1.470300035485$, where $J_0(x)$ is a Bessel function of the first kind given by

$$J_\alpha(x) = \sum_{m=0}^{\infty} \frac{(-1)^m}{m!\,\Gamma(m+\alpha+1)} \left(\frac{x}{2}\right)^{2m+\alpha}.$$

Programming Projects

1. Gauss–Lobatto Quadrature

One sometimes desires a quadrature rule to include the endpoints of the interval. The Gauss–Legendre quadrature rules do not include $x = \pm 1$. Gauss–Lobatto quadrature includes both of these points in the set.

1. Derive the $L = 2$ Gauss–Lobatto quadrature set. There is only one degree of freedom in this quadrature set, the weight, and it needs to integrate linear polynomials exactly. This quadrature rule will have the form

$$\int_{-1}^{1} f(x)\, dx = wf(-1) + wf(1).$$

2. Now derive the $L = 3$ Gauss–Lobatto quadrature set. Now there are two degrees of freedom because the x's must be ± 1 and 0. This rule will integrate cubics exactly and have the form:

$$\int_{-1}^{1} f(x)\, dx = w_1 f(-1) + w_2 f(0) + w_1 f(1).$$

3. Implement this quadrature rule and verify that it integrates the appropriate polynomials exactly.

2. Gauss–Hermite Quadrature

Other types of Gauss quadrature are possible. Consider integrals of the form

$$\int_0^\infty f(x)e^{-x}\,dx.$$

These integrals can be handled with Gauss–Hermite quadrature.

1. Derive an $L = 1$ quadrature rule (i.e., determine x_1 and w_1) such that

$$\int_0^\infty f(x)e^{-x}\,dx = w_1 f(x_1),$$

is exact for any linear function $f(x)$.
2. Derive an $L = 2$ quadrature rule such that

$$\int_0^\infty f(x)e^{-x}\,dx = w_1 f(x_1) + w_2 f(x_2),$$

is exact for any cubic (or lower degree) polynomial $f(x)$.
3. Implement these two quadrature rules and verify that they integrate linear and cubic polynomials exactly.

3. Integration and Root Finding

Consider a 1-D cylindrical reactor with geometric buckling 0.0203124 cm^{-1} and $D = 9.21$ cm, $\nu\Sigma_f = 0.1570$ cm^{-1}, and $\Sigma_a = 0.1532$ cm^{-1}. The geometric buckling for a cylinder is given by

$$B_g^2 = \left(\frac{2.405}{R}\right)^2.$$

1. Find the critical radius of this reactor, that is when $B_g^2 = B_m^2$ with

$$B_m^2 = \frac{\nu\Sigma_f - \Sigma_a}{D}.$$

2. Using the numerical integration method of your choice, find the peak scalar flux assuming that power per unit height is 2 MW/cm. Use 200 MeV/fission $= 3.204 \times 10^{-11}$ J.
3. Now assume the reactor has a height of 500 cm and a power of 1000 MW. What is the peak scalar flux? You will need a multi-dimensional integral in this case.

Maude: *Lord. You can imagine where it goes from here.*

Dude: *He fixes the cable?*

–from the film **The Big Lebowski**

CHAPTER POINTS

- Initial value problems require the application of an integration rule to update dependent variables.

- Explicit methods use information from previous and current times to march forward in time. These methods have a limited step size.

- Implicit methods define the update in terms of the state at a future time. These methods can take large steps, but may be limited in accuracy. Implicit methods also require the solution of linear or nonlinear equations for each update.

- To solve systems of equations, we can apply generalizations of the techniques developed for single ODEs.

Computational Nuclear Engineering and Radiological Science Using Python
DOI: 10.1016/B978-0-12-812253-2.00019-4

In this chapter we will use numerical integration techniques to solve ordinary differential equations (ODEs) where the value of the solution is specified at a single value of the independent variable. Because these problems often correspond to time-dependent problems with an initial condition, these are called initial value problems.

Consider the generic, initial value problem with a first-order derivative given by

$$y'(t) = f(y, t), \qquad y(0) = y_0,$$

where $f(y, t)$ is a function that in general depends on y and t. Typically, we call t the time variable and y our solution. For a problem of this sort we can simply integrate both sides of the equation from $t = 0$ to $t = \Delta t$, where Δt is called the time step. Doing this we get

$$y(\Delta t) - y(0) = \int_0^{\Delta t} f(y, t) \, dt. \tag{17.1}$$

17.1 FORWARD EULER

To proceed we will treat the integral in the previous equation using a numerical integration rule. One rule that is so basic we did not talk about in the chapters on numerical integration is the left-hand rectangle rule. Here we estimate the integral as

$$\int_0^{\Delta t} f(y, t) \, dt \approx \Delta t f(y(0), 0). \tag{17.2}$$

Using this relation in Eq. (17.1) gives us

$$y(\Delta t) = y(0) + \Delta t f(y(0), 0). \tag{17.3}$$

This will give an approximate value of the solution Δt after the initial condition. If we wanted to continue out to later times, we could apply the rule again repeatedly. To do this we define the value of y after n timesteps, each of width Δt as

$$y^n = y(t^n) = y(n \Delta t), \qquad \text{for } n = 0, \ldots, N. \tag{17.4}$$

Using this we can write the solution using the left-hand rectangle rule for integration as

$$y^{n+1} = y^n + \Delta t f(y^n, t^n).$$

This method is called the explicit Euler method or the forward Euler method after the Swiss mathematician whose name is commonly pronounced *oi-ler*, much like a hockey team from Edmonton. The method is said to be explicit, not because sometimes it will make you want to shout profanity, rather that the update is explicitly defined by the value of the solution at time t^n. Below we define a Python function that for a given right-hand side, initial condition, and time step and number of time steps, N, performs the forward Euler method. This function will take the name of the function on the right-hand side as an input.

BOX 17.1 NUMERICAL PRINCIPLE

An explicit numerical method formulates the update to time step $n+1$ in terms of quantities only at time step n or before. The classical example of an explicit method is the forward Euler method which writes the solution to $y'(t) = f(y, t)$ as

$$y^{n+1} = y^n + \Delta t f(y^n, t^n) \qquad \text{Forward Euler.}$$

```
In [1]: def forward_euler(f,y0,Delta_t,numsteps):
            """Perform numsteps of the forward Euler method starting at y0
            of the ODE y'(t) = f(y,t)
            Args:
                f: function to integrate takes arguments y,t
                y0: initial condition
                Delta_t: time step size
                numsteps: number of time steps

            Returns:
                a numpy array of the times and a numpy
                array of the solution at those times
            """
            numsteps = int(numsteps)
            y = np.zeros(numsteps+1)
            t = np.arange(numsteps+1)*Delta_t
            y[0] = y0
            for n in range(1,numsteps+1):
                y[n] = y[n-1] + Delta_t * f(y[n-1], t[n-1])
            return t, y
```

We will test this on a simple problem:

$$y'(t) = -y(t), \qquad y_0 = 1.$$

The solution to this problem is

$$y(t) = e^{-t}.$$

The following code compares the exact solution to the forward Euler solution with a time step of 0.1.

```
In [2]: RHS = lambda y,t: -y
        Delta_t = 0.1
        t_final = 2
        t,y = forward_euler(RHS,1,Delta_t,t_final/Delta_t)
        plt.plot(t,y,'o-',label="numerical solution")
        t_fine = np.linspace(0,t_final,100)
        plt.plot(t_fine,np.exp(-t_fine),label="Exact Solution")
```

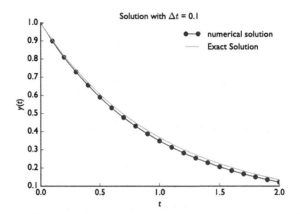

That looks pretty close. The numerical solution appears to be slightly below the exact solution, but the difference appears to be small. We could re-do this with different size time steps and compare the solutions as a function of Δt.

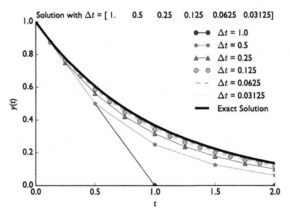

Also, we can compute the error as a function of Δt. On a log-log scale this will tell us about the convergence of this method in Δt.

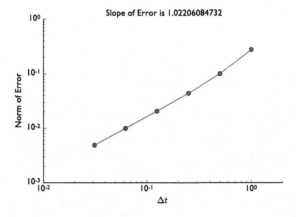

Notice that the error indicates that this is a first-order method in Δt: when we decrease Δt by a factor of 2, the error decreases by a factor of 2. In this case we measured the error with a slightly different error norm:

$$\text{Error} = \frac{1}{\sqrt{N}} \sqrt{\sum_{n=1}^{N} \left(y^n_{\text{approx}} - y^n_{\text{exact}} \right)^2},$$

where N is the number of steps the ODE is solved over.

One thing that can happen with the forward Euler method is that if the time step is too large, it can become unstable. What this means is the solution diverges to be plus or minus infinity (sometimes it goes to both). In our case, forward Euler is unstable if $\Delta t > 2$, as we will prove later and demonstrate numerically here.

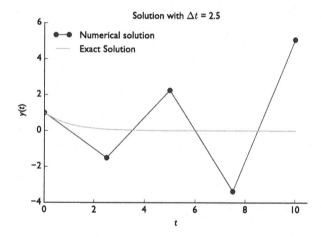

The solution grows over time, even though the true solution decays to 0. This happens because the magnitude of the solution grows, which makes the right-hand side for the next update larger. This makes the solution grow in magnitude each step.

Stability is an important consideration, and we will talk more about it later.

17.2 BACKWARD EULER

We could use a different method to integrate our original ODE rather than the left-hand rectangle rule. An obvious alternative is the right-hand rectangle rule:

$$y^{n+1} = y^n + \Delta t f(y^{n+1}, t^{n+1}).$$

This method is called he backward Euler method or the implicit Euler method.

It is implicit because the update is implicitly defined by evaluating f with the value of y we are trying to solve for. That means the update needs to solve the equation

$$y^{n+1} - y^n - \Delta t f(y^{n+1}, t^{n+1}) = 0, \tag{17.5}$$

using a nonlinear solver (unless f is linear in y). Therefore, this method is a bit harder to implement in code.

BOX 17.2 NUMERICAL PRINCIPLE

An implicit numerical method formulates the update to time step $n+1$ in terms of quantities at time step $n+1$ and possibly previous time steps. The classical example of an implicit method is the backward Euler method which write the solution to $y'(t) = f(y,t)$ as

$$y^{n+1} = y^n + \Delta t f(y^{n+1}, t^{n+1})$$

Backward Euler.

A well-known second-order implicit method is the Crank-Nicolson method. It uses values at time step $n+1$ and n:

$$y^{n+1} = y^n + \frac{1}{2}\Delta t \left[f(y^n, t^n) + f(y^{n+1}, t^{n+1}) \right]$$

Crank-Nicolson.

The function below uses the inexact Newton method we defined before to solve Eq. (17.5).

```
In [3]: def backward_euler(f,y0,Delta_t,numsteps):
            """Perform numsteps of the backward Euler method starting at y0
            of the ODE y'(t) = f(y,t)
            Args:
                f: function to integrate takes arguments y,t
                y0: initial condition
                Delta_t: time step size
                numsteps: number of time steps

            Returns:
                a numpy array of the times and a numpy
                array of the solution at those times
            """
            numsteps = int(numsteps)
            y = np.zeros(numsteps+1)
            t = np.arange(numsteps+1)*Delta_t
            y[0] = y0
            for n in range(1,numsteps+1):
                solve_func = lambda u: u-y[n-1] - Delta_t*f(u,t[n])
                y[n] = inexact_newton(solve_func,y[n-1])
            return t, y
```

Performing the test we did for forward Euler gives the following results.

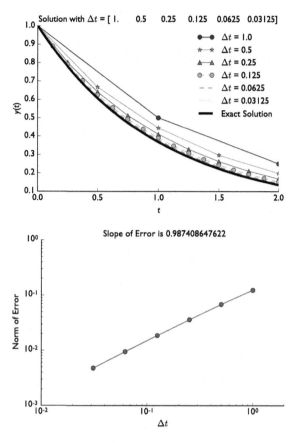

There are several differences between the results from backward Euler and forward Euler. The backward Euler solutions approach the exact solution from above, and the convergence of the error is at the same rate as forward Euler. It seems like we did not get much for the extra effort of solving a nonlinear equation at each step. What we do get is unconditional stability. We can see this by using a large time step.

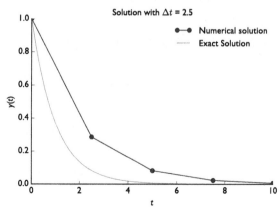

The solution, though not very accurate, still behaves reasonably well. The exact solution decays to 0 as $t \to \infty$, and the backward Euler solution behaves the same way. This behavior can be very useful on more complicated problems than this simple one.

17.3 CRANK–NICOLSON (TRAPEZOID RULE)

We could use the trapezoid rule to integrate the ODE over the time step. Doing this gives

$$y^{n+1} = y^n + \frac{\Delta t}{2}\left(f(y^n, t^n) + f(y^{n+1}, t^{n+1})\right).$$

This method, often called Crank–Nicolson, is also an implicit method because y^{n+1} is on the right-hand side of the equation. For this method the equation we have to solve at each time step is

$$y^{n+1} - y^n - \frac{\Delta t}{2}\left(f(y^n, t^n) + f(y^{n+1}, t^{n+1})\right) = 0.$$

Implementing this method is no more difficult than backward Euler. The only change is the function given to the inexact Newton method.

```
In [4]: def crank_nicolson(f,y0,Delta_t,numsteps):
            """Perform numsteps of the Crank--Nicolson method starting at y0
            of the ODE y'(t) = f(y,t)
            Args:
                f: function to integrate takes arguments y,t
                y0: initial condition
                Delta_t: time step size
                numsteps: number of time steps

            Returns:
                a numpy array of the times and a numpy
                array of the solution at those times
            """
            numsteps = int(numsteps)
            y = np.zeros(numsteps+1)
            t = np.arange(numsteps+1)*Delta_t
            y[0] = y0
            for n in range(1,numsteps+1):
                solve_func = lambda u:u-y[n-1]-0.5*Delta_t*(f(u,t[n])
                                               + f(y[n-1],t[n-1]))
                y[n] = inexact_newton(solve_func,y[n-1])
            return t, y
```

On our test where the solution is e^{-t}, we can see the benefit of having a second-order method:

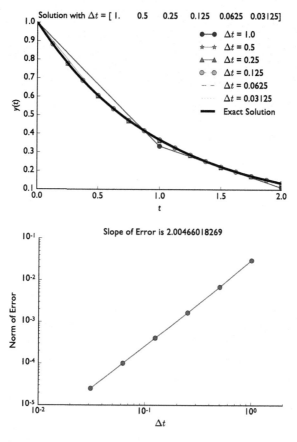

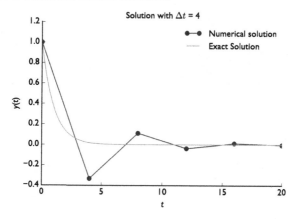

We do get the expected second-order convergence of the error as evidenced by the error plot. In the comparison of the solutions, except for the large value of Δt, it is hard to distinguish the approximate and the exact solutions.

In terms of stability, Crank–Nicolson is a mixed bag. The method will not diverge as $\Delta t \to \infty$, but it can oscillate around the correct solution:

Notice that the oscillation makes the numerical solution negative. This is the case even though the exact solution, e^{-t}, cannot be negative.

17.3.1 Comparison of the Methods

We will compare the methods on a problem that has a driving term as well as decay:

$$y'(t) = \left(\frac{1}{t+1/2} - 1 \right) y(t), \qquad y(0) = \frac{1}{2}. \tag{17.6}$$

The purpose of this exercise is to show that the issues of accuracy are even more important when the solution is not a simple exponential.

The solution to this problem is

$$y(t) = \left(t + \frac{1}{2} \right) e^{-t}.$$

We will start with a small value of Δt:

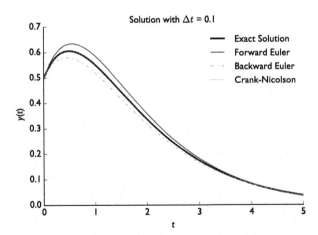

The two Euler methods are equally inaccurate, they just differ in how they are wrong (above or below). Crank–Nicolson does a good job of following the solution, and is indiscernible from the exact solution.

With a bigger time step we can see the warts of the methods that we have presented:

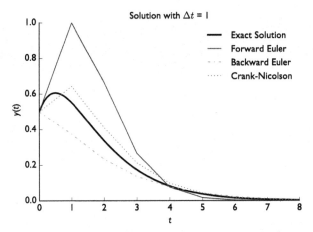
Solution with $\Delta t = 1$

In these results we see that for a large time step, forward Euler completely overshoots the initial growth, backward Euler just starts decaying, and Crank–Nicolson starts off too high before beginning to decay. It appears that even Crank-Nicolson is not accurate enough for this problem with this large step.

17.4 STABILITY

There is a formal definition of stability for a numerical method for integrating ODEs. To get to this definition, consider the ODE

$$y'(t) = -\alpha y(t).$$ (17.7)

For any single-step method we can write

$$y^{n+1} = g y^n.$$

A solution method is said to be stable if $|g| \leq 1$. Furthermore, a solution is said to be non-oscillatory if $0 \leq g \leq 1$. The quantity g is often called the growth rate.

We first look at forward Euler on this ODE:

$$y^{n+1} = y^n - \alpha \Delta t y^n = (1 - \alpha \Delta t) y^n,$$

this implies that the growth rate for forward Euler is

$$g_{FE} = 1 - \alpha \Delta t.$$

To make sure that $|g| \leq 1$ we need to have $\alpha \Delta t \leq 2$. To be non-oscillatory we need $\alpha \Delta t \leq 1$. This is why when we solved

$$y'(t) = -y(t)$$

with $\Delta t = 2.5$, the solution grew in an unstable manner. Because there is a restriction on the time step for stability, we call the forward Euler method conditionally stable.

BOX 17.3 NUMERICAL PRINCIPLE

There are several types of stability that are important. These are usually discussed in how the update for the ODE, $y'(t) = -\alpha y(t)$, is given. For any single-step method we can write the update for this problem as

$$y^{n+1} = gy^n.$$

Here g is the growth rate.

- A method is **conditionally stable** if the growth rate has a magnitude greater than 1 for certain positive values of the time step size.
- A method is **unconditionally stable** if the growth rate has a magnitude smaller than or equal to 1 for all positive values of the time step size.
- A method is **non-oscillatory** if the growth rate is between 0 and 1.

The value of the growth rate for backward Euler can be easily derived. We start with

$$y^{n+1} = y^n - \alpha \Delta t\, y^{n+1},$$

which when rearranged is

$$y^{n+1} = \frac{y^n}{1 + \alpha \Delta t}.$$

This makes

$$g_{BE} = \frac{1}{1 + \alpha \Delta t}.$$

For any $\Delta t > 0$, g will be between 0 and 1. Therefore, backward Euler is unconditionally stable and unconditionally non-oscillatory.

The Crank–Nicolson method has

$$g_{CN} = \frac{2 - \alpha \Delta t}{2 + \alpha \Delta t}.$$

This method will be unconditionally stable because

$$\lim_{\Delta t \to \infty} g_{CN} = -1.$$

It is conditionally non-oscillatory because $g_{CN} < 0$ for $\alpha \Delta t > 2$. In the original example, we had $\alpha \Delta t = 4$, and we saw noticeable oscillations.

In the contrast between the implicit methods, Crank–Nicolson and backward Euler, we see a common theme in numerical methods: to get better accuracy one often has to sacrifice robustness. In this case Crank–Nicolson allows oscillations in the solution, but its error decreases at second-order in the time step size. Backward Euler is non-oscillatory, but the error converges more slowly than Crank–Nicolson.

17.5 FOURTH-ORDER RUNGE–KUTTA METHOD

There is one more method that we need to consider at this point. It is an explicit method called a Runge–Kutta method. The particular one we will learn is fourth-order accurate in Δt. That means that if we decrease Δt by a factor of 2, the error will decrease by a factor of $2^4 = 16$. The method can be written as

$$y^{n+1} = y^n + \frac{1}{6}\left(\Delta y_1 + 2\Delta y_2 + 2\Delta y_3 + \Delta y_4\right), \qquad (17.8)$$

where

$$\Delta y_1 = \Delta t f(y^n, t^n),$$
$$\Delta y_2 = \Delta t f\left(y^n + \frac{\Delta y_1}{2}, t^n + \frac{\Delta t}{2}\right),$$
$$\Delta y_3 = \Delta t f\left(y^n + \frac{\Delta y_2}{2}, t^n + \frac{\Delta t}{2}\right),$$
$$\Delta y_4 = \Delta t f\left(y^n + \Delta y_3, t^n + \Delta t\right).$$

To get fourth-order accuracy, this method takes a different approach to integrating the right-hand side of the ODE. Basically, it makes several projections forward and combines them in such a way that the errors cancel to make the method fourth-order.

Implementing this method is not difficult either.

```
In [5]: def RK4(f,y0,Delta_t,numsteps):
            """Perform numsteps of the 4th-order Runge-Kutta
            method starting at y0 of the ODE y'(t) = f(y,t)
            Args:
                f: function to integrate takes arguments y,t
                y0: initial condition
                Delta_t: time step size
                numsteps: number of time steps

            Returns:
                a numpy array of the times and a numpy
                array of the solution at those times
            """
            numsteps = int(numsteps)
            y = np.zeros(numsteps+1)
            t = np.arange(numsteps+1)*Delta_t
            y[0] = y0
            for n in range(1,numsteps+1):
                dy1 = Delta_t * f(y[n-1], t[n-1])
                dy2 = Delta_t * f(y[n-1] + 0.5*dy1, t[n-1] + 0.5*Delta_t)
                dy3 = Delta_t * f(y[n-1] + 0.5*dy2, t[n-1] + 0.5*Delta_t)
                dy4 = Delta_t * f(y[n-1] + dy3, t[n-1] + Delta_t)
                y[n] = y[n-1] + 1.0/6.0*(dy1 + 2.0*dy2 + 2.0*dy3 + dy4)
            return t, y
```

We will first try this Runge–Kutta method on the problem with growth and decay given in Eq. (17.6) with a large time step.

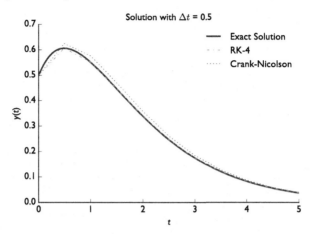

It seems to do better that Crank–Nicolson with a large time step, but since it is an explicit method, there are limits: with a large enough time step the solution will oscillate and can be unstable.

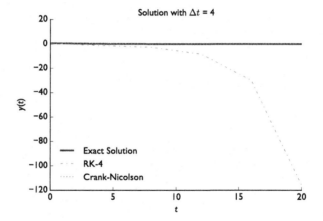

Therefore, we need to be careful when we use the RK-4 method that the time step is not too large.

17.5.1 Stability for RK4

For the fourth-order Runge–Kutta method presented above, we can look at the stability by examining how it performs on the problem $y'(t) = -\alpha y$. On this problem the intermediate values are

$$\Delta y_1 = -\alpha \Delta t y^n,$$

$$\Delta y_2 = -\alpha \Delta t \left(y^n + \frac{\Delta y_1}{2} \right)$$

$$= -\alpha \Delta t \left(1 - \frac{\alpha \Delta t}{2} \right) y^n,$$

$$\Delta y_3 = -\alpha \Delta t \left(y^n + \frac{\Delta y_2}{2} \right)$$

$$= -\alpha \Delta t \left(1 - \frac{\alpha \Delta t}{2} \left(1 - \frac{\alpha \Delta t}{2} \right) \right) y^n,$$

$$\Delta y_4 = -\alpha \Delta t \left(y^n + \Delta y_3 \right)$$

$$= -\alpha \Delta t \left(1 - \alpha \Delta t \left(1 - \frac{\alpha \Delta t}{2} \left(1 - \frac{\alpha \Delta t}{2} \right) \right) \right) y^n.$$

Using these results in the update for y^{n+1} in Eq. (17.8) gives

$$y^{n+1} = \left(\frac{\alpha^4 \Delta t^4}{24} - \frac{\alpha^3 \Delta t^3}{6} + \frac{\alpha^2 \Delta t^2}{2} - \alpha \Delta t + 1 \right) y^n.$$

This implies that the growth rate for RK4 is

$$g_{RK4} = \frac{\alpha^4 \Delta t^4}{24} - \frac{\alpha^3 \Delta t^3}{6} + \frac{\alpha^2 \Delta t^2}{2} - \alpha \Delta t + 1. \tag{17.9}$$

This is a fourth-degree polynomial in $\alpha \Delta t$. For $\alpha \Delta t > 0$ the value of g_{RK4} is positive. Additionally, the stability limit is

$$\alpha \Delta t \leq \frac{1}{3} \left(4 - 10 \sqrt[3]{\frac{2}{43 + 9\sqrt{29}}} + 2^{2/3} \sqrt[3]{43 + 9\sqrt{29}} \right) \approx 2.78529356.$$

This implies that RK4 is conditionally stable, and is non-oscillatory where it is stable. The stability criterion is less restrictive for RK4 than for forward Euler, and where Crank–Nicolson is non-oscillatory, RK4 will be as well.

17.6 SYSTEMS OF DIFFERENTIAL EQUATIONS

Often we will be concerned with solving a system of ODEs rather than a single ODE. The explicit methods translate to this scenario directly. However, we will restrict ourselves to systems that can be written in the form

$$\mathbf{y}'(t) = \mathbf{A}(t)\mathbf{y} + \mathbf{c}(t), \qquad \mathbf{y}(0) = \mathbf{y}_0.$$

In this equation $\mathbf{A}(t)$ is a matrix that can change over time, and $\mathbf{c}(t)$ is a function of t only. For systems of this type our methods are written as follows:

- Forward Euler

$$\mathbf{y}^{n+1} = \mathbf{y}^n + \Delta t \mathbf{A}(t^n)\mathbf{y}^n + \Delta t \mathbf{c}(t^n).$$

- Backward Euler

$$\mathbf{y}^{n+1} = \mathbf{y}^n + \Delta t \mathbf{A}(t^{n+1})\mathbf{y}^{n+1} + \Delta t \mathbf{c}(t^{n+1}),$$

which rearranges to

$$\left(\mathbf{I} - \Delta t \mathbf{A}(t^{n+1})\right)\mathbf{y}^{n+1} = \mathbf{y}^n + \Delta t \mathbf{c}(t^{n+1}).$$

Therefore, for backward Euler we will have to solve a linear system of equations at each time step.

- Crank-Nicolson

$$\left(\mathbf{I} - \frac{\Delta t}{2}\mathbf{A}(t^{n+1})\right)\mathbf{y}^{n+1} = \left(\mathbf{I} + \frac{\Delta t}{2}\mathbf{A}(t^n)\right)\mathbf{y}^n + \frac{\Delta t}{2}\left(\mathbf{c}(t^{n+1}) + \mathbf{c}(t^n)\right).$$

This will also involve a linear solve at each step.

- Fourth-order Runge–Kutta

$$\mathbf{y}^{n+1} = \mathbf{y}^n + \frac{1}{6}\left(\Delta\mathbf{y}_1 + 2\Delta\mathbf{y}_2 + 2\Delta\mathbf{y}_3 + \Delta\mathbf{y}_4\right),$$

$$\Delta y_1 = \Delta t \mathbf{A}(t^n)y^n + \mathbf{c}(t^n),$$

$$\Delta y_2 = \Delta t \mathbf{A}\left(t^n + \frac{\Delta t}{2}\right)\left(y^n + \frac{\Delta y_1}{2}\right) + \Delta t \mathbf{c}\left(t^n + \frac{\Delta t}{2}\right),$$

$$\Delta y_3 = \Delta t \mathbf{A}\left(t^n + \frac{\Delta t}{2}\right)\left(y^n + \frac{\Delta y_2}{2}\right) + \Delta t \mathbf{c}\left(t^n + \frac{\Delta t}{2}\right),$$

$$\Delta y_4 = \Delta t \mathbf{A}\left(t^n + \Delta t\right)\left(y^n + \Delta y_3\right) + \Delta t \mathbf{c}(t^n + \Delta t).$$

As noted above, the implicit methods require the solution of a linear system at each time step, while the explicit methods do not.

We first define a function for solving forward Euler on a system.

```
In [6]: def forward_euler_system(Afunc,c,y0,Delta_t,numsteps):
            """Perform numsteps of the backward euler method starting at y0
            of the ODE y'(t) = A(t) y(t) + c(t)
            Args:
                Afunc: function to compute A matrix
                c: nonlinear function of time
                Delta_t: time step size
                numsteps: number of time steps

            Returns:
                a numpy array of the times and a numpy
                array of the solution at those times
            """
            numsteps = int(numsteps)
            unknowns = y0.size
```

```
y = np.zeros((unknowns,numsteps+1))
t = np.arange(numsteps+1)*Delta_t
y[0:unknowns,0] = y0
for n in range(1,numsteps+1):
    yold = y[0:unknowns,n-1]
    A = Afunc(t[n-1])
    y[0:unknowns,n] = yold + Delta_t * (np.dot(A,yold) + c(t[n-1]))
return t, y
```

As a test we will solve the ODE:

$$y''(t) = -y(t), \qquad y(0) = 1, \quad y'(0) = 0.$$

At first blush this does not seem compatible with our method for solving a system of equations. Moreover, we have not covered how to solve ODEs with derivatives other than first derivatives. Nevertheless, we can write this as a system using the definition

$$u(t) = y'(t),$$

to get

$$\frac{d}{dt}\begin{pmatrix} u \\ y \end{pmatrix} = \begin{pmatrix} 0 & -1 \\ 1 & 0 \end{pmatrix}\begin{pmatrix} u \\ y \end{pmatrix}.$$

We will set this up in Python and solve it with forward Euler. The solution is $y(t) = \cos(t)$.

```
In [7]: #Set up A
        Afunc = lambda t: np.array([(0,-1),(1,0)])
        #set up c
        c = lambda t: np.zeros(2)
        #set up y
        y0 = np.array([0,1])
        Delta_t = 0.1
        t_final = 8*np.pi
        t,y = forward_euler_system(Afunc,c,y0,Delta_t,t_final/Delta_t)
```

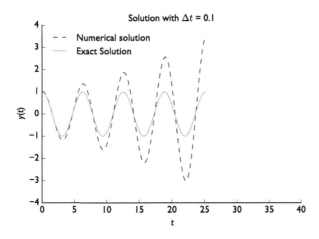

The error grows over time. What's happening here is that the numerical error builds over time and this causes the magnitude to grow over time. Using a smaller value of Δt can help, but not completely remove the problem.

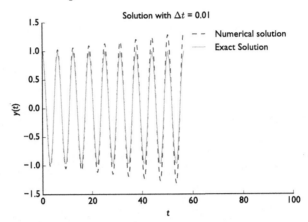

We have really just delayed the inevitable: the growth of the magnitude of the solution is increasing.

To understand what is going on we will look a phase field plot for this ODE. The phase field plot for this system tells us the direction of the derivatives of y and u given a value for each. Using the phase field and a starting point, we can follow the arrows to see how the true solution behaves. In the following phase figure, the solid black line shows the behavior of the solution when $y(0) = 1$ and $u(t) = y'(0) = 0$. The solution starts at the solid circle and goes around twice in the time plotted ($t = 0$ to 4π).

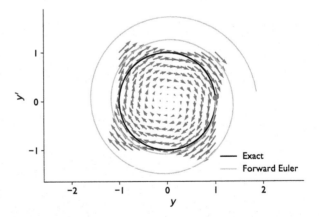

The solution looks like a circle in the phase field because of the repeating periodic nature of the solution. However, the forward Euler solution has errors that grow over time, so that each time around, the circle grows larger.

The backward Euler method is implemented next. To solve the linear system we will use a Gaussian elimination function that we defined previously.

```
In [8]: def backward_euler_system(Afunc,c,y0,Delta_t,numsteps):
            """Perform numsteps of the backward euler method starting at y0
            of the ODE y'(t) = A(t) y(t) + c(t)
            Args:
                Afunc: function to compute A matrix
                c: nonlinear function of time
                y0: initial condition
                Delta_t: time step size
                numsteps: number of time steps

            Returns:
                a numpy array of the times and a numpy
                array of the solution at those times
            """
            numsteps = int(numsteps)
            unknowns = y0.size
            y = np.zeros((unknowns,numsteps+1))
            t = np.arange(numsteps+1)*Delta_t
            y[0:unknowns,0] = y0
            for n in range(1,numsteps+1):
                yold = y[0:unknowns,n-1]
                A = Afunc(t[n])
                LHS = np.identity(unknowns) - Delta_t * A
                RHS = yold + c(t[n])*Delta_t
                y[0:unknowns,n] = GaussElimPivotSolve(LHS,RHS)
            return t, y
```

Results with $\Delta t = 0.1$ show that error builds over time, but in a different way than forward Euler.

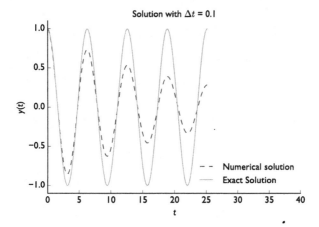

Now the numerical error builds over time, but the error causes the solution to damp out over time. As before, decreasing Δt only delays the onset of the error.

The phase field plot shows that instead of the path growing, backward Euler's path spirals down to zero.

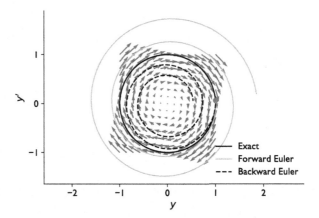

Larger time steps make the solution decay to zero faster. If we take a small time step, the decay is slower but still occurs.

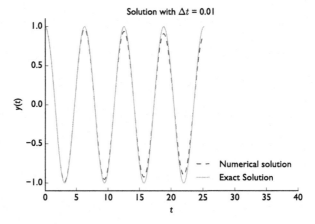

The takeaway here is that first-order accurate methods have errors that build over time. Forward Euler the errors cause the solution to grow, whereas, Backward Euler has the solution damp to zero over time.

17.6.1 Stability for Systems

For a system of initial value problems, the stability condition is derived by looking at the problem

$$\mathbf{y}' = -\mathbf{A}\mathbf{y}.$$

If we multiply both sides of the equation by a left-eigenvector of $\mathbf{A}$, $\mathbf{l}_k$ with associated eigenvector α_k we get

$$\mathbf{u}_k' = -\alpha_k \mathbf{u}_k, \tag{17.10}$$

where $\mathbf{u}_k = \mathbf{l}_k \cdot \mathbf{y}$.

Notice that each row of Eq. (17.10) is identical to the model equation we had for stability for single initial value problems in Eq. (17.7). Therefore, we can replace the α in our equations for the growth rate g by the eigenvalues of the matrix $\mathbf{A}$. Then we can determine under what conditions $|g| \leq 1$ to find a range of stability.

In the example above where

$$\mathbf{A} = \begin{pmatrix} 0 & -1 \\ 1 & 0 \end{pmatrix},$$

the eigenvalues are $\pm i$. Using the growth rate for forward Euler, we get $g_{\text{FE}} = 1 \pm i \Delta t$, which implies

$$|g_{\text{FE}}| = \sqrt{1 + \Delta t^2}.$$

This result means that for any Δt, forward Euler will not be stable because any positive Δt makes $|g_{\text{FE}}| > 1$. This is what we saw in the example. Backward Euler, however, is stable because

$$|g_{\text{BE}}| = \frac{1}{\sqrt{1 + \Delta t^2}},$$

which is less than one for any positive Δt.

17.6.2 Crank–Nicolson for Systems

Now we will look at Crank–Nicolson to see how it behaves. A Python implementation of Crank–Nicolson is given next:

```
In [9]: def cn_system(Afunc,c,y0,Delta_t,numsteps):
            """Perform numsteps of the Crank--Nicolson method starting at y0
            of the ODE y'(t) = A(t) y(t) + c(t)
            Args:
                Afunc: function to compute A matrix
                c: nonlinear function of time
                y0: initial condition
                Delta_t: time step size
                numsteps: number of time steps

            Returns:
                a numpy array of the times and a numpy
                array of the solution at those times
            """
            numsteps = int(numsteps)
            unknowns = y0.size
            y = np.zeros((unknowns,numsteps+1))
            t = np.arange(numsteps+1)*Delta_t
            y[0:unknowns,0] = y0
            for n in range(1,numsteps+1):
                yold = y[0:unknowns,n-1]
                A = Afunc(t[n])
                LHS = np.identity(unknowns) - 0.5*Delta_t * A
                A = Afunc(t[n-1])
```

```
RHS = yold +
        0.5*Delta_t * np.dot(A,yold) + 0.5*(c(t[n-1]) +
        c(t[n]))*Delta_t
    y[0:unknowns,n] = GaussElimPivotSolve(LHS,RHS)
return t, y
```

On our test problem Crank–Nicolson solution does not display the significant error that the first-order methods did.

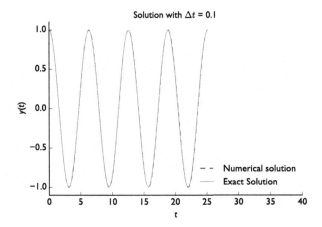

The error build up is not nearly as large of a problem. Even if we look at the solution over a much longer time, the error is not significant:

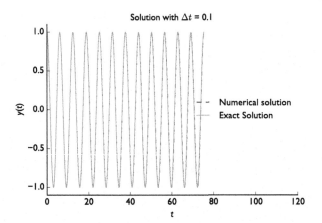

The phase field plot demonstrates that the accuracy of Crank-Nicolson allows it to not have any strange behavior in the phase field. The numerical solution appears as a circle in the phase field:

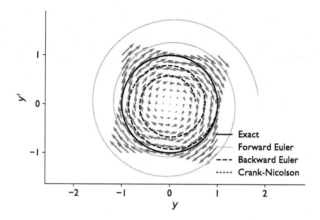

One can show that for this problem, Crank–Nicolson is unconditionally stable because $|g_{CN}| = 1$ for any Δt. This also explains why the solution did not damp or grow over time.

17.6.3 RK4 for Systems

Finally, we generalize our implementation of fourth-order Runge–Kutta to handle systems of equations.

```
In [10]: def RK4_system(Afunc,c,y0,Delta_t,numsteps):
             """Perform numsteps of the 4th-order Runge--Kutta method starting at y0
             of the ODE y'(t) = f(y,t)
             Args:
                 f: function to integrate takes arguments y,t
                 y0: initial condition
                 Delta_t: time step size
                 numsteps: number of time steps

             Returns:
                 a numpy array of the times and a numpy
                 array of the solution at those times
             """
             numsteps = int(numsteps)
             unknowns = y0.size
             y = np.zeros((unknowns,numsteps+1))
             t = np.arange(numsteps+1)*Delta_t
             y[0:unknowns,0] = y0
             for n in range(1,numsteps+1):
                 yold = y[0:unknowns,n-1]
                 A = Afunc(t[n-1])
                 dy1 = Delta_t * (np.dot(A,yold) + c(t[n-1]))
                 A = Afunc(t[n-1] + 0.5*Delta_t)
                 dy2 = Delta_t * (np.dot(A,y[0:unknowns,n-1] + 0.5*dy1)
                             + c(t[n-1] + 0.5*Delta_t))
                 dy3 = Delta_t * (np.dot(A,y[0:unknowns,n-1] + 0.5*dy2)
                             + c(t[n-1] + 0.5*Delta_t))
                 A = Afunc(t[n] + Delta_t)
```

```
        dy4 = Delta_t * (np.dot(A,y[0:unknowns,n-1] + dy3) + c(t[n]))
        y[0:unknowns,n] = y[0:unknowns,n-1] +
                            1.0/6.0*(dy1 + 2.0*dy2 + 2.0*dy3 + dy4)
    return t, y
```

Like Crank–Nicolson, RK4 does not display noticeable error on our test problem.

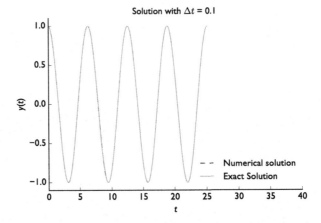

On this simple problem, RK4 appears to be as good as Crank-Nicolson. The stability limit for RK4 can be derived by replacing α with $\pm i$ in Eq. (17.9). The result is that $\Delta t < 2.78529356$ will give a stable, if inaccurate, solution.

17.7 POINT-REACTOR KINETICS EQUATIONS

The simplest model to describe the behavior of the power in a nuclear reactor are the point-reactor kinetics equations (PRKEs) with one delayed neutron precursor group. This is an important model for the time-dependent behavior of nuclear reactors, and other nuclear systems. See the further reading section below for suggestions on background references for this model.

The PRKEs are a system of two ODEs:

$$p'(t) = \frac{\rho - \beta}{\Lambda} p(t) + \lambda C(t),$$

$$C'(t) = \frac{\beta}{\Lambda} p(t) - \lambda C(t).$$

Here p is the number of free neutrons in the reactor, ρ is the reactivity

$$\rho = \frac{k - 1}{k},$$

for k the effective multiplication factor, β is the fraction of neutrons produced by fission that are delayed, Λ is the mean time between neutron generations in the reactor, λ is the mean number of delayed neutrons produced per unit time, and C is the number of delayed neutron precursors (nuclides that will emit a neutron).

In our notation

$$\mathbf{y}(t) = (p(t), C(t))^t,$$

$$\mathbf{A}(t) = \begin{pmatrix} \frac{\rho - \beta}{\Lambda} & \lambda \\ \frac{\beta}{\Lambda} & -\lambda \end{pmatrix},$$

and

$$\mathbf{c}(t) = \mathbf{0}.$$

Recall that at steady-state, (i.e. $p'(t) = C'(t) = 0$), to have a non-trivial solution $\rho = 0$, i.e., the solution is critical. In this case, the solution to this system is

$$p = \frac{\lambda \Lambda}{\beta}.$$

We will need to know this for some of our test cases.

Also, some typical values for the constants are $\beta = 750 \times 10^{-5}$, $\Lambda = 2 \times 10^{-5}$ s, $\lambda = 10^{-1}$ s^{-1}.

17.7.1 Rod-Drop

Now we simulate the situation where we have a large amount of negative reactivity (a dollar ($) is a reactivity unit, where one dollar of reactivity is equal to the value of β for the system) inserted into a critical reactor, $\rho = -2\$ = -2\beta$. This is the scenario that occurs when a control is rapidly inserted into the reactor. The tricky part here is setting up the function for $\mathbf{A}(t)$. We will start with RK4.

```
In [32]: #Set up A
         beta = 705.0e-5
         Lambda = 2.0e-5 #s
         l = 1.0e-1 #s**-1
         rho = -2.0*beta
         Afunc = lambda t: np.array([((rho-beta)/Lambda,l),
                                     (beta/Lambda,-1)])
         #set up c
         c = lambda t: np.zeros(2)
         #set up inital vector
         y0 = np.array([1,beta/(l*Lambda)])
         Delta_t = 0.001 #1 millisecond
         t_final = 5
         t,y = RK4_system(Afunc,c,y0,Delta_t,t_final/Delta_t)
```

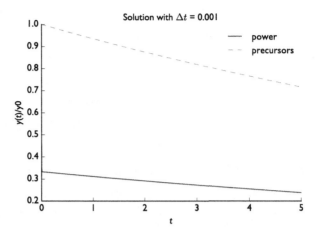

We look at the results on a semilog scale so we can see the prompt jump: the rapid decrease in the neutron population after the control rod is inserted.

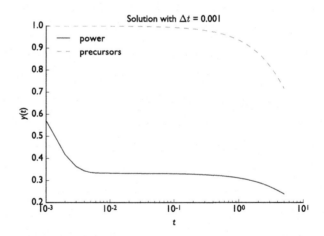

Notice that our time step is smaller than the prompt jump (as can be seen on the semilog scale). With a larger time step, RK4 has problems. In fact with a time step of $\Delta t = 0.01$ s, the power is 10^{303} at 12 s. It seems that this time step is beyond the stability limit for RK4. This can be checked by evaluating the eigenvalues of the $\mathbf{A}$ matrix for this problem and using the formula for g_{RK4}. The stability limit is $\Delta t \leq 0.00247574$ s.

Using backward Euler, we can take this larger time step safely:

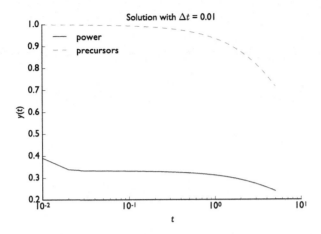

Yes, backward Euler works with this time step size, where RK4 could not. It is here that we see the benefit of implicit methods: with backward Euler we need 10 times fewer time steps to get the solution. The price of this was that we had to solve a linear system at each step.

17.7.2 Linear Reactivity Ramp

In this case we will insert positive reactivity linearly, this could be done by slowly removing a control rod, and see what happens. We will have a linear ramp between $\rho = 0$ and $\rho = 0.5\$$ over 1 second.

```
In [33]: #reactivity function
         rho = lambda t: (t<1)*beta*0.5*t + beta*0.5*(t>=1)
```

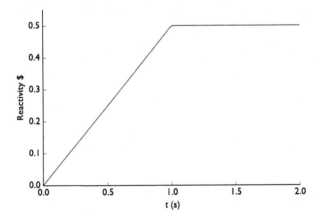

The solution with backward Euler demonstrates two different growth rates: one corresponding to the ramp, and another once the reactivity stops increasing.

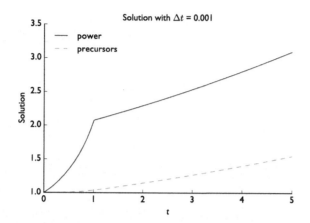

To make the scenario more complicated we will now look at a positive reactivity ramp that is then has the control rod completely re-inserted at $t = 1$, and then instantly brought back to critical at $t = 2$. The reactivity for this scenario looks like

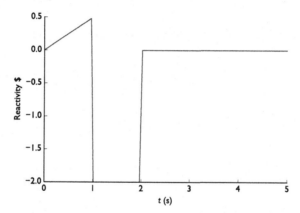

We will compare backward Euler and RK4 on this problem with the same time step.

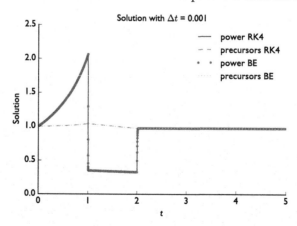

The prompt jump is a wild phenomenon: the solution changes very rapidly during it and the methods still keep up with a small enough time step. In this example, we cannot see on the graph a difference between the solutions. Additionally, the precursors change very slowly over the entire simulation.

CODA

We have discussed 4 classical techniques for integrating initial value problems: forward and backward Euler, Crank–Nicolson, and fourth-order Runge Kutta. Through these methods we discussed the issues of order of accuracy, stability, oscillatory solutions, and the building of error.

Initial value problems are only the first of the type of ODEs that we can solve using the techniques we have covered up to this point. To solve initial value problems we appealed to methods of doing numerical integration. In the next chapter we will solve boundary value problems and use the techniques to evaluate derivatives numerically we discussed previously.

FURTHER READING

Although we have covered the majority of methods used in practice for the solution of initial value problems, there are some modifications that are made to the implementation that makes these methods more powerful. For example, adaptive Runge–Kutta methods take time steps that are smaller when the solution changes rapidly, and larger when the solution is slowly varying. Also there are other methods that we did not cover, including the backward differentiation formulas, known as BDF formulas. The BDF-2 method is a second-order implicit method, but it is less oscillatory than Crank-Nicolson.

For background on the PRKEs, see the reactor physics texts previously mentioned by Lewis [7] and Stacey [10]. There has been much work on developing the best numerical methods for the PRKEs. The recent method developed by Ganapol may be the most effective [21].

PROBLEMS

Short Exercises

Using the five methods we have presented in this chapter to compute the solution to the following problems using the step sizes $\Delta t = 0.5, 0.1, 0.05, 0.01$. Give an explanation for the behavior of each solution.

17.1. $y'(t) = -\left[y(t)\right]^2$ for $t \in [1, 10]$, and $y(1) = 0.5$. The solution to this problem is $y(t) = 1/(t+1)$. *Hint: You can make the substitution $\hat{t} = t - 1$ and reformulate the initial condition in terms of $\hat{t} = 0$.*

17.2. $y'(t) = \sin(t)/t$ for $t \in [0, 10]$, and $y(0) = 0$. The solution to this problem is $y(t) = \mathrm{Si}(t)$, where $\mathrm{Si}(t)$ is the Sine integral function.

17.3. $y''(t) = (t - 20)y$ for $t \in [0, 24]$ and $y(0) = -0.176406127078$, and $y'(0) = 0.892862856736$. The true solution to this problem is $y(t) = \mathrm{Ai}(t - 20)$, where $\mathrm{Ai}(t)$ is the Airy function.

Programming Projects

1. Point Reactor Kinetics

Assume one group of delayed neutrons and an initially critical reactor with no extraneous source. The point reactor kinetics equations under these assumptions are:

$$\frac{dP}{dt} = \frac{\rho - \beta}{\Lambda}P + \lambda C,$$

$$\frac{dC}{dt} = \frac{\beta}{\Lambda}P - \lambda C,$$

with initial conditions: $P(0) = P_0$ and $C(0) = C_0 = \frac{\beta P_0}{\lambda \Lambda}$. Other useful values are $\beta = 750$ pcm, $\Lambda = 2 \times 10^{-5}$ s, $\lambda = 10^{-1}$ s^{-1}.

At time $t = 0$ a control rod is instantaneously withdrawn and the inserted reactivity is $\rho = 50$ pcm $= 50 \times 10^{-5}$. Write a Python code to solve the PRKEs. Do not assume that reactivity is a constant, rather allow it to be a function of t so that you can do a ramp or sinusoidal reactivity.

Your code will utilize the following methods:

- Forward Euler
- Backward Euler
- Fourth-order Runge Kutta
- Crank–Nicolson.

For each method provide plots for $t = 0 \ldots 5$ is using $\Delta t = \{1, 3, 5, 6\}$ ms.

- One plot with all the methods per time step size
- One plot with all time step sizes per method.

Make the axes reasonable—if a method diverges do not have the scale go to $\pm\infty$.

Comment your results and explain their behavior using what you know about the accuracy and stability of the methods, including their stability limits. The explanation should include justifications for convergence/divergence or the presence of oscillations. To answer these questions you should look at the analytical solutions of the equations.

REACTIVITY RAMP

A positive reactivity ramp is inserted as follows:

$$\rho(t) = \min(\rho_{\max}t, \rho_{\max}).$$

Pick a method and a time step size of your choice. Solve this problem for $\rho_{\max} = 0.5\beta$ and $\rho_{\max} = 1.1\beta$. Explain your choice and your findings.

SINUSOIDAL REACTIVITY

Try

$$\rho(t) = \rho_0 \sin(\omega t).$$

See if the reactor remains stable for any amplitude ρ_0 and/or frequency ω.

2. Iodine Ingestion

The diagram below shows the metabolic compartment model for the uptake of iodine in the thyroid gland of an adult human. The goal is to calculate the fraction of the total ingested activity present in the blood (compartment 2) and the thyroid gland (compartment 3) as a function of time for an ingestion, at time 0, of 1 Bq of ^{131}I (half-life of 8.04 days). In the diagram below λ_{21} represents the biological transfer into compartment 2 from compartment 1. All λ's are expressed in days^{-1}.

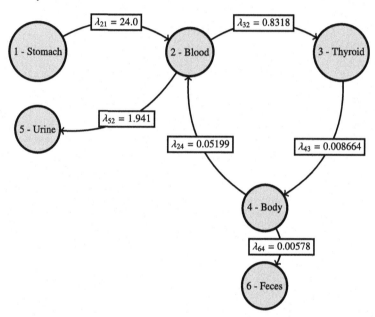

- Write down the physical equations governing the conservation of iodine in each of the following compartments: stomach, blood, thyroid, and body. This will form a system of coupled ODEs. Remember that the substance is removed from a compartment by both radioactive decay and transfer between biological compartments.
- Why is it enough to look at the conservation of iodine in stomach, blood, thyroid, and body to get the answer?
- Your system of ODEs should have the following form:

$$Y' = AY + B(t),$$

where $B(t)$ represents the external source terms in your conservation laws. Only the component corresponding to the stomach of B is nonzero. This component is the following function:

$$h(t) = \begin{cases} 1 & \text{if } t = 0 \\ 0 & \text{otherwise} \end{cases}.$$

In practice you will let h be equal to $1/\Delta t$ in the first time step so that $\int_0^{\Delta t} h(t)\, dt = 1$. After that h will always be zero.

- Write a code to solve this system of ODEs. The function B should be a function of t in your code. Your code will utilize the following methods:
  - Forward Euler
  - Backward Euler
  - Fourth-order Runge–Kutta
  - Crank–Nicolson.

For each method provide plots for $t = 0 \ldots 1$ day using the following step sizes:
  - 0.5/24 hr
  - 1/24 hr
  - 1.4/24 hr
  - 2/24 hr.

Note: all λ's are in days^{-1} so be careful about how you express Δt in your code.
  - One plot with all the methods per time step size
  - One plot with all time step sizes per method.

Make the axes reasonable—if a method diverges don't have the scale go to $\pm\infty$.

Comment your results and explain their behavior using what you know about the accuracy and stability of the methods, including their stability limits. The explanation should include justifications for convergence/divergence or the presence of oscillations. To answer these questions you should look at the analytical solutions of the equations.

There is no real direction here, neither lines of power nor cooperation.

−Thomas Pynchon, **Gravity's Rainbow**

CHAPTER POINTS

- Boundary value problems specify the value of the solution to the differential equations at multiple points.

- We develop a method for solving the diffusion equation for neutrons in slab, spherical, and cylindrical geometries.

- Time-dependent problems can be written as the solution to a succession of modified steady state problems.

In this lecture we will solve the one-dimensional (1-D), one-group neutron diffusion equation. This is an example of a boundary-value problem. A boundary value problem is a differential equation or a system of differential equations that specifies the value of the solution or its derivatives at more than one point.

In the previous chapter we dealt with initial value problems where the conditions for the solution were prescribed at a single point that we called $t = 0$. In boundary value problems we specify the solution at multiple points as we shall see. This means we cannot start at one point and move the solution in a particular direction. Rather, we must solve a system of equations to find the solution that satisfies the boundary values.

The general neutron diffusion equation for the scalar flux of neutrons, $\phi(r, t)$, is given by

$$\frac{1}{v}\frac{\partial \phi}{\partial t} - \nabla \cdot D(r)\nabla\phi(r, t) + \Sigma_a(r)\phi(r, t) = \nu\Sigma_f(r)\phi(r, t) + Q(r, t).$$

Our notation is standard: v is the neutron speed, $D(r)$ is the diffusion coefficient, Σ_a is the macroscopic absorption cross-section, Σ_f is the macroscopic fission cross section, ν is the number of neutrons per fission, and $Q(r, t)$ is a prescribed source. The boundary conditions we will consider for this equation are generic conditions,

$$\mathcal{A}(r)\phi(r, t) + \mathcal{B}(r)\frac{d\phi}{dr} = \mathcal{C}(r) \qquad \text{for } r \in \partial V.$$

The initial condition is

$$\phi(r, 0) = \phi^0(r).$$

We can eliminate the time variable from this equation by integrating over a time step from $t^n = n\Delta t$ to $t^{n+1} = (n + 1)\Delta t$:

$$\frac{1}{v}\left(\phi^{n+1}(r) - \phi^n(r)\right) - \int_{t^n}^{t^{n+1}} dt \, [\nabla \cdot D(r)\nabla\phi(r, t) + \Sigma_a(r)\phi(r, t)]$$

$$= \int_{t^n}^{t^{n+1}} dt \, [\nu\Sigma_f(r)\phi(r, t) + Q(r, t)], \tag{18.1}$$

where $\phi^n(r) = \phi(r, n\Delta t)$.

Using the backward Euler approach to the integrals (that is the right-hand rectangle rule) we get

$$\frac{1}{v\Delta t}\left(\phi^{n+1}(r) - \phi^n(r)\right) - \nabla \cdot D(r)\nabla\phi^{n+1}(r) + \Sigma_a(r)\phi^{n+1}(r) = \nu\Sigma_f(r)\phi^{n+1}(r) + Q^{n+1}(r).$$

The next step is to define a new absorption cross-section and source as

$$\Sigma_a^* = \Sigma_a(r) + \frac{1}{v\Delta t}, \qquad Q^{n+1,*}(r) = Q^{n+1}(r) + \frac{1}{v\Delta t}\phi^n(r).$$

With these definitions we get the following equation

$$-\nabla \cdot D(r)\nabla\phi^{n+1}(r) + \Sigma_a^*(r)\phi^{n+1}(r) = \nu\Sigma_f(r)\phi^{n+1}(r) + Q^{n+1,*}(r).$$

This is a steady-state diffusion equation for the scalar flux at time $n+1$. Therefore, we can focus our efforts on solving steady-state problems, knowing that to solve time-dependent equations, we just have to redefine the source and absorption cross-sections appropriately.

18.1 DISCRETIZING THE STEADY-STATE DIFFUSION EQUATION

We begin with the steady-state diffusion equation with a reflecting boundary condition at $r = 0$ and a general boundary condition at $r = R$:

$$-\nabla \cdot D(r)\nabla\phi(r) + \Sigma_a(r)\phi(r) = \nu\Sigma_f(r)\phi(r) + Q(r),$$

$$\left.\frac{d\phi}{dr}\right|_{r=0} = 0,$$

$$\mathcal{A}(R)\phi(R) + \mathcal{B}(R)\left.\frac{d\phi}{dr}\right|_{r=R} = \mathcal{C}(R).$$

We seek to solve this numerically on a grid of spatial cells

In our notation there will be I cells and $I + 1$ edges. The cell centers are given by

$$r_i = i\Delta r + \frac{\Delta r}{2}, \qquad i = 0, 1, \ldots, I - 1,$$

and the left and right edges are given by the formulas

$$r_{i-1/2} = i\Delta r, \qquad r_{i+1/2} = (i+1)\Delta r, \qquad i = 0, 1, \ldots, I - 1.$$

We can construct a Python function that creates the cell edges and cell centers given R and I.

```
In [1]: def create_grid(R,I):
            """Create the cell edges and centers for a
            domain of size R and I cells
            Args:
                R: size of domain
                I: number of cells

            Returns:
                Delta_r: the width of each cell
                centers: the cell centers of the grid
                edges: the cell edges of the grid
```

```
"""
Delta_r = float(R)/I
centers = np.arange(I)*Delta_r + 0.5*Delta_r
edges = np.arange(I+1)*Delta_r
return Delta_r, centers, edges
```

Our problem will be set up such that inside each cell the material properties are constant so that inside cell i,

$$D(r) = D_i, \quad \Sigma_a(r) = \Sigma_{a,i}, \quad \Sigma_f(r) = \Sigma_{f,i}, \quad Q(r) = Q_i, \qquad r \in (r_{i-1/2}, r_{i+1/2}).$$

18.1.1 The Diffusion Operator in Different Geometries

The definition of $\nabla \cdot D(r)\nabla$ in several geometries is given below:

$$\nabla \cdot D(r)\nabla = \begin{cases} \frac{d}{dr} D(r) \frac{d}{dr} & \text{1-D Slab} \\ \frac{1}{r^2} \frac{d}{dr} r^2 D(r) \frac{d}{dr} & \text{1-D Sphere} \\ \frac{1}{r} \frac{d}{dr} r D(r) \frac{d}{dr} & \text{1-D Cylinder} \end{cases} .$$

Also, the differential volume element in each geometry is

$$dV = \begin{cases} dr & \text{1-D Slab} \\ 4\pi r^2 dr & \text{1-D Sphere} \\ 2\pi r dr & \text{1-D Cylinder} \end{cases} .$$

We define ϕ_i as the average value of the scalar flux in cell i. This quantity is given by

$$\phi_i = \frac{1}{V_i} \int_{r_{i-1/2}}^{r_{i+1/2}} dV\, \phi(r),$$

where V_i is the cell volume

$$V_i = \begin{cases} \Delta r & \text{1-D Slab} \\ \frac{4}{3}\pi (r_{i+1/2}^3 - r_{i-1/2}^3) & \text{1-D Sphere} \\ \pi (r_{i+1/2}^2 - r_{i-1/2}^2) & \text{1-D Cylinder} \end{cases} .$$

To develop the discrete equations we will integrate the diffusion equation term by term. We first integrate the absorption term in the diffusion equation over cell i and divide by V_i,

$$\frac{1}{V_i} \int_{r_{i-1/2}}^{r_{i+1/2}} dV\, \Sigma_a(r)\phi(r) = \frac{\Sigma_{a,i}}{V_i} \int_{r_{i-1/2}}^{r_{i+1/2}} dV\, \phi(r) = \Sigma_{a,i}\phi_i.$$

The fission term is handled in a similar manner

$$\frac{1}{V_i} \int_{r_{i-1/2}}^{r_{i+1/2}} dV\, \nu \Sigma_f(r)\phi(r) = \nu \Sigma_{f,i}\phi_i.$$

The source term is similarly straightforward because $Q(r)$ is constant in cell i,

$$\frac{1}{V_i} \int_{r_{i-1/2}}^{r_{i+1/2}} dV\, Q(r) = Q_i.$$

The diffusion term is a bit trickier,

$$-\frac{1}{V_i} \int_{r_{i-1/2}}^{r_{i+1/2}} dV\, \nabla \cdot D(r) \nabla \phi(r) = \begin{cases} -\frac{1}{V_i} \int_{r_{i-1/2}}^{r_{i+1/2}} dr\, \frac{d}{dr} D(r) \phi(r) \frac{d}{dr} & \text{1-D Slab} \\ -\frac{4\pi}{V_i} \int_{r_{i-1/2}}^{r_{i+1/2}} dr\, \frac{d}{dr} r^2 D(r) \frac{d}{dr} \phi(r) & \text{1-D Sphere} \\ -\frac{2\pi}{V_i} \int_{r_{i-1/2}}^{r_{i+1/2}} dr\, \frac{d}{dr} r D(r) \frac{d}{dr} \phi(r) & \text{1-D Cylinder} \end{cases}.$$

To simplify these we can use the fundamental theorem of calculus to get

$$-\frac{1}{V_i} \int_{r_{i-1/2}}^{r_{i+1/2}} dV\, \nabla \cdot D(r) \nabla \phi(r) = \begin{cases} -\frac{1}{V_i} \left[D_{i+1/2} \frac{d}{dr} \phi(r_{i+1/2}) - D_{i-1/2} \frac{d}{dr} \phi(r_{i-1/2}) \right] \\ \text{1-D Slab} \\ -\frac{4\pi}{V_i} \left[D_{i+1/2} r_{i+1/2}^2 \frac{d}{dr} \phi(r_{i+1/2}) - D_{i-1/2} r_{i-1/2}^2 \frac{d}{dr} \phi(r_{i-1/2}) \right] \\ \text{1-D Sphere} \\ -\frac{2\pi}{V_i} \left[D_{i+1/2} r_{i+1/2} \frac{d}{dr} \phi(r_{i+1/2}) - D_{i-1/2} r_{i-1/2} \frac{d}{dr} \phi(r_{i-1/2}) \right] \\ \text{1-D Cylinder} \end{cases}.$$

$$(18.2)$$

Defining the surface area, $S_{i\pm1/2}$ at the cell edge for each geometry will allow us to simplify Eq. (18.2) further:

$$S_{i\pm1/2} = \begin{cases} 1 & \text{1-D Slab} \\ 4\pi r_{i\pm1/2}^2 & \text{1-D Sphere} \\ 2\pi r_{i\pm1/2} & \text{1-D Cylinder} \end{cases}.$$

Using this definition in Eq. (18.2) we get

$$-\frac{1}{V_i} \int_{r_{i-1/2}}^{r_{i+1/2}} dV\, \nabla \cdot D(r) \nabla \phi(r) = -\frac{1}{V_i} \left[D_{i+1/2} S_{i+1/2} \frac{d}{dr} \phi(r_{i+1/2}) - D_{i-1/2} S_{i-1/2} \frac{d}{dr} \phi(r_{i-1/2}) \right].$$

The final piece of the derivation is to write the value of the derivative at the cell edge using the central difference formula. To do this we will have to interpret the cell average scalar flux as the value of the scalar flux at the cell center:

$$\phi_i \approx \phi(r_i).$$

With this definition we can write

$$\frac{d}{dr} \phi(r_{i+1/2}) = \frac{\phi_{i+1} - \phi_i}{\Delta r} + O(\Delta r^2).$$

Therefore, the fully discrete diffusion term is

$$-\frac{1}{V_i} \int_{r_{i-1/2}}^{r_{i+1/2}} dV \, \nabla^2 \phi(r) = -\frac{1}{V_i} \left[D_{i+1/2} S_{i+1/2} \frac{\phi_{i+1} - \phi_i}{\Delta r} - D_{i-1/2} S_{i-1/2} \frac{\phi_i - \phi_{i-1}}{\Delta r} \right].$$

Putting all of this together gives us the diffusion equation integrated over a cell volume:

$$-\frac{1}{V_i} \left[D_{i+1/2} S_{i+1/2} \frac{\phi_{i+1} - \phi_i}{\Delta r} - D_{i-1/2} S_{i-1/2} \frac{\phi_i - \phi_{i-1}}{\Delta r} \right] + \left(\Sigma_{\mathrm{a},i} - \nu \Sigma_{\mathrm{f},i} \right) \phi_i = Q_i. \qquad (18.3)$$

18.1.2 Interface Diffusion Coefficient

We need to define what we mean by $D_{i+1/2}$ and $D_{i-1/2}$ because the diffusion coefficient is only constant inside a cell. For these terms will define a diffusion coefficient so that the neutron current is continuous at a cell face. In particular we define $\phi_{i+1/2}$ so that

$$2D_{i+1} \frac{\phi_{i+1} - \phi_{i+1/2}}{\Delta r} = D_{i+1/2} \frac{\phi_{i+1} - \phi_i}{\Delta r},$$

and

$$2D_i \frac{\phi_{i+1/2} - \phi_i}{\Delta r} = D_{i+1/2} \frac{\phi_{i+1} - \phi_i}{\Delta r}.$$

These two equations state that we want to define $D_{i+1/2}$ and $\phi_{i+1/2}$ so that the current is the same if we calculate it from the left or the right. We also get a similar system defining $D_{i-1/2}$. Solving these equations for $D_{i+1/2}$ and $D_{i-1/2}$ we get

$$D_{i\pm1/2} = \frac{2D_i D_{i\pm1}}{D_i + D_{i\pm1}}. \qquad (18.4)$$

This definition for the diffusion coefficient at the interface takes the harmonic mean of the diffusion coefficient on each side and assures that the neutron current density is continuous at the interface.

18.1.3 Boundary Conditions

We need to enforce our general boundary condition at $r = R$, that is

$$\mathcal{A}(R)\phi(R) + \mathcal{B}(R) \frac{d\phi}{dr}\bigg|_{r=R} = \mathcal{C}(R).$$

To see how this is going to come into the equation, we examine at the equation for $i = I - 1$,

$$-\frac{1}{V_I} \left[D_{I-1/2} S_{I-1/2} \frac{\phi_I - \phi_{I-1}}{\Delta r} - D_{I-3/2} S_{I-3/2} \frac{\phi_{I-1} - \phi_{I-2}}{\Delta r} \right] + \left(\Sigma_{\mathrm{a},I-1} - \nu \Sigma_{\mathrm{f},I-1} \right) \phi_I = Q_{I-1}.$$

Notice that this equation has ϕ_I, which is the undefined value of the scalar flux outside the domain. We need to create a value for this parameter that is consistent with the boundary condition. To do this we will make the following approximations:

$$\phi(R) \approx \frac{1}{2}(\phi_{I-1} + \phi_I),$$

and

$$\left.\frac{d\phi}{dr}\right|_{r=R} \approx \frac{\phi_I - \phi_{I-1}}{\Delta r}.$$

Using these in our boundary condition, we get the final equation that we need

$$\left(\frac{A}{2} - \frac{B}{\Delta r}\right)\phi_{I-1} + \left(\frac{A}{2} + \frac{B}{\Delta r}\right)\phi_I = C.$$

Here we have dropped the spatial dependence of A, B, and C in the boundary condition.

18.1.3.1 Types of Boundary Conditions on the Outer Surface

Using the general boundary condition, we can enforce a variety of boundary conditions [22]. A Dirichlet boundary condition of the form

$$\phi(R) = c,$$

can be enforced by setting $A = 1$, $B = 0$, and $C = c$.

An albedo boundary condition is used for the case where some fraction of the neutrons that leave the system are reflected back. If we call this fraction that is reflected α, the albedo boundary condition can be written as

$$\frac{(1-\alpha)}{4(1+\alpha)}\phi(R) + \left.\frac{D}{2}\frac{d\phi}{dr}\right|_{r=R} = 0,$$

where the diffusion coefficient is evaluated at $r = R$. This boundary condition is obtained by setting $C = 0$, and

$$A = \frac{(1-\alpha)}{4(1+\alpha)}, \qquad B = \frac{D}{2}.$$

The reflecting boundary is a special case of the albedo condition with $\alpha = 1$. For this boundary condition we set $A = C = 0$ and $B = 1$.

The final boundary condition that we consider is the partial current boundary condition, also called a Marshak boundary condition. These allow us to specify the amount of neutrons that enter the system from the edge, rather than just specifying the scalar flux on the boundary as in the Dirichlet condition. At $R = r$ the rate at which neutrons enter the domain per unit area, (i.e., the partial current), is given by

$$\text{incoming partial current} = \frac{1}{4}\phi(R) + \left.\frac{D}{2}\frac{d\phi}{dr}\right|_{r=R}.$$

Therefore, if we wish to set the incoming partial current into the system at a particular value, J_{in}, we set our boundary constants to be

$$A = \frac{1}{4}, \qquad B = \frac{D}{2}, \qquad C = J_{\text{in}}.$$

A vacuum boundary condition can be obtained by setting $J_{\text{in}} = 0$.

18.1.3.2 *Reflecting Boundary Condition at* $r = 0$

At the $r = 0$ boundary we require a reflecting boundary in the curvilinear geometries (spherical and cylindrical), and we want to specify the same for the slab. A reflecting boundary has

$$\frac{d}{dr}\phi(0) = 0.$$

It turns out that this is automatically enforced in the curvilinear geometries because $S_{-1/2} = 0$, and the derivative at the inner edge of cell 0 is effectively zero. To enforce this in slab geometry we just need to force $S_{-1/2} = 0$ to make the equation for $i = 0$

$$-\frac{D_{1/2}S_{1/2}}{V_0}\frac{\phi_1 - \phi_0}{\Delta r} + \left(\Sigma_{\text{a},0} - \nu\Sigma_{\text{f},0}\right)\phi_0 = Q_0.$$

18.2 PYTHON CODE FOR THE DIFFUSION EQUATION

We have now completely specified the discrete equations for our diffusion problem. This section will detail how to build the matrices and vectors.

There are $I + 1$ equations in our system:

$$-\frac{1}{V_i}\left[D_{i+1/2}S_{i+1/2}\frac{\phi_{i+1} - \phi_i}{\Delta r} - D_{i-1/2}S_{i-1/2}\frac{\phi_i - \phi_{i-1}}{\Delta r}\right] + \left(\Sigma_{\text{a},i} - \nu\Sigma_{\text{f},i}\right)\phi_i = Q_i,$$

$$i = 0, \ldots, I - 1,$$

and

$$\left(\frac{A}{2} - \frac{B}{\Delta r}\right)\phi_{I-1} + \left(\frac{A}{2} + \frac{B}{\Delta r}\right)\phi_I = C.$$

This is a system of equations to solve, to do this we first define our solution vector and righthand side

$$\boldsymbol{\phi} = \begin{pmatrix} \phi_0 \\ \phi_1 \\ \vdots \\ \phi_I \end{pmatrix}, \qquad \mathbf{b} = \begin{pmatrix} Q_0 \\ Q_1 \\ \vdots \\ C \end{pmatrix}.$$

Our system will be written as

$$\mathbf{A}\boldsymbol{\phi} = \mathbf{b}.$$

To define $\mathbf{A}$, we will factor our equations to be

$$-\frac{D_{i+1/2}S_{i+1/2}}{\Delta r\,V_i}\phi_{i+1} + \left[\frac{1}{\Delta r\,V_i}\left(D_{i+1/2}S_{i+1/2} + D_{i-1/2}S_{i-1/2}\right) + \Sigma_{a,i} - \nu\Sigma_{f,i}\right]\phi_i$$

$$-\frac{D_{i-1/2}S_{i-1/2}}{\Delta r\,V_i}\phi_{i-1} = Q_i, \qquad i = 0,\dots,I-1,$$

and

$$\left(\frac{A}{2} - \frac{B}{\Delta r}\right)\phi_{I-1} + \left(\frac{A}{2} + \frac{B}{\Delta r}\right)\phi_I = C.$$

From these equations we get that the element of $\mathbf{A}$ in row i and column j is

$$A_{ij} = \begin{cases} \left[\frac{1}{\Delta r V_i}\left(D_{i+1/2}S_{i+1/2} + D_{i-1/2}S_{i-1/2}\right) + \Sigma_{a,i} - \nu\Sigma_{f,i}\right] & i = j \text{ and } i = 0,1,\dots I-1 \\ -\frac{1}{\Delta r V_i}D_{i+1/2}S_{i+1/2} & i+1 = j \text{ and } i = 0,1,\dots I-2 \\ -\frac{1}{\Delta r V_i}D_{i-1/2}S_{i-1/2} & i-1 = j \text{ and } i = 1,2,\dots I-1 \\ \left(\frac{A}{2} - \frac{B}{\Delta r}\right) & j = I-1 \text{ and } i = I \\ \left(\frac{A}{2} + \frac{B}{\Delta r}\right) & j = I \text{ and } i = I \\ 0 & \text{otherwise} \end{cases}$$

We will now set up a function that

- builds the matrix $\mathbf{A}$,
- builds the vector $\mathbf{b}$,
- uses Gauss elimination to solve the system for the scalar fluxes $\boldsymbol{\phi}$.

The code will call the grid function that we defined before. Additionally, our function will take as arguments the name of a function that defines each of the material properties.

```
In [2]: def DiffusionSolver(R,I,D,Sig_a,nuSig_f, Q,BC, geometry):
        """Solve the neutron diffusion equation in a 1-D geometry
        using cell-averaged unknowns
        Args:
            R: size of domain
            I: number of cells
            D: name of function that returns diffusion coefficient for a given r
            Sig_a: name of function that returns Sigma_a for a given r
            nuSig_f: name of function that returns nu Sigma_f for a given r
            Q: name of function that returns Q for a given r
            BC: Boundary Condition at r=R in form [A,B,C]
            geometry: shape of problem 0 for slab
                1 for cylindrical
                2 for spherical
```

```
Returns:
    centers: the cell centers of the grid
    phi:  cell-average value of the scalar flux

"""
#create the grid
Delta_r, centers, edges = create_grid(R,I)
A = np.zeros((I+1,I+1))
b = np.zeros(I+1)
#define surface areas and volumes
assert( (geometry==0) or (geometry == 1) or (geometry == 2))
if (geometry == 0):
    #in slab it's 1 everywhere except at the left edge
    S = 0.0*edges+1
    S[0] = 0.0 #this will enforce reflecting BC
    #in slab its dr
    V = 0.0*centers + Delta_r
elif (geometry == 1):
    #in cylinder it is 2 pi r
    S = 2.0*np.pi*edges
    #in cylinder its pi (r^2 - r^2)
    V = np.pi*( edges[1:(I+1)]**2
                - edges[0:I]**2 )
elif (geometry == 2):
    #in sphere it is 4 pi r^2
    S = 4.0*np.pi*edges**2
    #in sphere its 4/3 pi (r^3 - r^3)
    V = 4.0/3.0*np.pi*( edges[1:(I+1)]**3
                - edges[0:I]**3 )

#Set up BC at R
A[I,I] = (BC[0]*0.5 + BC[1]/Delta_r)
A[I,I-1] = (BC[0]*0.5 - BC[1]/Delta_r)
b[I] = BC[2]
r = centers[0]
DPlus = 0
#fill in rest of matrix
for i in range(I):
    r = centers[i]
    DMinus = DPlus
    DPlus = 2*(D(r)*D(r+Delta_r))/(D(r)+D(r+Delta_r))
    A[i,i] = (1.0/(Delta_r * V[i])*DPlus*S[i+1] +
              Sig_a(r) - nuSig_f(r))
    if (i>0):
        A[i,i-1] = -1.0*DMinus/(Delta_r * V[i])*S[i]
        A[i,i] += 1.0/(Delta_r * V[i])*(DMinus*S[i])
    A[i,i+1] = -DPlus/(Delta_r * V[i])*S[i+1]
    b[i] = Q(r)

#solve system
phi = GaussElimPivotSolve(A,b)
#remove last element of phi because it is outside the domain
phi = phi[0:I]
return centers, phi
```

18.3 A TEST PROBLEM FOR EACH GEOMETRY

Before we proceed, recall that the steady state diffusion equation will only have a solution for subcritical problems with a source. If the system is critical or supercritical, there is no finite steady state solution. We will discuss in future lectures how to solve these problems.

In an infinite, homogeneous medium all the spatial derivatives go to zero and the diffusion equation reads

$$\phi(r) = \frac{Q}{\Sigma_a - \nu \Sigma_f}.$$

Therefore, we expect that if we make our problem have a reflecting boundary condition it should reproduce this infinite medium solution.

To try this out we define functions for each of D, Σ_a, $\nu \Sigma_f$, and Q:

```
In [3]: #in this case all three are constant
        def D(r):
            return 0.04;
        def Sigma_a(r):
            return 1;
        def nuSigma_f(r):
            return 0;
        def Q(r):
            return 1
        print("For this problem the diffusion length is", np.sqrt(D(1)/Sigma_a(1)))
        inf_med = Q(1)/(Sigma_a(1) - nuSigma_f(1))
        print("The infinite medium solution is",inf_med)
```

```
For this problem the diffusion length is 0.2
The infinite medium solution is 1.0
```

To compute the solution we call our `DiffusionSolver` function. In this case we set $R = 10$ and $I = 10$; this makes $\Delta r = 1$. We set the boundary condition parameter to be `[0,1,0]` to make it correspond to the reflecting boundary condition discussed above.

```
In [4]: R = 10
        I = 10
        #Solve Diffusion Problem in Slab geometry
        x, phi_slab = DiffusionSolver(R, I,D, Sigma_a, nuSigma_f, Q,[0,1,0],0)
        #Solve Diffusion Problem in cylindrical geometry
        rc, phi_cyl = DiffusionSolver(R, I,D, Sigma_a, nuSigma_f, Q,[0,1,0],1)
        #Solve Diffusion Problem in spherical geometry
        rs, phi_sphere = DiffusionSolver(R, I,D, Sigma_a, nuSigma_f, Q,[0,1,0],2)
```

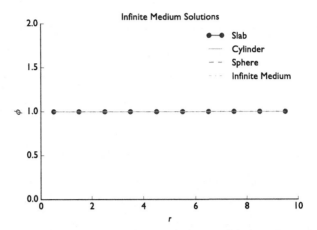

These results demonstrate that our method for solving the diffusion equation can solve the simplest possible problem. This is an important test, however, because if our method cannot solve this problem correctly, it is not likely to be able to solve more difficult problems.

Next, we test our implementation on a heterogeneous problem where the material properties are discontinuous at $r = 5$ with fuel from 0 to 5 and moderator from 5 to 10. If we are solving for the thermal flux, there will be a source in the moderator.

```
In [5]: def D(r):
            value = 5.0*(r<=5) + 1.0*(r>5)
            return value;
        def Sigma_a(r):
            value = 1.0*(r<=5) + 0.1*(r>5)
            return value;
        def nuSigma_f(r):
            value = 0.4*(r<=5) + 0.0*(r>5)
            return value;
        def Q(r):
            value = 0*(r<=5) + 1.0*(r>5)
            return value
```

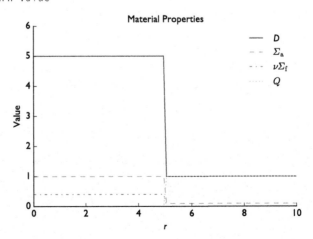

Now, we use the `DiffusionSolver` function to solve this problem with a zero-Dirichlet boundary conditions at $r = R$ and a reflecting boundary at $r = 0$:

```
In [6]:  R = 10
         I = 30
         #Solve Diffusion Problem in Slab geometry
         x, phi_slab = DiffusionSolver(R, I,D, Sigma_a, nuSigma_f, Q,[1,0,0],0)
         #Solve Diffusion Problem in cylindrical geometry
         rc, phi_cyl = DiffusionSolver(R, I,D, Sigma_a, nuSigma_f, Q,[1,0,0],1)
         #Solve Diffusion Problem in cylindrical geometry
         rs, phi_sphere = DiffusionSolver(R, I,D, Sigma_a, nuSigma_f, Q,[1,0,0],2)
```

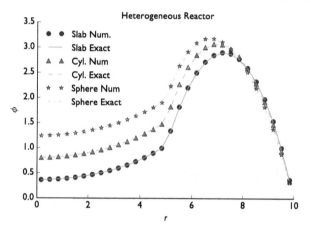

In this figure the exact solutions are also shown. These can be found by solving the diffusion equation in each region and joining the solutions by making the scalar flux and neutron current continuous at $r = 5$ [10]. For example, the slab solution is

$$\phi(r) = \begin{cases} 0.181299e^{-0.34641r}\left(e^{0.69282r} + 1\right) & r \le 5 \\ 34.5868\sinh(0.316228r) - 35.3082\cosh(0.316228r) + 10 & r \ge 5 \end{cases}.$$

The numerical solutions seem to agree with the exact solutions, and indeed the difference between the two decreases as I gets larger.

18.3.1 Reed's Problem

Reed's problem is a common test problem for numerical methods for solving neutron diffusion and transport problems. It is a heterogeneous reactor problem that has several regions. The material properties for our geometry (reflecting at $r = 0$ and vacuum at $r = R$) is defined below.

```
In [7]:  #in this case all three are constant
         def D(r):
             value = (1.0/3.0*(r>5) +
                      1.0/3.0/0.001 *((r<=5) * (r>3)) +
                      1.0/3.0/5.0 *((r<=3) * (r>2)) +
```

```
                1.0/3.0/50.0 * (r<=2))
        return value;
def Sigma_a(r):
    value = 0+(0.1*(r>5) +
              5.0 * ((r<3) * (r>2))+
              50.0 * (r<=2))
    return value;
def nuSigma_f(r):
    return 0*r;
def Q(r):
    value = 0 + 1.0*((r<7) * (r>5)) + 50.0*(r<=2)
    return value;
```

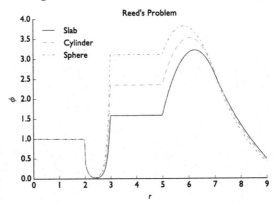

This problem is set up so that there is a

- strong absorber with a strong source from $r = 0$ to 2,
- strong absorber without a source from $r = 2$ to 3,
- void from $r = 3$ to 4,
- scatterer with source from $r = 5$ to 7, and
- scatterer without source from $r = 7$ to 9.

We approximate the void by having a very large diffusion coefficient and set $\Sigma_a = 0$. The solution to this problem using our DiffusionSolver function is given below.

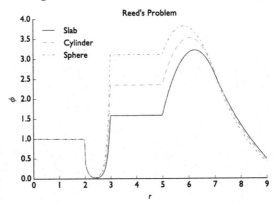

Notice that the scalar flux for the sphere is highest, followed by the cylinder, then the slab. This is due to the fact that the leakage from the sphere is the smallest because it has the smallest ratio of surface area to volume. The solutions we obtain are consistent with previous solutions to Reed's problem: the solution is flat in the void region, peaks in the scattering region with source, and goes to a constant in the strong source and absorber region.

We can also solve Reed's problem in time dependent mode. In this case we will set the initial condition to have zero scalar flux everywhere, and solve a series of steady-state problems as we indicated at the beginning of the chapter. In the code below we solve this time dependent problem in spherical geometry with $\Delta t = 0.5$ and $v = 1$. We then plot the solution at $t = 1, 10, 100$.

```
In [8]: dt = 0.5
        tfinal = 100
        v = 1
        steps = np.linspace(dt,tfinal,tfinal/dt)
        R = 9
        I = 100
        dx = R/I
        phi_old = np.zeros(I)
        #define sigma_a star function
        Sigma_a_star = lambda r: Sigma_a(r) + 1/(v*dt)
        for step in steps:
            #construct Q star function, needs to convert r to the cell index
            Q_star = lambda r: Q(r) + phi_old[int((r-0.5*dx)/dx-1)]/(v*dt)
            #Solve Diffusion Problem in cylindrical geometry
            rs, phi_sphere = DiffusionSolver(R, I,D, Sigma_a_star,
                                    nuSigma_f, Q_star,
                                    [0.25,0.5*D(R),0],2)
            #update old solution
            phi_old = phi_sphere.copy()
            if (math.fabs(step-1) < dt):
                plt.plot(rs,phi_sphere,label="t = " + str(step))
            elif (math.fabs(step-10) < dt):
                plt.plot(rs,phi_sphere,'-',label="t = " + str(step))
            elif (math.fabs(step-100) < dt):
                plt.plot(rs,phi_sphere,'-.',label="t = " + str(step))
```

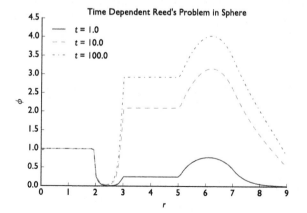

In this problem, steady state is reached relatively quickly near the center of the sphere, but the scattering region takes on the order of 100 seconds to reach steady state. I ran this problem with a coarser spatial mesh because we are solving 1000 steady state problems (instead of just one) to do the time dependent simulation.

CODA

In this chapter we have solved our first boundary value problem, and it is an important one for nuclear engineering: the neutron diffusion equation with a source. To do this we used a finite difference approximation to the second-derivative. Moreover, we have shown how to solve the diffusion equation in multiple geometries and in steady-state and time dependent modes.

The source-driven problems we solved here are important and can address many different applications: from shielding analyses to reactor accident scenarios. Nevertheless, there is a more important type of calculation we can perform: k-eigenvalue calculations to determine the criticality of a nuclear system. We will demonstrate how to do this in the next chapter.

PROBLEMS

Programming Projects

1. Code Testing

In this exercise you will test the implementation of the slab geometry diffusion equation solver. The first step is to develop two analytic solutions to the slab geometry diffusion equation with constant material properties.

QUADRATIC SOLUTION

The first problem we solve has $\Sigma_a = \nu\Sigma_f = 0$ with a zero Dirichlet boundary condition at $r = 1$. The specific equation you need to solve is

$$-D\frac{d^2\phi}{dr^2} = Q,$$

with boundary conditions

$$\left.\frac{d\phi}{dr}\right|_{r=0} = 0, \qquad \phi(1) = 0.$$

Solve this problem for $\phi(r)$.

HYPERBOLIC COSINE SOLUTION

The second problem we solve has $\Sigma_a - \nu \Sigma_f = 1$ with a zero boundary condition at $r = 1$. The specific equation you need to solve is

$$-D\frac{d^2\phi}{dr^2} + \phi(r) = Q,$$

with boundary conditions

$$\left.\frac{d\phi}{dr}\right|_{r=0} = 0, \qquad \phi(1) = 0.$$

Solve this problem for $\phi(r)$.

VERIFYING THE IMPLEMENTATION

Using the analytic solutions you calculated, compare numerical solutions to the exact answer using several values of Δr. Demonstrate that the method is working correctly by

- Stating what the expected behavior of the error should be as Δr changes on each of the two problems.
- Demonstrating that the observed behavior of the error as Δr changes is indeed what you see.

2. Time-Dependent, Super-Critical Excursion

A burst reactor is constructed by building a sphere of plutonium-239 that has a cylindrical hole inside which a slug of plutonium can be inserted. You will model this reactor as a solid sphere with a radius of 7 cm when the slug is in the reactor and as a hollow shell with the same outer radius and an inner radius of 2 cm when the slug is not present. For the plutonium use the following cross-sections from the report *Reactor Physics Constants*, ANL-5800:

Quantity	Value
σ_f [b]	1.85
σ_a [b]	2.11
σ_{tr} [b]	6.8
ν	2.98

For the density of plutonium use 19.74 g/cm^3; the diffusion coefficient is $D = 1/3\Sigma_{tr}$. In the hollow area of the shell, use $\Sigma_a = \nu \Sigma_f = 0$ and $D = 100$ cm. The average neutron speed in this problem is $v = 10^4$ cm/s. Recall that the macroscopic cross-section for reaction i is $\Sigma_i = N\sigma_i$, where N is the number density of nuclei.

An experiment is performed where there is initially 1000 neutrons uniformly distributed in the system at time $t = 0$ when the solid sphere is assembled. The sphere is solid until $t = 0.1$ s at which time you can assume the reactor is a hollow shell. Use a vacuum boundary condition on the edge of the sphere.

Your task is to compute the total number of neutrons that leak out of the sphere from time $t = 0$ to $t = 1$ s as well as the peak fission rate density in the reactor during the experiment.

Finally, determine the maximum value of

$$\frac{d}{dt} \ln \phi(r) \approx \frac{\ln \phi^{n+1}(r) - \ln \phi^n(r)}{\Delta t}.$$

Note: The leakage rate from the sphere per unit surface area is

$$-D\frac{d\phi}{dr}\Big|_{r=7},$$

and the fission rate density at any point in the system is $\Sigma_f(r)\phi(r)$.

3. Shielding Problem

You are tasked with shielding a spherical source of fast neutrons of radius 2.3 cm where the source emits 10^6 neutrons per second per cubic centimeter. You are constructing the shield out of iron and for fast neutrons the cross-sections for iron are (again from ANL-5800)

Quantity	Value
σ_a [b]	0.006
D [cm]	0.1234567

The fission cross-section for iron is 0. Assume the source has the absorption cross-section and diffusion coefficient of pure plutonium-239; ignore fission inside the plutonium (see the previous problem for the cross-sections). Set the boundary condition outside of the shield to have a zero incoming partial current.

How thick does the shield need to be so that no more than 10^4 neutrons leak out per second? The density of iron is 7.874 g/cm^3. *Hint: This might be a good problem for a nonlinear solver: you want to know when the leakage rate out of the reactor equals 10^4 as a function of the thickness of the shield.*

Ni vis humana quot annis
maxima quaeque manu legeret.

Put forth his hand with power, and year by year
Choose out the largest.

–Virgil, **Georgics**

CHAPTER POINTS

- We can adapt our fixed-source diffusion method to find the criticality of a nuclear system.

- To do this we use inverse power iteration, which requires repeated solution of a diffusion system.

- This method gives us the fundamental mode and k_{eff} for a nuclear system.

Computational Nuclear Engineering and Radiological Science Using Python
DOI: 10.1016/B978-0-12-812253-2.00021-2

19.1 NUCLEAR SYSTEM CRITICALITY

In the previous chapter we solved steady-state problems in source-driven, subcritical nuclear systems. A different, and perhaps more common, situation is that we want to know the degree of criticality, often quoted as k_{eff} or reactivity, of a system. In 1-D geometry with a single energy group the k-eigenvalue problem is written as

$$-\nabla \cdot D(r)\nabla\phi(r) + \Sigma_a(r)\phi(r) = \frac{\nu\Sigma_f(r)}{k}\phi(r),$$

with a vacuum, reflecting, or albedo boundary condition at the outer surface

$$\mathcal{A}\phi(r) + \mathcal{B}\frac{d\phi}{dr} = 0 \qquad \text{for } r = R,$$

and a reflecting boundary condition at $r = 0$

$$\frac{d}{dr}\phi(r) = 0 \qquad \text{for } r = 0.$$

Notice that at the outer surface, the boundary condition form is the same as that from the previous chapter, except that $\mathcal{C}$ must be zero.

Note the differences from the equation we solved in the last lecture:

- We added the eigenvalue, $1/k$, to the fission term,
- Eigenvalue problems never have sources ($Q(r) = 0$), and always have vacuum, albedo, or reflecting boundary conditions (i.e., $\mathcal{C} = 0$).

It is also useful to recall what the eigenvalue is doing physically in this equation. If the system is supercritical, then $k > 1$ which depresses the rate at which fission neutrons are produced in order to get a steady solution. On the other hand, if $k < 1$ the system is subcritical and more fission neutrons are needed to make the system not decay to 0 at steady-state. A critical system does not need more or fewer fission neutrons (i.e., $k = 1$).

Upon performing the integration over a cell inside a grid of cells, as we did last time, we get the system

$$-\frac{1}{V_i}\left[D_{i+1/2}S_{i+1/2}\frac{\phi_{i+1} - \phi_i}{\Delta r} - D_{i-1/2}S_{i-1/2}\frac{\phi_i - \phi_{i-1}}{\Delta r}\right] + \Sigma_{a,i}\phi_i = \frac{\nu\Sigma_{f,i}}{k}\phi_i,$$

$$i = 0, \ldots, I - 1, \tag{19.1}$$

and

$$\left(\frac{\mathcal{A}}{2} - \frac{\mathcal{B}}{\Delta r}\right)\phi_{I-1} + \left(\frac{\mathcal{A}}{2} + \frac{\mathcal{B}}{\Delta r}\right)\phi_I = 0.$$

We can write this as a generalized eigenvalue problem

$$\mathbf{A}\vec{\phi} = \lambda\mathbf{B}\vec{\phi},$$

where, in our case, $\lambda = 1/k$, and the matrices $\mathbf{A}$ and $\mathbf{B}$ are defined by the discrete equations above. In this case

$$\mathbf{B} = \begin{pmatrix} \nu \Sigma_{f,1} & & & & \\ 0 & \nu \Sigma_{f,2} & & & \\ 0 & 0 & \ddots & & \\ 0 & 0 & \cdots & \nu \Sigma_{f,I} & \\ 0 & 0 & 0 & \cdots & 0 \end{pmatrix}.$$

The matrix $\mathbf{A}$ is defined in the previous chapter, without the fission terms.

The fundamental mode for the system is the largest value of k and its corresponding eigenvector. This eigenvalue has a special name: k_{eff}. We will now discuss a method for finding this eigenvalue/eigenvector pair.

19.2 INVERSE POWER METHOD

If we consider the generalized eigenvalue problem,

$$\mathbf{A}\mathbf{x} = \lambda \mathbf{B}\mathbf{x},$$

we can make it look like a standard eigenvalue problem by multiplying both sides by the inverse of $\mathbf{A}$ and then rearranging a bit to get

$$\mathbf{A}^{-1}\mathbf{B}\mathbf{x} = \frac{1}{\lambda}\mathbf{x}.$$

In other words, the eigenvalue of the generalized eigenvalue problem is the reciprocal of the eigenvalue to the eigenvalue problem

$$\mathbf{C}\mathbf{x} = l\mathbf{x},$$

where

$$\mathbf{C} = \mathbf{A}^{-1}\mathbf{B}, \qquad l = \frac{1}{\lambda}.$$

We can sketch out a simple algorithm, and then we will show that it should give the largest eigenvalue of C. This will be the largest value of k, which is the value of k_{eff} for the system. Here are the steps in the algorithm.

1. Start with a random initial guess for $\mathbf{x}$ of unit norm, called $\mathbf{x}_0$, set $i = 0$.
2. Compute the product $\mathbf{b}_{i+1} = \mathbf{C}\mathbf{x}_i$,
3. Let $l_{i+1} = ||\mathbf{b}_{i+1}||$,
4. Set

$$\mathbf{x}_{i+1} = \frac{\mathbf{b}_{i+1}}{||\mathbf{b}_{i+1}||},$$

and set $i = i + 1$ (this step normalizes $\mathbf{x}_{i+1}$ to be a unit vector),
5. if $|l_{i+1} - l_i| < \epsilon$, then stop. Otherwise, go to step 2.

The final value of l_{i+1} is an approximation to the maximum magnitude eigenvalue of $\mathbf{C}$. To decide if the eigenvalue is positive or negative we can look at the first index of $\mathbf{b}_{i+1}$ and $\mathbf{b}_i$. The sign of the eigenvalue is equal to the sign of $b_{1,i+1}/b_{1,i}$. Also, $\mathbf{x}_{i+1}$ is the eigenvector associated with eigenvalue l_{i+1}.

Why might this algorithm work? To show that it is reasonable we will assume we know the N eigenvalues l_n and N eigenvectors $\mathbf{u}_n$ of $\mathbf{C}$, an $N \times N$ matrix. Using this information we can write

$$\mathbf{x}_i = \sum_{n=1}^{N} \alpha_{n,i} \mathbf{u}_n, \qquad \mathbf{b}_i = \sum_{n=1}^{N} \beta_{n,i} \mathbf{u}_n,$$

because we can decompose vectors in the range of $\mathbf{C}$ into a linear combination of eigenvectors, which is what these relations say.

Using these relations we can write

$$\mathbf{b}_{i+1} = \sum_{n=1}^{N} \beta_{n,i+1} \mathbf{u}_n = \mathbf{C}\mathbf{x}_i = \mathbf{C} \sum_{n=1}^{N} \alpha_{n,i} \mathbf{u}_n.$$

Of course, $\mathbf{C}\mathbf{u}_n = l_n \mathbf{u}_n$, so we get

$$\mathbf{b}_{i+1} = \sum_{n=1}^{N} \alpha_{n,i} l_n \mathbf{u}_n.$$

If we assume, that the eigenvalues are numbered in decreasing magnitude, $|l_{n-1}| > |l_n|$, then we notice that $\mathbf{b}_{i+1}$ is larger in the $\mathbf{u}_1$ component than any of the others because $\mathbf{x}_i$ is a unit vector. After many iterations of the algorithm, we can expect that

$$\mathbf{b}_{i+1} \approx \alpha_{1,i} l_1 \mathbf{u}_1, \qquad i \gg 1,$$

because this component grows each iteration and grows faster than the others. Therefore

$$||\mathbf{b}_{i+1}|| \approx l_1,$$

because $\mathbf{x}_i$ is a unit-vector by construction.

Therefore, we have the steps we need to find the largest magnitude eigenvector and its associated eigenvector for the matrix $\mathbf{C}$. Recall that to compute $\mathbf{b}_{i+1} = \mathbf{C}\mathbf{x}_i$ we are actually calculating

$$\mathbf{b}_{i+1} = \mathbf{C}\mathbf{x}_i = \mathbf{A}^{-1}\mathbf{B}\mathbf{x}_i.$$

We do not want to explicitly compute the inverse of $\mathbf{A}$, therefore we can solve the following problem to compute $\mathbf{b}_{i+1}$:

$$\mathbf{A}\mathbf{b}_{i+1} = \mathbf{B}\mathbf{x}_i.$$

Notice that at each iteration we are solving a problem with the same matrix and a changing right-hand side. This indicates LU factorization is the correct strategy. When we are solving a diffusion k-eigenvalue problem, this means we will have to solve a steady-state diffusion equation with a known source at each power iteration.

Finally, we relate the eigenvalue l to the k-eigenvalue of the original problem. Recall that $l = 1/\lambda = k$ is the largest eigenvalue of C which is the largest value of k, and therefore the fundamental mode of the system. Therefore, using inverse power iteration gives us the fundamental eigenvalue of the system, k_{eff}, and the fundamental mode eigenvector.

19.3 FUNCTION FOR INVERSE POWER ITERATION

We can translate our simple algorithm above and translate it into python.

```
In [1]: def inversePower(A,B,epsilon=1.0e-6,LOUD=False):
            """Solve the generalized eigenvalue problem
            Ax = 1 B x using inverse power iteration
            Inputs
            A: The LHS matrix (must be invertible)
            B: The RHS matrix
            epsilon: tolerance on eigenvalue
            Outputs:
            1: the smallest eigenvalue of the problem
            x: the associated eigenvector
            """
            N,M = A.shape
            assert(N==M)
            #generate initial guess
            x = np.random.random((N))
            x = x / np.linalg.norm(x) #make norm(x)==1
            l_old = 0
            converged = 0
            #compute LU factorization of A
            row_order = LU_factor(A,LOUD=False)
            iteration = 1;
            while not(converged):
                b = LU_solve(A,np.dot(B,x),row_order)
                l = np.linalg.norm(b)
                sign = b[0]/x[0]/l
                x = b/l
                converged = (np.fabs(1-l_old) < epsilon)
                l_old = l
                if (LOUD):
                    print("Iteration:",iteration,"\tMagnitude of l =",1.0/l)
                iteration += 1
            return sign/l, x
```

To test this method, we can solve a very simple eigenproblem:

$$\begin{pmatrix} 1 & 0 \\ 0 & 0.1 \end{pmatrix} \mathbf{x} = l\mathbf{x}.$$

The smallest eigenvalue is 0.1, and we will solve this using our inverse power iteration method.

```
In [2]: #define A
        A = np.identity(2)
        A[1,1] = 0.1
        #define B
        B = np.identity(2)
        l, x = inversePower(A,B,LOUD=True)
```

```
Iteration: 1    Magnitude of l = 0.168384028973
Iteration: 2    Magnitude of l = 0.100930486877
Iteration: 3    Magnitude of l = 0.10000934947
Iteration: 4    Magnitude of l = 0.100000093499
Iteration: 5    Magnitude of l = 0.100000000935
Iteration: 6    Magnitude of l = 0.100000000009
```

That test worked. Now we can use our method to solve for the eigenvalue of a 1-D reactor.

19.4 SOLVING 1-D DIFFUSION EIGENVALUE PROBLEMS

We will now modify our code from the previous lecture to deal with the differences in eigenvalue problems. In particular we need to remove the sources, and move the fission terms to the **B** matrix.

```
In [3]: def DiffusionEigenvalue(R,I,D,Sig_a,nuSig_f,BC, geometry,epsilon = 1.0e-8):
            """Solve a neutron diffusion eigenvalue problem in a 1-D geometry
            using cell-averaged unknowns
            Args:
                R: size of domain
                I: number of cells
                D: name of function that returns diffusion coefficient for a given r
                Sig_a: name of function that returns Sigma_a for a given r
                nuSig_f: name of function that returns nu Sigma_f for a given r
                BC: Boundary Condition at r=R in form [A,B]
                geometry: shape of problem
                        0 for slab
                        1 for cylindrical
                        2 for spherical

            Returns:
                k: the multiplication factor of the system
                phi:  the fundamental mode with norm 1
                centers: position at cell centers

            """
            #create the grid
            Delta_r, centers, edges = create_grid(R,I)
            A = np.zeros((I+1,I+1))
            B = np.zeros((I+1,I+1))
            #define surface areas and volumes
            assert( (geometry==0) or (geometry == 1) or (geometry == 2))
            if (geometry == 0):
                #in slab it's 1 everywhere except at the left edge
```

```
        S = 0.0*edges+1
        S[0] = 0.0 #to enforce Refl BC
        #in slab its dr
        V = 0.0*centers + Delta_r
    elif (geometry == 1):
        #in cylinder it is 2 pi r
        S = 2.0*np.pi*edges
        #in cylinder its pi (r^2 - r^2)
        V = np.pi*( edges[1:(I+1)]**2
                    - edges[0:I]**2 )
    elif (geometry == 2):
        #in sphere it is 4 pi r^2
        S = 4.0*np.pi*edges**2
        #in sphere its 4/3 pi (r^3 - r^3)
        V = 4.0/3.0*np.pi*( edges[1:(I+1)]**3
                    - edges[0:I]**3 )

    #Set up BC at R
    A[I,I] = (BC[0]*0.5 + BC[1]/Delta_r)
    A[I,I-1] = (BC[0]*0.5 - BC[1]/Delta_r)


    #fill in rest of matrix
    for i in range(I):
        r = centers[i]
        A[i,i] = (0.5/(Delta_r * V[i])*((D(r)+D(r+Delta_r))*S[i+1]) +
                    Sig_a(r))
        B[i,i] = nuSig_f(r)
        if (i>0):
            A[i,i-1] = -0.5*(D(r)+D(r-Delta_r))/(Delta_r * V[i])*S[i]
            A[i,i] += 0.5/(Delta_r * V[i])*((D(r)+D(r-Delta_r))*S[i])
        A[i,i+1] = -0.5*(D(r)+D(r+Delta_r))/(Delta_r * V[i])*S[i+1]

    #find eigenvalue
    l,phi = inversePower(A,B,epsilon)
    k = 1.0/l
    #remove last element of phi because it is outside the domain
    phi = phi[0:I]
    return k, phi, centers
```

To test this code we should compute the eigenvalue and fundamental mode for a homogeneous 1-D reactor. We will set up the problem with $D = 3.850204978408833$ cm, $\nu\Sigma_f = 0.1570$ cm^{-1}, and $\Sigma_a = 0.1532$ cm^{-1}. We also know that in spherical geometry the critical size can be found from

$$\frac{\nu\Sigma_f - \Sigma_a}{D} = \left(\frac{\pi}{R}\right)^2,$$

which leads to

$$R^2 = \frac{\pi^2 D}{\nu\Sigma_f - \Sigma_a}.$$

For this system the critical size is 100 cm.

When we run our eigenvalue solver on this problem, we should see that the eigenvalue converges to 1 as the number of mesh points is refined, also the eigenvector should be the fundamental mode, which has the shape

$$\phi(r) = \frac{A}{r} \sin\left(\frac{\pi r}{R}\right).$$

```
In [4]: nuSigmaf_func = lambda r: 0.1570
        D_func = lambda r: 3.850204978408833
        Sigmaa_func = lambda r: 0.1532
        R = 100
        I = 20
        #solution in spherical geometry with 100 cells
        k, phi,centers = DiffusionEigenvalue(R,I,D_func,
                                    Sigmaa_func,nuSigmaf_func,
                                    [1,0],
                                    2, epsilon=1.0e-10)
```

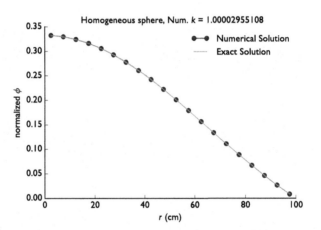

The flux shape looks correct, but the eigenvalue is slightly off. We can see how the eigenvalue solution converges to the right answer.

```
In [5]: Points = np.array((10,20,40,80,160,320))
        for I in Points:
            k, phi,centers = DiffusionEigenvalue(R,I,D_func,Sigmaa_func,
                                    nuSigmaf_func,[1,0],
                                    2,epsilon=1.0e-9)
            print("I =", I, "\t\tk =",k,"\tError =",np.fabs(k-1))
```

```
I = 10        k = 1.00011764407     Error = 0.000117644067299
I = 20        k = 1.00002956354     Error = 2.95635412424e-05
I = 40        k = 1.00000740964     Error = 7.40964406831e-06
I = 80        k = 1.00000186269     Error = 1.86269399705e-06
I = 160       k = 1.00000047537     Error = 4.75371085829e-07
I = 320       k = 1.00000012886     Error = 1.28862679416e-07
```

We can do the same check for slab geometry. In this case, the critical half-size of the reactor is

$$4R^2 = \frac{\pi^2 D}{\nu \Sigma_f - \Sigma_a},$$

giving a critical size of 50 cm.

The shape in this case is

$$\phi(r) = A \cos\left(\frac{\pi r}{R}\right).$$

The solution with 20 cells is

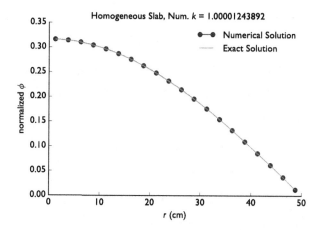

Again, we match the exact solution pretty well with only 20 cells. The final check is the cylinder. For a cylinder

$$R^2 = \frac{2.405^2 D}{\nu \Sigma_f - \Sigma_a},$$

and the fundamental mode is

$$\phi(r) = C J_0\left(\frac{2.405 r}{R}\right),$$

where J_0 is a Bessel function of the first kind. The critical size is about 76.5535 cm.

Once again, 20 cells gives us a reasonable approximation to the exact solution:

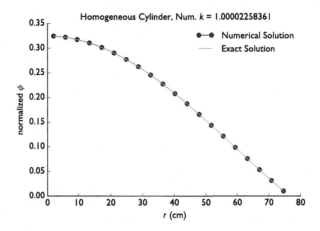

We have not done enough homogeneous problems to call this a thorough verification of the code, but we do have some confidence that the code is working correctly.

19.4.1 Heterogeneous Problems

Previously, we looked at a problem with fuel on the interior and a reflector on the outside. The fuel extends from $r = 0$ to $r = 5$ cm and the reflector extends out to $R = 10$.

```
In [6]: def D(r):
            value = 5.0*(r<=5) + 1.0*(r>5)
            return value;
        def Sigma_a(r):
            value = 0.5*(r<=5) + 0.01*(r>5)
            return value;
        def nuSigma_f(r):
            value = 0.7*(r<=5) + 0.0*(r>5)
            return value;
```

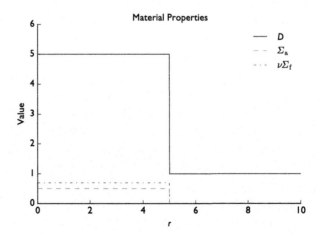

For this problem the eigenvalue in each geometry is found in the next code snippet. We then plot the solution and compare it with the exact solution and its eigenvalue.

```
In [7]:  R = 10
         I = 200
         #Solve Diffusion Problem in slab geometry
         k, phi_slab,centers = DiffusionEigenvalue(R, I,D, Sigma_a,
                                                  nuSigma_f,[1,0], 0)
         #Solve Diffusion Problem in cylindrical geometry
         kc, phi_cyl,centers_cyl = DiffusionEigenvalue(R, I,D, Sigma_a,
                                                  nuSigma_f,[1,0], 1)
         #Solve Diffusion Problem in spherical geometry
         ks, phi_sphere,centers_sp = DiffusionEigenvalue(R, I,D,Sigma_a,
                                                  nuSigma_f,[1,0], 2)
```

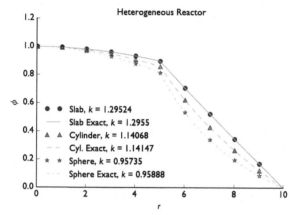

With 200 cells the eigenvalues are approximated to about 4 digits of accuracy. We see that the eigenvalue goes down as we go from slab to cylinder to sphere. This is not a surprise because the slab geometry is infinite in 2 directions, cylinder is infinite in 1, and the sphere is a finite size. Therefore, the leakage goes up as I make the problem "smaller".

One thing we should see is that the eigenvalue goes up as we make the moderator larger. Doubling the thickness of the reflector should increase k_{eff}:

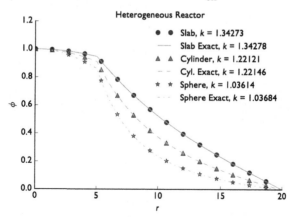

This does depress the leakage and increases the multiplication factor, as expected. With 600 cells we get 5 digits of accuracy in the slab and 4 digits in cylinders and spheres for the eigenvalue.

CODA

In this chapter we adapted our diffusion code for source driven problems to solve *k*-eigenvalue problems to find the multiplication factor for a system. The basis for this is inverse power iteration. At each step of the iterative process we have to solve a steady-state diffusion problem where the source is the fission neutrons created by the previous iteration's solution.

Eigenvalue calculations are important in reactor design, but the one-group approximation is a bit dubious for thermal, light water reactors. A better model for these systems is a two-group diffusion model where we solve for the scalar flux of fast and thermal neutrons separately. This is done in the next chapter.

PROBLEMS

Programming Projects

1. Reflector Effective Albedo

For the heterogeneous reactor example used above, determine the effective albedo of the reflector in the slab case. To do this you will solve for the eigenvalue of a slab reactor with $R = 5$ cm with an albedo boundary condition. Then your task is to find a value of α, the fraction of neutrons reflected back, that produces a value of k_{eff} that matches that of the reactor with a reflector. You can use a nonlinear solver to accomplish this by defining a function that calls the eigenvalue solver with a particular value of α in the albedo boundary condition and returns the difference between k_{eff} and the eigenvalue for the reactor with a reflector.

2. Spherical Plutonium Reactor k-Eigenvalue

For the burst reactor made of plutonium in the defined in the "super-critical excursion" problem of Chapter 18, compute the value of k_{eff} with and without the plug inserted.

3. Criticality for 1-D Heterogeneous System Using the 1-Group Neutron Diffusion Equation

Consider a 1-D heterogeneous cylinder of thickness R. The medium is made of 5 regions 10 cm thick (total domain size is $R = 50$ cm):

Region	1	2	3	4	5
Medium	fuel 1	fuel 3	fuel 1	fuel 2	reflector 1

Use a uniform mesh size of $\Delta r = 1$ cm (i.e., 10 cells per region, i.e., a total of 50 cells). Find the value of k_{eff} for this reactor. Plot the solution for $\phi(r)$ in the reactor and comment on the shape of the solution.

The data you will need:

Medium	D (cm)	Σ_a (cm^{-1})	$\nu\Sigma_f$ (cm^{-1})
reflector 1	2	0.020	0
fuel 1	1.1	0.070	0.095
fuel 2	1.2	0.065	0.095
fuel 3	1.5	0.085	0.095

You, precious birds: your nests, your houses are in the trees, in the bushes. Multiply there, scatter there, in the branches of trees, the branches of bushes

–Popul Vuh, *as translated by Dennis Tedlock*

CHAPTER POINTS

- Two-group eigenvalue problems form a larger system of equations to apply inverse power iteration.

- The structure of two-group problems for thermal nuclear reactor systems allows inverse power iteration to be applied by solving two systems the size of a one-group problem at each iteration.

20.1 TWO-GROUP CRITICALITY PROBLEMS

To this point we have only solved single-group problems. Next we will extend our capabilities to two-group criticality problems. In this case we have two coupled diffusion equations of the form

$$-\nabla \cdot D_1(r)\nabla\phi_1(r) + \Sigma_{t1}(r)\phi_1(r) = \Sigma_{s1\to1}\phi_1(r) +$$

$$\Sigma_{s2\to1}\phi_2(r) + \frac{\chi_1}{k}\left[\nu\Sigma_{f1}(r)\phi_1(r) + \nu\Sigma_{f2}(r)\phi_2(r)\right], \quad (20.1a)$$

$$-\nabla \cdot D_2(r)\nabla\phi_2(r) + \Sigma_{t2}(r)\phi_2(r) = \Sigma_{s1\to2}\phi_1(r) + \Sigma_{s2\to2}\phi_2(r) +$$

$$\frac{\chi_2}{k}\left[\nu\Sigma_{f1}(r)\phi_1(r) + \nu\Sigma_{f2}(r)\phi_2(r)\right], \quad (20.1b)$$

with a vacuum, reflecting, or albedo boundary condition at the outer surface

$$A_g\phi_g(r) + B_g\frac{d\phi_g}{dr} = 0 \qquad \text{for } r = R, \quad g = 1, 2,$$

and a reflecting boundary condition at $r = 0$

$$\frac{d}{dr}\phi_g(r) = 0 \qquad \text{for } r = 0, \quad g = 1, 2.$$

Note that the outer surface boundary conditions can be different for the different groups. In these equations ϕ_1 is the "fast" scalar flux and ϕ_2 is the thermal scalar flux. The fraction of fission neutrons born in group g is denoted by χ_g, Σ_{tg} is the total macroscopic cross-section in group g, $\nu\Sigma_{fg}$ is the product of the mean number of fission neutrons times the macroscopic fission cross-section for group g, and $\Sigma_{sg'\to g}$ is the macroscopic cross-section for scattering from group g' to group g.

Two-group problems are usually set up so that the following simplifications can be made

- There is no upscattering: $\Sigma_{s2\to1} = 0$;
- All fission neutrons are born fast: $\chi_2 = 0$;
- We define the removal cross-section for group 1: $\Sigma_{r1} = \Sigma_{t1} - \Sigma_{s1\to1} = \Sigma_{a1} + \Sigma_{s1\to2}$;
- We define the removal cross-section for group 2: $\Sigma_{r2} = \Sigma_{t2} - \Sigma_{s2\to2} = \Sigma_{a2}$.

Upon making these simplifications system (20.1) becomes

$$-\nabla \cdot D_1(r)\nabla\phi_1(r) + \Sigma_{r1}(r)\phi_1(r) = \frac{1}{k}\left[\nu\Sigma_{f1}(r)\phi_1(r) + \nu\Sigma_{f2}(r)\phi_2(r)\right],$$

$$-\nabla \cdot D_2(r)\nabla\phi_2(r) + \Sigma_{r2}(r)\phi_2(r) = \Sigma_{s1\to2}\phi_1(r).$$

These equations can be discretized use the same procedure we used for one-group equations. The only difference is that now we will have coupling between each equation in the form of the fission terms and the downscattering.

If we apply our cell-centered discretization from the previous two lectures to the two-group equations, we get for the fast-group equations

$$-\frac{1}{V_i}\left[D_{1,i+1/2}S_{i+1/2}\frac{\phi_{1,i+1} - \phi_{1,i}}{\Delta r} - D_{1,i-1/2}S_{i-1/2}\frac{\phi_{1,i} - \phi_{1,i-1}}{\Delta r}\right] + \Sigma_{r1,i}\phi_{1,i}$$

$$= \frac{1}{k}\left[\nu\Sigma_{f1,i}\phi_{1,i} + \nu\Sigma_{f2,i}\phi_{2,i}\right], \qquad i = 0, \dots, I - 1,$$

and

$$\left(\frac{\mathcal{A}_1}{2} - \frac{\mathcal{B}_1}{\Delta r}\right)\phi_{1,I-1} + \left(\frac{\mathcal{A}_1}{2} + \frac{\mathcal{B}_1}{\Delta r}\right)\phi_{1,I} = 0.$$

The thermal group equations are

$$-\frac{1}{V_i}\left[D_{2,i+1/2}S_{i+1/2}\frac{\phi_{2,i+1} - \phi_{2,i}}{\Delta r} - D_{2,i-1/2}S_{i-1/2}\frac{\phi_{2,i} - \phi_{2,i-1}}{\Delta r}\right] + \Sigma_{\mathrm{r}2,i}\phi_{2,i}$$
$$= \Sigma_{\mathrm{s}1\to 2,i}\phi_{1,i}, \qquad i = 0, \dots, I-1,$$

and

$$\left(\frac{\mathcal{A}_2}{2} - \frac{\mathcal{B}_2}{\Delta r}\right)\phi_{2,I-1} + \left(\frac{\mathcal{A}_2}{2} + \frac{\mathcal{B}_2}{\Delta r}\right)\phi_{2,I} = 0.$$

These equations define our eigenvalue problem. Next, we discuss the particulars of this problem and how it can be solved.

20.2 GENERALIZED EIGENVALUE PROBLEM

We can write the $2(I + 1)$ equations as a generalized eigenvalue problem of the form

$$\mathbf{A}\boldsymbol{\Phi} = \frac{1}{k}\mathbf{B}\boldsymbol{\Phi},$$

as before. The difference is that in this case we will write the system in a bit of a different form. First we define the solution vector $\boldsymbol{\Phi}$ as

$$\boldsymbol{\Phi} = \begin{pmatrix} \phi_{1,1} \\ \phi_{1,2} \\ \vdots \\ \phi_{1,I+1} \\ \phi_{2,1} \\ \phi_{2,2} \\ \vdots \\ \phi_{2,I+1} \end{pmatrix}.$$

The matrix $\mathbf{A}$ is what we call a block-matrix. That is a matrix that we define in terms of other, smaller matrices. In particular,

$$\mathbf{A} = \begin{pmatrix} \mathbf{M}_{11} & \mathbf{0} \\ \mathbf{M}_{21} & \mathbf{M}_{22} \end{pmatrix}.$$

The matrix $\mathbf{A}$ is a $2(I + 1) \times 2(I + 1)$ matrix with the $\mathbf{M}_{ij}$ each being $(I + 1) \times (I + 1)$ matrices. The $\mathbf{M}_{11}$ matrix is the left-hand side of the fast-group equation and $\mathbf{M}_{22}$ is the left-hand side of the thermal flux equation. We can write out the non-zero entries of these matrices explicitly as

$$(\mathbf{M}_{gg})_{ii} = \begin{cases} \frac{2}{V_i \Delta r}\left[D_{g,i+1/2}S_{i+1/2} - D_{g,i-1/2}S_{i-1/2}\right] + \Sigma_{\mathrm{r}g,i} & i = 0\ldots I-1 \\ \left(\frac{A_g}{2} + \frac{B_g}{\Delta r}\right) & i = I \end{cases},$$

$$(\mathbf{M}_{gg})_{i,i+1} = \left\{ -\frac{D_{g,i+1/2}S_{i+1/2}}{V_i \Delta r} \quad i = 0\ldots I-1 \right.,$$

$$(\mathbf{M}_{gg})_{i,i-1} = \begin{cases} -\frac{D_{g,i-1/2}S_{i-1/2}}{V_i \Delta r} & i = 1\ldots I-1 \\ \left(\frac{A_g}{2} - \frac{B_g}{\Delta r}\right) & i = I \end{cases}.$$

The matrix $\mathbf{M}_{21}$ is a diagonal matrix for the downscattering terms:

$$(\mathbf{M}_{21})_{ii} = \begin{cases} -\Sigma_{\mathrm{s}1\rightarrow 2,i} & i = 0\ldots I-1 \\ 0 & i = I \end{cases}.$$

With these definitions the larger matrix $\mathbf{A}$ is defined. Now to define the right-hand side matrix $\mathbf{B}$. This matrix is written in block form as

$$\mathbf{B} = \begin{pmatrix} \mathbf{P}_{11} & \mathbf{P}_{12} \\ \mathbf{0} & \mathbf{0} \end{pmatrix}.$$

The fission matrices, $\mathbf{P}_{1g}$ are $(I+1) \times (I+1)$ diagonal matrices of the form:

$$(\mathbf{P}_{1g})_{ii} = \nu \Sigma_{\mathrm{f}g,i}.$$

With these definitions we have now completely specified the generalized eigenvalue problem

$$\mathbf{A}\Phi = \frac{1}{k}\mathbf{B}\Phi.$$

20.3 INVERSE POWER METHOD FOR THE TWO GROUP PROBLEM

This eigenvalue problem has a particular structure that we can take advantage of. To use the inverse power iteration, recall that we have to solve systems of equations of the form

$$\mathbf{A}x_{i+1} = \mathbf{B}x_i.$$

In our case we can solve the matrix using block-forward substitution. What this means is that since our matrix $\mathbf{A}$ is of the form

$$\mathbf{A} = \begin{pmatrix} \mathbf{M}_{11} & \mathbf{0} \\ \mathbf{M}_{21} & \mathbf{M}_{22} \end{pmatrix},$$

we can solve the generic system

$$\begin{pmatrix} \mathbf{M}_{11} & \mathbf{0} \\ \mathbf{M}_{21} & \mathbf{M}_{22} \end{pmatrix} \begin{pmatrix} \mathbf{y}_1 \\ \mathbf{y}_2 \end{pmatrix} = \begin{pmatrix} \mathbf{z}_1 \\ \mathbf{z}_2 \end{pmatrix},$$

as

$$\mathbf{y}_1 = \mathbf{M}_{11}^{-1}\mathbf{z}_1,$$

and

$$\mathbf{y}_2 = \mathbf{M}_{22}^{-1}\left(\mathbf{z}_2 - \mathbf{M}_{21}\mathbf{y}_1\right).$$

Therefore, instead of solving a large system involving $\mathbf{A}$, we solve

$$\mathbf{M}_{11}\mathbf{y}_1 = \mathbf{z}_1,$$

and

$$\mathbf{M}_{22}\mathbf{y}_2 = \left(\mathbf{z}_2 - \mathbf{M}_{21}\mathbf{y}_1\right).$$

That is, we solve two smaller systems. In particular we solve two, 1-group steady state diffusion equations in each iteration. Therefore, we can use our 1-group steady-state diffusion code from before, by modifying how it is called and what are the sources.

The benefit of solving a smaller system twice can be seen when we look at the scaling for the time to solution for LU factorization. As previously mentioned, LU factorization scales as the number of equations, n, to the third power: $O(n^3)$. This means that doubling the number of equations increases the time to solution by a factor of $2^3 = 8$. Solving the smaller system twice takes twice as long. Therefore, we save a factor of 4 in time to solution by solving two smaller systems. Also, the memory required is smaller because we do not form the matrix $\mathbf{A}$ and the $(I + 1)^2$ zeros in the upper right block.

20.3.1 Inverse Power Iteration Function

Before looking at two-group diffusion problems, we will first show how to compute the eigenvalue of a block matrix system like the system above.

We now take our simple algorithm above and translate it into Python. This function will use the LU factorization functions discussed previously.

```
In [1]: def inversePowerBlock(M11, M21, M22, P11, P12,epsilon=1.0e-6,LOUD=False):
            """Solve the generalized eigenvalue problem
            (M11  0 ) (phi_1) = 1 (P11 P12) using inverse power iteration
            (M21 M22) (phi_2)   (0   0 )
            Inputs
            Mij: An LHS matrix (must be invertible)
            Plj: A fission matrix
            epsilon: tolerance on eigenvalue
            Outputs:
            l: the smallest eigenvalue of the problem
            x1: the associated eigenvector for the first block
            x2: the associated eigenvector for the second block
```

```
"""
N,M = M11.shape
assert(N==M)
#generate initial guess
x1 = np.random.random((N))
x2 = np.random.random((N))
l_old = np.linalg.norm(np.concatenate((x1,x2)))
x1 = x1/l_old
x2 = x2/l_old
converged = 0
#compute LU factorization of M11
row_order11 = LU_factor(M11,LOUD=False)
#compute LU factorization of M22
row_order22 = LU_factor(M22,LOUD=False)
iteration = 1;
while not(converged):
    #solve for b1
    b1 = LU_solve(M11,np.dot(P11,x1) + np.dot(P12,x2),row_order11)
    #solve for b2
    b2 = LU_solve(M22,np.dot(-M21,b1),row_order22)
    #eigenvalue estimate is norm of combined vectors
    l = np.linalg.norm(np.concatenate((b1,b2)))
    x1 = b1/l
    x2 = b2/l
    converged = (np.fabs(1-l_old) < epsilon)
    l_old = l
    if (LOUD):
        print("Iteration:",iteration,"\tMagnitude of l =",1.0/l)
    iteration += 1
return 1.0/l, x1, x2
```

To test this method, we can solve a very simple eigenproblem:

$$
\begin{pmatrix} 10 & 0 & 0 & 0 \\ 0 & 0.5 & 0 & 0 \\ -1 & 0 & 1 & 0 \\ 0 & -1 & 0 & 0.1 \end{pmatrix} \mathbf{x} = l \begin{pmatrix} 1 & 0 & 1 & 0 \\ 0 & 1 & 0 & 1 \\ 0 & 0 & 0 & 0 \\ 0 & 0 & 0 & 0 \end{pmatrix} \mathbf{x}.
$$

The smallest eigenvalue is $\frac{1}{22} \approx 0.0454545\ldots$, and we will solve this using our inverse power iteration method.

```
In [2]: #define A
        M11 = np.identity(2)
        M11[0,0] = 10.0
        M11[1,1] = 0.5
        M22 = np.identity(2)
        M22[1,1] = 0.1
        M21 = -np.identity(2)
        #define P
        P11 = np.identity(2)
        P12 = np.identity(2)
        l, x1, x2  = inversePowerBlock(M11,M21,M22,P11,P12,epsilon=1.0e-8,LOUD=True)
```

```
Iteration: 1    Magnitude of l = 0.0887392840225
Iteration: 2    Magnitude of l = 0.0454591935894
Iteration: 3    Magnitude of l = 0.0454545458387
Iteration: 4    Magnitude of l = 0.0454545454546
Iteration: 5    Magnitude of l = 0.0454545454545
```

Now that our test passed, we will use this function to solve for the eigenvalue of a 1-D reactor.

20.4 SOLVING 1-D, TWO-GROUP DIFFUSION EIGENVALUE PROBLEMS

We will now modify our code from the previous lecture to deal with two-group eigenvalue problems. Now we will need to define more matrices and call our InversePowerBlock function.

```
In [3]: def TwoGroupEigenvalue(R,I,D1,D2,Sig_r1,Sig_r2,
                               nu_Sigf1, nu_Sigf2,Sig_s12,
                               BC1,BC2,
                               geometry,epsilon = 1.0e-8):
        """Solve a neutron diffusion eigenvalue problem in a 1-D geometry
        using cell-averaged unknowns
        Args:
            R: size of domain
            I: number of cells
            Dg: name of function that returns diffusion coefficient
                for a given r
            Sig_rg: name of function that returns Sigma_rg for a given r
            nuSig_fg: name of function that returns nu Sigma_fg for a given r
            Sig_s12: name of function that returns Sigma_s12 for a given r
            BC1: Boundary Value of fast phi at r=R in form [A,B]
            BC2: Boundary Value of thermal phi at r=R in form [A,B]
            geometry: shape of problem
                    0 for slab
                    1 for cylindrical
                    2 for spherical

        Returns:
            k: the multiplication factor of the system
            phi_fast:  the fast flux fundamental mode with norm 1
            phi_thermal:  the thermal flux fundamental mode with norm 1
            centers: position at cell centers

        """
        #create the grid
        Delta_r, centers, edges = create_grid(R,I)
        M11 = np.zeros((I+1,I+1))
        M21 = np.zeros((I+1,I+1))
        M22 = np.zeros((I+1,I+1))
        P11 = np.zeros((I+1,I+1))
```

```python
P12 = np.zeros((I+1,I+1))
#define surface areas and volumes
assert( (geometry==0) or (geometry == 1) or (geometry == 2))
if (geometry == 0):
    #in slab it's 1 everywhere except at the left edge
    S = 0.0*edges+1
    S[0] = 0.0 #to enforce Refl BC
    #in slab its dr
    V = 0.0*centers + Delta_r
elif (geometry == 1):
    #in cylinder it is 2 pi r
    S = 2.0*np.pi*edges
    #in cylinder its pi (r^2 - r^2)
    V = np.pi*( edges[1:(I+1)]**2
                - edges[0:I]**2 )
elif (geometry == 2):
    #in sphere it is 4 pi r^2
    S = 4.0*np.pi*edges**2
    #in sphere its 4/3 pi (r^3 - r^3)
    V = 4.0/3.0*np.pi*( edges[1:(I+1)]**3
                - edges[0:I]**3 )

#Set up BC at R
M11[I,I] = (BC1[0]*0.5 + BC1[1]/Delta_r)
M11[I,I-1] = (BC1[0]*0.5 - BC1[1]/Delta_r)
M22[I,I] = (BC2[0]*0.5 + BC2[1]/Delta_r)
M22[I,I-1] = (BC2[0]*0.5 - BC2[1]/Delta_r)

#fill in rest of matrix
for i in range(I):
    r = centers[i]
    M11[i,i] = (0.5/(Delta_r * V[i])*((D1(r)+D1(r+Delta_r))*S[i+1]) +
                Sig_r1(r))
    M22[i,i] = (0.5/(Delta_r * V[i])*((D2(r)+D2(r+Delta_r))*S[i+1]) +
                Sig_r2(r))
    M21[i,i] = -Sig_s12(r)
    P11[i,i] = nu_Sigf1(r)
    P12[i,i] = nu_Sigf2(r)
    if (i>0):
        M11[i,i-1] = -0.5*(D1(r)+D1(r-Delta_r))/(Delta_r * V[i])*S[i]
        M11[i,i] += 0.5/(Delta_r * V[i])*((D1(r)+D1(r-Delta_r))*S[i])
        M22[i,i-1] = -0.5*(D2(r)+D2(r-Delta_r))/(Delta_r * V[i])*S[i]
        M22[i,i] += 0.5/(Delta_r * V[i])*((D2(r)+D2(r-Delta_r))*S[i])
    M11[i,i+1] = -0.5*(D1(r)+D1(r+Delta_r))/(Delta_r * V[i])*S[i+1]
    M22[i,i+1] = -0.5*(D2(r)+D2(r+Delta_r))/(Delta_r * V[i])*S[i+1]

#find eigenvalue
l,phi1,phi2 = inversePowerBlock(M11,M21,M22,P11,P12,epsilon)
k = 1.0/l
#remove last element of phi because it is outside the domain
phi1 = phi1[0:I]
phi2 = phi2[0:I]
return k, phi1, phi2, centers
```

To test this code we will solve a 1-group eigenvalue problem and pretend it has two groups. We will use the same case that we used for the fundamental mode for a homogeneous 1-group, 1-D reactor. We will set up the problem with $D_1 = D_2 = 3.850204978408833$ cm, $\nu \Sigma_{f,1} = \nu \Sigma_{f,2} = 0.1570$ cm^{-1}, and $\Sigma_{a,1} = \Sigma_{a,2} = 0.1532$ cm^{-1}. If we set $\Sigma_{s1 \to 2} = 0$, there will be no coupling from group 1 to group 2, therefore the scalar flux in group 2 will be zero everywhere and group 1 will be the same as that from the one group problem. Recall that for 1-group in spherical geometry the critical size can be found from

$$\frac{\nu \Sigma_f - \Sigma_a}{D} = \left(\frac{\pi}{R}\right)^2,$$

which leads to

$$R^2 = \frac{\pi^2 D}{\nu \Sigma_f - \Sigma_a}.$$

For this system the critical size is 100 cm.

We will run our eigenvalue solver on this problem. We should see that the eigenvalue converges to 1 as the number of mesh points is refined, also the eigenvector should be the fundamental mode.

```
In [4]: nuSigmaf_func = lambda r: 0.1570
        D_func = lambda r: 3.850204978408833
        Sigmaa_func = lambda r: 0.1532
        Sigmas_func = lambda r: 0.0
        R = 100
        I = 20
        #solution in spherical geometry with 100 cells
        k, phi_f,phi_t,centers = TwoGroupEigenvalue(R,I,D_func,D_func,
                                         Sigmaa_func,Sigmaa_func,
                                         nuSigmaf_func,nuSigmaf_func,
                                         Sigmas_func,[1,0,0],[1,0,0],
                                         2, epsilon=1.0e-10)
```

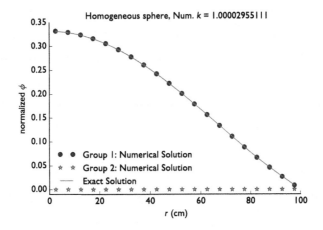

The flux shape looks correct, but the eigenvalue is slightly off as we saw last chapter; refining the mesh will improve the eigenvalue estimate. The thermal-flux for this problem is zero everywhere as expected.

Another way to check a two group problem is to solve an infinite medium problem. The formula for k_∞ is

$$k_\infty = \frac{\nu\Sigma_{f1} + \frac{\Sigma_{s1\to2}}{\Sigma_{r2}}\nu\Sigma_{f2}}{\Sigma_{r1}}.$$

Also, the ratio of the scalar fluxes will be

$$\frac{\phi_1}{\phi_2} = \frac{\Sigma_{r2}}{\Sigma_{s1\to2}}.$$

For our test we will use the following values for the cross-sections (all values in cm^{-1})

- $\nu\Sigma_{f1} = 0.0085$
- $\Sigma_{s1\to2} = 0.0241$
- $\Sigma_{a1} = 0.0121$
- $\Sigma_{r1} = \Sigma_{s1\to2} + \Sigma_{a1} = 0.0362$
- $\nu\Sigma_{f2} = 0.185$
- $\Sigma_{r2} = \Sigma_{a2} = 0.121$

Using these values we get that $k_\infty = 1.25268$ and

$$\frac{\phi_1}{\phi_2} = 5.021.$$

we will test our code using this solution. We will set $D_1 = D_2 = 0.1$ cm and we expect that with a reflecting boundary condition at the outer surface that $k_\infty \to 1.25268$ and the scalar flux ratio goes to 5.021.

```
In [5]: nuSigmaf1_func = lambda r: 0.0085
        nuSigmaf2_func = lambda r: 0.185
        D_func = lambda r: 0.1
        Sigmar1_func = lambda r: 0.0362
        Sigmar2_func = lambda r: 0.121
        Sigmas12_func = lambda r: 0.0241
        R = 5
        I = 50
        k, phi_f,phi_t,centers = TwoGroupEigenvalue(R,I,D_func,D_func,
                                    Sigmar1_func,Sigmar2_func,
                                    nuSigmaf1_func,nuSigmaf2_func,
                                    Sigmas12_func,[0,1,0],[0,1,0],
                                    2, epsilon=1.0e-8)
```

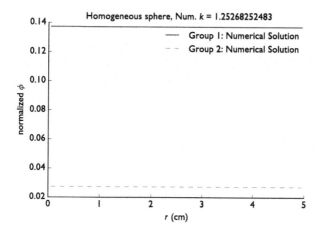

The eigenvalue is the value we expect, to within the iterative tolerance. We can also check the ratios of the scalar fluxes. The ratio should be 5.021.

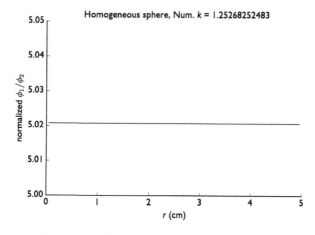

This is indeed what we see in the solution.

20.5 TWO-GROUP REFLECTED REACTOR

Bare homogeneous reactor calculations are not where the usefulness of a numerical method is really needed as these problems can be solved by hand. A more complicated problem that we can use our new tool for is to analyze the increase in k_{eff} by surround the reactor with a reflector. That is, a material that can scatter neutrons back into the reactor. We consider a spherical reactor that has a reflector around it. As an example we set, for the reactor

- $D_1 = D_2 = 1$ cm
- $\nu \Sigma_{f1} = 0.00085$ cm^{-1}

- $\Sigma_{s1\to2} = 0.001 \text{ cm}^{-1}$
- $\Sigma_{a1} = 0.009 \text{ cm}^{-1}$
- $\Sigma_{r1} = \Sigma_{s1\to2} + \Sigma_{a1} = 0.01 \text{ cm}^{-1}$
- $\nu\Sigma_{f2} = 0.057 \text{ cm}^{-1}$
- $\Sigma_{r2} = \Sigma_{a2} = 0.05 \text{ cm}^{-1}$,

and for the reflector

- $D_1 = D_2 = 1 \text{ cm}$
- $\nu\Sigma_{f1} = 0.0 \text{ cm}^{-1}$
- $\Sigma_{s1\to2} = 0.009 \text{ cm}^{-1}$
- $\Sigma_{a1} = 0.001 \text{ cm}^{-1}$
- $\Sigma_{r1} = \Sigma_{s1\to2} + \Sigma_{a1} = 0.01 \text{ cm}^{-1}$
- $\nu\Sigma_{f2} = 0.0 \text{ cm}^{-1}$
- $\Sigma_{r2} = \Sigma_{a2} = 0.00049 \text{ cm}^{-1}$.

The code to set up and solve this problem follows.

```
In [6]: R_reac = 50.0
        nuSigmaf1_func = lambda r: 0.00085*(r<=R_reac)  + 0.0
        nuSigmaf2_func = lambda r: 0.057*(r<=R_reac)  + 0.0
        D_func = lambda r: 1.0
        Sigmar1_func = lambda r: 0.01
        Sigmar2_func = lambda r: 0.01*(r<=R_reac)  + 0.00049*(r>R_reac)
        Sigmas12_func = lambda r: 0.001*(r<=R_reac)  + 0.009*(r>R_reac)
        R = 100
        I = 100
        k, phi_f,phi_t,centers = TwoGroupEigenvalue(R,I,D_func,D_func,
                                        Sigmar1_func,Sigmar2_func,
                                        nuSigmaf1_func,nuSigmaf2_func,
                                        Sigmas12_func,
                                        [0.25,0.5*D_func(R)],
                                        [0.25,0.5*D_func(R)],
                                        2, epsilon=1.0e-6)
```

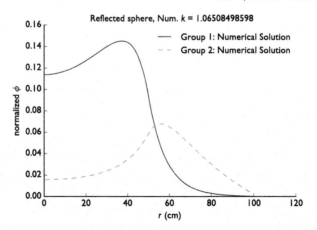

The scalar fluxes in this problem have several noticeable features:

- The thermal scalar flux has a peak in the reflector. This is from fast neutrons leaking out of the reactor and then slowing down in the reflector.
- The fast scalar flux has a peak toward the edge of the reactor. This peak is caused by thermal neutrons leaking back into the reactor from the reflector. These neutrons then cause fission in the reactor, producing fast fissions.

We can compare these results to an unreflected reactor of the same size.

```
In [7]:   R_reac = 500.0
          R = 50
          k, phi_f,phi_t,centers = TwoGroupEigenvalue(R,I,D_func,D_func,
                                        Sigmar1_func,Sigmar2_func,
                                        nuSigmaf1_func,nuSigmaf2_func,
                                        Sigmas12_func,
                                        [0.25,0.5*D_func(R)],
                                        [0.25,0.5*D_func(R)],
                                        2, epsilon=1.0e-6)
```

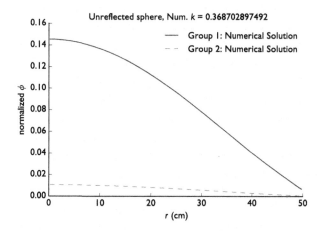

The eigenvalue decreased dramatically. This is due to the fact that the number of thermal neutrons in the system has been severely depressed, and the fission cross-section is higher for thermal neutrons. The implications of this phenomenon are explored in an exercise.

CODA

In this chapter we extended our reactor analysis capabilities to include two-group eigenvalue problems. We were able to apply this to a reflected reactor problem and observe some important reactor physics phenomenon. This marks the end of our foray into numerical solutions to the diffusion model of neutron transport. There is much more we could do: more groups, more dimensions, etc.

Rather than going deeper into diffusion models, we will pivot now to solving the neutron transport equation without making the diffusion approximation. We are going to investigate a tool that is crucial in many nuclear engineering and radiological health applications: Monte Carlo calculations.

PROBLEMS

Programming Projects

1. Effective Albedo for Reflected Two-Group Reactor

The reflected two-group example discussed in the text above had the thermal flux peak inside the reflector. Replace this reflector with an albedo boundary condition at $r = 50$ cm. You can define a different value of α for the thermal and fast boundary condition. Using numerical experimentation, determine values for these alphas that result in a match for the eigenvalue of the reflected reactor, has the thermal flux have a maximum at $r = 50$ cm, and has the fast flux peak near the edge of the reactor. **Hint:** Think about what it means physically to have a reflector if a majority of the neutrons that leak out of the fuel are fast, and a majority of the neutrons that return are thermal.

Discuss your findings and compare the arrived at eigenvalue with the infinite medium eigenvalue of the reactor material. Does this explain why a vein of natural uranium in a mine (something very large) is subcritical, but natural uranium surrounded by heavy water (as in CANDU reactor) or graphite (as in the Chicago Pile) can be made critical?

2. 2-Group Heterogeneous Reactor Multiplication Factor

Consider the following 1-D cylindrical core consisting of 10 fuel regions + 1 reflector region (each region is of width 20 cm, total domain size is $R = 200$ cm).

$$U_{\text{rodded}} \quad M \quad U \quad M \quad U \quad M \quad U \quad M \quad U \quad R$$

In this table, R= reflector, U = UO_2, M = MOX, U_{rodded} = UO_2 + absorber. In this reactor we assume that our groups are set up so that only down scattering can be considered and all fission neutrons are born fast.

Write a Python code to solve this problem. Plot the solution for $\phi(r)$ for each group and comment on the behavior. Also give the value of the multiplication factor, k_{eff}. For this reactor give the ratio of the peak fission rate density to the average fission rate density over the fuel regions (i.e., not including the reflector). This ratio is called the power peaking factor.

Then, consider the same reactor where the control rods are removed and the "U_{rodded}" region becomes a "U" region. Discuss the change in the eigenvalue and the shape of the flux. The data you will need is below:

Material	D_1	D_2	Σ_{r1}	Σ_{r2}	$\nu\Sigma_{f1}$	$\nu\Sigma_{f2}$	$\Sigma_{s1\to2}$
U	1.2	0.4	0.029653979	0.093079585	0.004567474	0.114186862	0.020432526
M	1.2	0.4	0.029653979	0.23633218	0.006851211	0.351903125	0.015865052
R	1.2	0.2	0.051	0.04	0	0	0.05
U_{Rodded}	1.2	0.4	0.029820069	0.098477509	0.004567474	0.114186862	0.02032872

MONTE CARLO METHODS

"Everything is composed of small particles of itself and they are flying around in concentric circles and arcs and segments and innumerable other geometrical figures too numerous to mention collectively, never standing still or resting but spinning away and darting hither and thither and back again, all the time on the go. These diminutive gentlemen are called atoms. Do you follow me intelligently?"

–Flann O'Brien, **The Third Policeman**

CHAPTER POINTS

- Monte Carlo methods give us a way to simulate the behavior of radiation without having to resort to discretized differential equations.

- Using sampling and tracking of individual, simulated neutrons we can solve shielding problems using realistic materials.

- Rejection sampling is a technique for drawing random numbers from a distribution where we do not know the cumulative density function.

Computational Nuclear Engineering and Radiological Science Using Python
DOI: 10.1016/B978-0-12-812253-2.00024-8

21.1 ANALOG PHYSICS

In this chapter we will use Python to create synthetic neutrons. We will then follow these neutrons to see what will happen in a system. The process of using simulated, synthetic particles that behave in a similar manner as actual neutrons (or other particles) is an example of analog physics. This gives a retro-sounding ring to the process, like it is something that goes along with vinyl records and vacuum-tube amplifiers. We mean here that our synthetic neutrons are analogs of true neutrons.

The way we will deal with these particles is using random sampling of their interactions, just as quantum mechanics governs the behavior of particles using probabilities. This is different than what we did in diffusion methods for simulating neutron behavior: in those models we derived differential equations for the expected behavior of a collection of neutrons. In analog physics we simulate many particles and then we compute the mean behavior (or other quantities that interest us).

These methods are called Monte Carlo methods. The first modern Monte Carlo methods were developed by Stanislaw Ulam during the Manhattan Project, and Nicholas Metropolis coined the name after a famous casino. One benefit of Monte Carlo methods is that they require much less mathematics in the algorithms. Once we know how to draw random numbers appropriately, we just need to "roll dice" many, many times to get the answer.

21.2 PROBABILITY PRELIMINARIES

We will need to know a few things about probabilities before we begin. A cumulative distribution function (CDF) is defined as a function $F(x)$ that is the probability that a random variable c, from a particular distribution, is less than x. In mathematical form we write this as

$$F(x) = P(c < x).$$

Because probabilities are always in the range [0, 1], the function $F(x) \in [0, 1]$. As an example, consider the random variable defined by the value of a roll of a single die. In this case, the CDF is given by

$$F(x) = \begin{cases} 0 & x \leq 1 \\ \frac{1}{6} & 1 < x \leq 2 \\ \frac{1}{3} & 2 < x \leq 3 \\ \frac{1}{2} & 3 < x \leq 4 \\ \frac{2}{3} & 4 < x \leq 5 \\ \frac{5}{6} & 5 < x \leq 6 \\ 1 & 6 < x \end{cases}.$$

Using this definition, $F(6.1)$ is 1, because it is certain that the roll will give a number less than 6.1.

The way that the CDF is defined leads to two important limits:

$$\lim_{x \to \infty} F(x) = 1, \quad \text{and} \quad \lim_{x \to -\infty} F(x) = 0.$$

Along with the CDF we will use the probability density function (PDF) , written as $f(x)$. The PDF is defined such that

$$f(x)dx = \text{The probability that the random variable is in } dx \text{ about } x.$$

The PDF is the derivative of the CDF and they are related by

$$f(x) = \frac{dF}{dx}, \quad \text{and} \quad F(x) = \int_{-\infty}^{x} f(x')\,dx'.$$

Also, from these relations we get

$$\int_{-\infty}^{\infty} f(x)\,dx = 1.$$

This relation shows that probability densities are normalized to 1 and can be interpreted as the probability of the random variable being between $-\infty$ and ∞ is 1.

Going back to the roll of a single die, the PDF for this random variable is the sum of Dirac delta functions because the value of a roll can only be the integers 1–6. This PDF is

$$f(x) = \frac{1}{6}\left[\delta(x-1) + \delta(x-2) + \delta(x-3) + \delta(x-4) + \delta(x-5) + \delta(x-6)\right].$$

Notice that the factor of one-sixth is required to satisfy the normalization condition.

Using the PDF we can find the expected value of some function of the random variable. Consider the function $g(x)$, the expected value of this function, $E[g(x)]$, is defined by the integral

$$E[g(x)] = \int_{-\infty}^{\infty} g(x) f(x)\,dx.$$

An important expected value is the mean of the random variable. The mean is found by determining the expected value of the function $g(x) = x$. The mean is sometimes written as $\bar{x}$ and it is defined as

$$\bar{x} = \int_{-\infty}^{\infty} x f(x)\,dx.$$

21.3 THE EXPONENTIAL DISTRIBUTION

The most important distribution for Monte Carlo methods for neutron transport is the exponential distribution. This is because the probability that a neutron travels a number of mean-free paths in $d\lambda$ about λ before having a collision is given by

$$f(\lambda)\,d\lambda = e^{-\lambda}\,d\lambda, \text{ for } \lambda > 0.$$

This distribution tells us that if N neutrons travel λ mean-free paths without a collision, then we expect Ne^{-1} neutrons to travel $\lambda + 1$ mean-free paths. The average distance traveled by a neutron without having a collision, in units of mean-free paths, based on this distribution can be found via the integral

$$\bar{\lambda} = \int_0^\infty \lambda e^{-\lambda}\,d\lambda = 1.$$

This shows why we call λ the mean-free path: it is the expected distance a neutron will travel without having a collision.

Usually, we want to work in units of distance rather than mean-free paths. To convert the exponential distribution to be a function of distance, rather than mean-free paths, we use the total macroscopic cross-section for the material times a distance to write

$$\Sigma_t x = \lambda, \qquad d\lambda = \Sigma_t dx.$$

Making this substitution we get the PDF for the exponential distribution as

$$f(x) = \Sigma_t e^{-\Sigma_t x}.$$

We can check that this is a proper PDF by integrating over all x and seeing that the integral equals one

$$\int_0^\infty dx\,\Sigma_t e^{-\Sigma_t x} = -\left. e^{-\Sigma_t x}\right|_0^\infty = 1.$$

From this we can also define the CDF:

$$F(x) = \int_0^x dx\,\Sigma_t e^{-\Sigma_t x} = -\left. e^{-\Sigma_t x}\right|_0^x = 1 - e^{-\Sigma_t x}.$$

Below is a plot of the CDF and PDF:

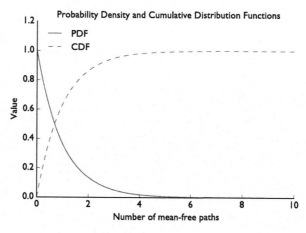

21.4 A FIRST MONTE CARLO PROGRAM

Now we would like to use Monte Carlo solve the following problem: a beam of neutrons strikes a 3 cm thick slab of material with $\Sigma_t = 2.0$ cm^{-1}. What fraction of the neutrons get through the slab without a collision? Using the results from the previous section, we know that the answer to this problem is defined by the integral

$$\int_3^\infty \Sigma_t e^{-\Sigma_t x}\, dx = e^{-2\times 3} \approx 0.002478752177.$$

We would like to solve this problem with Monte Carlo using the following procedure

1. Create neutron.
2. Sample randomly a distance to collision from the exponential distribution.
3. Check to see if the distance to collision is greater than 3.
4. Go back to 1 until we've run "enough" neutrons.

At the end of this procedure the ratio of the number of neutrons that went through the slab to those that we created is the fraction that we are looking for.

The hard part is that we do not know how to generate a random sample from the exponential distribution. We do, however, know how to get a random number between 0 and 1 using NumPy's random function or the functions from the random module. We also know that the CDF, $F(x)$, is always between 0 and 1. Therefore, the following procedure can give me a random number from the exponential distribution:

1. Pick a random number between 0 and 1, call it θ
2. Invert the CDF to solve for x in $F(x) = \theta$; this value of x is my random sample.

In our case we need to solve for x in

$$\theta = 1 - e^{-\Sigma_t x},$$

which gives us

$$x = \frac{-\log(1-\theta)}{\Sigma_t}.$$

We can translate this algorithm to Python in a few short steps. The algorithm will require the user to enter the number of neutrons requested, N, and the thickness of the slab and the macroscopic cross-section for the material.

```
In [2]: def slab_transmission(Sig_t,thickness,N):
            """Compute the fraction of neutrons that leak through a slab
            Inputs:
            Sig_t:      The total macroscopic x-section
            thickness:  Width of the slab
            N:          Number of neutrons to simulate

            Returns:
            transmission:  The fraction of neutrons that made it through
```

```
"""
thetas = np.random.random(N)
x = -np.log(1-thetas)/Sig_t
transmission = np.sum(x>thickness)/N

#for a small number of neutrons we'll output a little more
if (N<=1000):
    plt.scatter(x,np.arange(N))
    plt.xlabel("Distance to collision")
    plt.ylabel("Neutron Number")
return transmission
```

BOX 21.1 PYTHON PRINCIPLE

NumPy can generate random numbers from a variety of distributions. The most common two we use for Monte Carlo simulations are np.random.random(N), which gives N random numbers between 0 and 1, and np.random.uniform(lower, upper, N), which gives N random numbers between lower and upper. For both of these there are single-value versions in the random library: random.random() and random.uniform(lower, upper). In the np.random and random libraries there are more exotic distributions built-in as well.

To test this function we will execute it with a small number of neutrons and look at where the collisions take place. The function will make a graph showing where neutrons had a collision if the number of neutrons is less than or equal to 1000. This initial run will use 1000 neutrons to test this feature.

```
In [3]: #test the functionwith a small number of neutrons
        Sigma_t = 2.0
        thickness = 3.0
        N = 1000
        transmission = slab_transmission(Sigma_t, thickness, N)
        print("Out of",N,"neutrons only",int(transmission*N),
            "made it through.\n The fraction that made it through was",
            transmission)
```

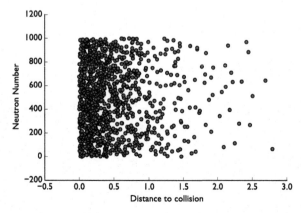

```
Out of 1000 neutrons only 4 made it through.
 The fraction that made it through was 0.004
```

Notice that most of the neutrons have a collision very close to the edge of the slab. Only 4 out of 1000 made it all the way through the slab (that is, only 4 had a distance to collision greater than 3). In this case we get a pretty good answer to our question (0.004 is off by less than a factor of two), but we can check that it converges to the correct answer as $N \to \infty$.

```
In [4]:   neuts = np.array([2000,4000,8000,16000,32000,
                            64000,128000,256e3,512e3,1024e3,2056e3])
          for N in neuts:
              transmission = slab_transmission(Sigma_t, thickness, N)
              print("Out of",N,"neutrons only",int(transmission*N),
                    "made it through.\n The fraction that made it through was",
                    transmission)
```

```
Out of 2000.0 neutrons only 6 made it through.
 The fraction that made it through was 0.003
Out of 4000.0 neutrons only 13 made it through.
 The fraction that made it through was 0.00325
Out of 8000.0 neutrons only 29 made it through.
 The fraction that made it through was 0.003625
Out of 16000.0 neutrons only 47 made it through.
 The fraction that made it through was 0.0029375
Out of 32000.0 neutrons only 73 made it through.
 The fraction that made it through was 0.00228125
Out of 64000.0 neutrons only 165 made it through.
 The fraction that made it through was 0.002578125
Out of 128000.0 neutrons only 341 made it through.
 The fraction that made it through was 0.0026640625
Out of 256000.0 neutrons only 618 made it through.
 The fraction that made it through was 0.0024140625
Out of 512000.0 neutrons only 1285 made it through.
 The fraction that made it through was 0.002509765625
Out of 1024000.0 neutrons only 2528 made it through.
 The fraction that made it through was 0.00246875
Out of 2056000.0 neutrons only 5176 made it through.
 The fraction that made it through was 0.00251750972763
```

We need about 1 million simulated neutrons to get to two digits of the correct answer. Also, if we ran this again we would get different answers because we are using random numbers. The run-to-run variability in the answers is often quantified using the standard deviation. This variability can be thought of as the error in the Monte Carlo calculation. Practitioners often call this variability noise. The error will go down slowly because the way that Monte Carlo works. The standard deviation of the estimate when we look at the run-to-run variability will decrease at a rate proportional to $1/\sqrt{N}$. This means that to cut the error in half we need to quadruple the number of neutrons. Another way to think about it is to say that Monte Carlo methods are one-half order accurate. The $1/\sqrt{N}$ convergence is valid for large values of N and is a result of the central limit theorem.

BOX 21.2 NUMERICAL PRINCIPLE

Running a Monte Carlo calculation multiple times, even with the same number of neutrons, will give you different answers. The standard deviation of this variability is proportional to $N^{-\frac{1}{2}}$, where N is the number of neutrons used in the calculation. This variability manifests itself as errors in the solution and, therefore, we want the standard deviation to be small. The standard deviation goes to zero slowly as $N \to \infty$: to cut the standard deviation in half, one needs to quadruple the number of neutrons.

21.5 ISOTROPIC NEUTRONS ON A SLAB

Even though it takes a lot of particles to make the error in Monte Carlo small, that does not mean it is a bad method. In fact, the nice thing about Monte Carlo is that you have to know very little about the mathematics of the system, all you have to do is be able to push particles around and roll dice. To demonstrate this we will make our problem a little harder. Now say that the neutrons hitting the slab are not a beam but a distribution of neutrons where a neutron's path of flight relative to the normal direction to the slab is measured by the angle ϕ. We say that the distribution of neutrons in angle relative to the slab is uniform in the cosine of the angle ϕ, i.e., the neutrons are isotropic in the cosine of the angle. The angles $\phi \in [-\pi/2, \pi/2]$ point into the slab, this means that the quantity $\cos \phi$ is uniformly distributed between 0 and 1.

For a neutron traveling in direction ϕ the slab can look thicker than 3.0 cm, because if the neutron is traveling at a grazing angle to the slab it will have to travel through more of the slab to get to the other side. We can express this as

$$\text{thickness}(\phi) = \frac{3}{\cos \phi}.$$

A quick check reveals that this gives us what we want: when $\phi = 0$ the neutron is traveling straight through the slab and the thickness to that neutron is 3 cm. Also, when $\phi = \pm\pi/2$ the thickness of the slab is infinite because the neutron is traveling parallel to the slab.

We can make the math easier by defining $\mu = \cos \phi$ and noticing that $\mu \in [0, 1]$. To handle our more complicated problem we make a small change to our Monte Carlo method to have each neutron have its own angle of flight relative to the slab.

1. Create neutron with μ sampled from the uniform distribution $\mu \in [0, 1]$.
2. Sample randomly a distance to collision from the exponential distribution.
3. Check to see if the distance to collision is greater than $3/\mu$.
4. Go back to 1 until we have run "enough" neutrons.

The only change is that now we pick μ uniformly between 0 and 1 (recall that each value of the cosine of the angle was equally likely. That is why this is a random distribution). Furthermore, we check to see if the distance to collision is greater than $3/\mu$. Those are the only changes.

The solution to this problem is more complicated to find mathematically, but the answer is expressed by the exponential integral function:

$$\text{transmission} = \int_0^1 \frac{d\mu}{\Sigma_t} e^{-\Sigma_t x/\mu} = E_2(\Sigma_t x).$$

For our case of $x = 3.0$ and $\Sigma_t = 2$ we get

$$E_2(6) \approx 0.000318257.$$

Notice how the transmission went down because most particles will not have $\mu = 1$.

We can simply modify our function above to handle this case. We do this by modifying the previous function to take as an input parameter whether the neutrons are isotropic in the cosine of the incident angle.

```
In [5]: def slab_transmission(Sig_t,thickness,N,isotropic=False):
            """Compute the fraction of neutrons that leak through a slab
            Inputs:
            Sig_t:      The total macroscopic x-section
            thickness: Width of the slab
            N:          Number of neutrons to simulate
            isotropic: Are the neutrons isotropic or a beam

            Returns:
            transmission:  The fraction of neutrons that made it through
            """
            if (isotropic):
                mu = np.random.random(N)
            else:
                mu = np.ones(N)
            thetas = np.random.random(N)
            x = -np.log(1-thetas)/Sig_t
            transmission = np.sum(x>thickness/mu)/N

            #for a small number of neutrons we'll output a little more
            if (N<=1000):
                plt.scatter(x*mu,np.arange(N))
                plt.xlabel("Distance traveled into slab")
                plt.ylabel("Neutron Number")
            return transmission
```

As before we will run the algorithm with a small number of neutrons and visualize where the interactions take place. The figure will now show the distance the neutron travels before a collision as measured from the face the of slab.

```
In [6]: ###testthe function with a small number of neutrons
        Sigma_t = 2.0
        thickness = 3.0
```

```
N = 1000
transmission = slab_transmission(Sigma_t, thickness, N, isotropic=True)
print("Out of",N,"neutrons only",int(transmission*N),
        "made it through.\n The fraction that made it through was",
        transmission)
```

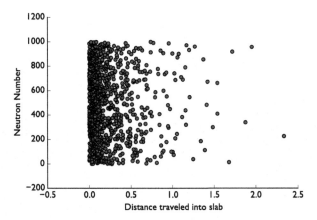

```
Out of 1000 neutrons only 0 made it through.
 The fraction that made it through was 0.0
```

When the neutrons enter the slab at different angles, fewer transmit through the slab without a collision. In this case, none made it through the slab. Also, notice that the scale of the figure changed because no neutron made it more the 2.5 cm into the slab. We can do the same convergence study as before by increasing the number of simulated neutrons and looking at the accuracy of the calculations.

```
In [7]:  neuts = np.array([2000,4000,8000,16000,32000,
                    64000,128000,256e3,512e3,1024e3,2056e3])
         for N in neuts:
             transmission = slab_transmission(Sigma_t, thickness, N, isotropic=True)
             print("Out of",N,"neutrons only",int(transmission*N),
                    "made it through.\n The fraction that made it through was",
                    transmission)
```

```
Out of 2000.0 neutrons only 0 made it through.
 The fraction that made it through was 0.0
Out of 4000.0 neutrons only 0 made it through.
 The fraction that made it through was 0.0
Out of 8000.0 neutrons only 3 made it through.
 The fraction that made it through was 0.000375
Out of 16000.0 neutrons only 3 made it through.
 The fraction that made it through was 0.0001875
Out of 32000.0 neutrons only 11 made it through.
 The fraction that made it through was 0.00034375
Out of 64000.0 neutrons only 20 made it through.
 The fraction that made it through was 0.0003125
Out of 128000.0 neutrons only 30 made it through.
 The fraction that made it through was 0.000234375
```

```
Out of 256000.0 neutrons only 73 made it through.
 The fraction that made it through was 0.00028515625
Out of 512000.0 neutrons only 140 made it through.
 The fraction that made it through was 0.0002734375
Out of 1024000.0 neutrons only 324 made it through.
 The fraction that made it through was 0.00031640625
Out of 2056000.0 neutrons only 717 made it through.
 The fraction that made it through was 0.00034873540856
```

We do start to get to the correct answer, but since so few neutrons get through we have to simulate a lot of them. Here is the result if we try 10 million:

```
In [8]:  N = 1e7
         transmission = slab_transmission(Sigma_t, thickness, N, isotropic=True)
         print("Out of",N,"neutrons only",int(transmission*N),
             "made it through.\n The fraction that made it through was",
             transmission)
```

```
Out of 10000000.0 neutrons only 3164 made it through.
 The fraction that made it through was 0.0003164
```

We are getting several digits of accuracy, but it took a lot of neutrons. Of course in real life 10 million neutrons is not very many. We typically talk about neutrons in numbers like 10^{10} or greater. This is the price to pay with analog physics: the number of our pretend neutrons are always going to be smaller that the actual neutrons.

21.6 A FIRST MONTE CARLO SHIELDING CALCULATION

Both problems that we solved above can be solved pretty easily by hand. To make the problem more difficult we can add some scattering. We want to know what fraction of the neutrons get through the slab before being absorbed. This is a typical question in radiation shielding.

In particular, say that the slab is made up of a material that has $\Sigma_t = 2.0$ cm^{-1}, $\Sigma_s = 0.75$ cm^{-1}, and $\Sigma_a = 1.25$ cm^{-1}. Also, say that the neutrons are scattered isotropically when they scatter, that is the direction can change to any other direction upon scattering. This problem cannot be solved very well by diffusion (remember diffusion is an approximation). We can solve it by modifying our procedure from before.

We will need to add the fact that a collision can be a scatter or an absorption. We will still sample a distance to collision using the exponential distribution and the total macroscopic cross-section, $\Sigma_t = \Sigma_s + \Sigma_a$, as before. The difference is that when the neutron collides, we sample whether it is absorbed or scattered based on the scattering ratio: Σ_s/Σ_t. If it scatters, we sample another μ for it and keep following it. Otherwise, we stop following the neutron because it has been absorbed.

The algorithm for this problem just builds on what we did before.

1. Create a counter, $t = 0$ to track the number of neutrons that get through.
2. Create neutron with μ sampled from the uniform distribution $\mu \in [0, 1]$. Set $x = 0$.

3. Sample randomly a distance to collision, l, from the exponential distribution.

4. Move the particle to $x = x + l\mu$.

5. Check to see if $x > 3$. If so $t = t + 1$. Check if $x < 0$, if so go to 2.

6. Sample a random number s in $[0,1]$, if $s < \Sigma_s/\Sigma_t$, the particle scatters and sample $\mu \in$ $[-1, 1]$ and go to step 3. Otherwise, continue.

7. Go back to 2 until we have run "enough" neutrons.

In this case we need to check to make sure that the neutron does not exit the slab at $x = 0$. This is now possible because a scattered neutron can travel backwards toward the face of the slab. If that happens, to our mind that is the same as absorption because that neutron is not going to transmit through the slab.

Our algorithm is going to have to change a lot in this case. For each created neutron we have to follow it until it leaks out of the slab or is absorbed. This could be many collisions if the scattering ratio is high and Σ_t is large. Nevertheless, we can modify the steps of the simple algorithm to simulate this more complicated scenario.

```
In [9]:  def slab_transmission(Sig_s,Sig_a,thickness,N,isotropic=False):
             """Compute the fraction of neutrons that leak through a slab
             Inputs:
             Sig_s:      The scattering macroscopic x-section
             Sig_a:      The absorption macroscopic x-section
             thickness:  Width of the slab
             N:          Number of neutrons to simulate
             isotropic:  Are the neutrons isotropic or a beam

             Returns:
             transmission:  The fraction of neutrons that made it through
             """
             Sig_t = Sig_a + Sig_s
             iSig_t = 1/Sig_t
             transmission = 0.0
             N = int(N)
             for i in range(N):
                 if (isotropic):
                     mu = random.random()
                 else:
                     mu = 1.0
                 x = 0
                 alive = 1
                 while (alive):
                     #get distance to collision
                     l = -math.log(1-random.random())*iSig_t
                     #move particle
                     x += l*mu
                     #still in the slab?
                     if (x>thickness):
                         transmission += 1
                         alive = 0
                     elif (x<0):
                         alive = 0
                     else:
                         #scatter or absorb
```

```
        if (random.random() < Sig_s*iSig_t):
            #scatter, pick new mu
            mu = random.uniform(-1,1)
        else: #absorbed
            alive = 0
    transmission /= N
    return transmission
```

As a test, this should do the same thing as the previous example, if we set $\Sigma_s = 0$ and $\Sigma_a = 2$. The scattering ratio in this case is 0 so that all the collisions are absorption, and as result, we just need to track distance to collision.

```
In [10]: N = 100000
         Sigma_s = 0.0
         Sigma_a = 2.0
         transmission = slab_transmission(Sigma_s,Sigma_a, thickness, N,
                                         isotropic=True)
         print("Out of",N,"neutrons only",int(transmission*N),
             "made it through.\n The fraction that made it through was",
             transmission)

Out of 100000 neutrons only 25 made it through.
 The fraction that made it through was 0.00026
```

That seems to be working in that it is close to the answer we saw before. It would be a good idea to run this with larger values of N to show we converge to the correct answer. Forgoing that for now, we will try the problem with $\Sigma_s = 0.75$ and $\Sigma_a = 1.25$. In this case we would expect the transmission rate to go up because the total macroscopic cross-section is the same, but the scattering ratio is greater than zero.

```
In [11]: N = 1000000
         Sigma_s = 0.75
         Sigma_a = 2.0 - Sigma_s
         transmission = slab_transmission(Sigma_s,Sigma_a, thickness,
                                         N, isotropic=True)
         print("Out of",N,"neutrons only",int(transmission*N),
             "made it through.\n The fraction that made it through was",
             transmission)

Out of 1000000 neutrons only 861 made it through.
 The fraction that made it through was 0.000861
```

In this result we see about a factor of three increase. With scattering it takes much longer to do the simulation because we might have to follow each neutron for several steps when we follow it until it is absorbed or leaves the slab. An exercise at the end of the chapter explores this further.

Slab transmission with scattering is a problem where we cannot write down the answer easily. As I mentioned, diffusion cannot solve this problem accurately because it is a boundary driven problem with, potentially, a small amount of scattering. To derive the full solution to the transport equation is beyond the scope of this class, and requires sophisticated mathematics such as singular eigenfunction expansions [23,24]. It is not a stretch to say in this case that the Monte Carlo approach is much easier.

21.7 TRACKING IN A SPHERE

We would like to be able to track in geometries other than slab geometry. One common geometry is a sphere. We will consider a spherical shell around an isotropic source and compute the number of neutrons that escape through the outer radius. We have to be careful because in this geometry a neutron that exits the inner radius will strike the shell on the other side.

Consider a spherical shell of inner radius, R_i, and outer radius R_o. Neutrons initially strike R_i with a direction given by the vector $\Omega = (\theta, \varphi)$. These directions are defined so that

$$\Omega \cdot \hat{x} = \sin\theta\cos\varphi, \qquad \Omega \cdot \hat{y} = \sin\theta\sin\varphi,$$
$$\Omega \cdot \hat{z} = \cos\theta.$$

The ranges of the angles are $\theta \in [0, \pi]$ and $\varphi \in [0, 2\pi]$ and $\hat{x}, \hat{y}, \hat{z}$ are the unit vectors in the x, y, and z directions. Using these definitions we know that if a neutron travels a distance s, its position changes by

$$\Delta x = s\sin\theta\cos\varphi, \qquad \Delta y = s\sin\theta\sin\varphi,$$
$$\Delta z = s\cos\theta.$$

This coordinate system is shown in the figure below.

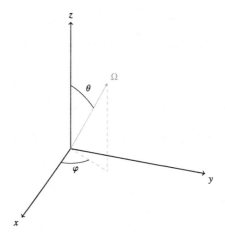

One more geometric idiosyncrasy with spherical shells that we have to deal with is the neutrons that cross the inner radius of the shell. They will strike the inner radius again at the other side of the shell. To find where it will strike the inner radius again we need to find the value of s such that

$$(x_i + s\sin\theta\cos\varphi)^2 + (y_i + s\sin\theta\sin\varphi)^2 + (z_i + s\cos\theta)^2 = R_i^2,$$

where (x_i, y_i, z_i) is the point where the neutron is. We could solve this quadratic equation for s, but this is an opportunity to use a root finding method (such as Ridder's method) to find s. I will choose a closed root finding method, like Ridder's method, in this case because the quadratic could have two solutions and I want to make sure that I get an answer between $s = 0$ and $2R_i$. I have saved the `ridder` function that we previously saw in the file `ridder.py`, and I will import that function below.

We now have everything we need for our Monte Carlo program. We will create neutrons initially at R_i. Given the symmetry of the sphere we can set $z = R_i$ and $y = x = 0$ initially. This will mean that the neutrons will have $\theta \in [0, \pi/2]$ initially (otherwise the neutron will not enter the shell). We then sample a distance to collision, s, much like we did before and follow the neutron around. Now at each step we need to update x, y, and z, and check the radius that the neutron is at to make sure it is still in the shell.

The code to do this is below.

```
In [12]: from ridder import ridder
         def shell_transmission(Sig_s,Sig_a,Ri,Ro,N):
             """Compute the fraction of neutrons that leak through a slab
             Inputs:
             Sig_s:      The scattering macroscopic x-section
             Sig_a:      The absorption macroscopic x-section
             Ri:         Inner radius of the shell
             Ro:         Outer radius of the shell
             N:          Number of neutrons to simulate

             Returns:
             transmission:  The fraction of neutrons that made it through
             """
             Sig_t = Sig_a + Sig_s
             iSig_t = 1/Sig_t
             transmission = 0.0
             N = int(N)

             for i in range(N):
                 #get initial direction
                 theta = random.uniform(0,0.5*np.pi)
                 phi = random.uniform(0,2*np.pi)
                 r = Ri
                 z = Ri
                 x = 0
                 y = 0
                 alive = 1
                 #vector to keep track of positions
                 xvec = x*np.ones([1])
                 yvec = y*np.ones([1])
                 zvec = z*np.ones([1])
                 while (alive):
                     #get distance to collision
                     s = -math.log(1.0-random.random())*iSig_t
                     #move particle
                     z += s*math.cos(theta)
                     y += s*math.sin(theta)*math.sin(phi)
                     x += s*math.sin(theta)*math.cos(phi)
                     xvec = np.append(xvec,x)
                     yvec = np.append(yvec,y)
                     zvec = np.append(zvec,z)
```

```python
            r = math.sqrt(z**2 + y**2 + x**2)
            #still in the shell?
            if (r>Ro):
                transmission += 1
                alive = 0
            elif (r<Ri):
                #find s so that the neutron is on the other side of the shell
                f = lambda s: ((x + s*math.sin(theta)*math.cos(phi))**2 +
                               (y+s*math.sin(theta)*math.sin(phi))**2 +
                               (z + s*math.cos(theta))**2 - Ri**2)
                s = ridder(f,1e-10,2*Ri,1.0e-10)
                z += s*math.cos(theta)
                y += s*math.sin(theta)*math.sin(phi)
                x += s*math.sin(theta)*math.cos(phi)
                r = Ri

                #check that we are on the inner radius
                assert(math.fabs(x**2+y**2+z**2 - Ri**2) < 1e-6)

        else:
            #scatter or absorb
            if (random.random() < Sig_s*iSig_t):
                #scatter, pick new angles
                theta = random.uniform(0,math.pi)
                phi = random.uniform(0,2*math.pi)
            else: #absorbed
                alive = 0
    transmission /= N
    return transmission
```

We will simulate transmission through a shell of thickness 3 cm and inner radius 2 cm. Also, we will visualize the tracks that the neutrons take through the shell.

```python
In [13]: N = 100
         Sigma_s = 1.0
         Sigma_a = 1.0
         Ri = 2
         Ro = Ri + 3
         transmission = shell_transmission(Sigma_s,Sigma_a,Ri,Ro,N)
         print("Out of",N,"neutrons only",int(transmission*N),
               "made it through.\n The fraction that made it through was",
               transmission)
```

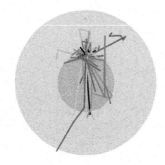

```
Out of 100 neutrons only 1 made it through.
 The fraction that made it through was 0.01
```

Notice that all the neutrons start at $z = R_i$ and $x = y = 0$ as prescribed in the code. We can also see the fact that when a neutron re-enters the hollow center, it streams across to the other side. This figure is also a way to check that the streaming through the hollow part of the shell is handled correctly: we should not see any neutron tracks end in the hollow part of the sphere (though this is hard to tell with a 2-D projection of the sphere). We can also see the one neutron that escaped the sphere.

Upon increasing the number of neutrons, we expect to get a more accurate answer, though we have not said what that answer is. With $N = 10^5$ we get

```
In [14]: N = 10**5
         transmission = shell_transmission(Sigma_s,Sigma_a,Ri,Ro,N)
         print("Out of",N,"neutrons only",int(transmission*N),
               "made it through.\n The fraction that made it through was",
               transmission)
```

```
Out of 100000 neutrons only 602 made it through.
 The fraction that made it through was 0.00602
```

21.8 A REAL SHIELDING PROBLEM

We will now take a large step forward. We will try to solve the problem of designing a lead-208 shield for a bare reactor made of uranium-235. We will use the full energy dependent cross-sections for lead and the actual fission spectrum of U-235. To do this we will have to read in the data for the lead microscopic cross-sections and the fission spectrum. We covered how to do this in Section 5.1. Using that knowledge we can read in a csv file of the format: incident neutron energy, cross-section in barns (10^{-24} cm^2). The code below does this. A plot of the cross-sections follows.

```
In [15]: import csv
         lead_s = [] #create a blank list for the x-sects
         lead_s_energy = [] #create a blank list for the x-sects energies
         #this loop will only execute if the file opens
         with open('pb_scat.csv') as csvfile:
             pbScat = csv.reader(csvfile)
             for row in pbScat: #have for loop that loops over each line
                 lead_s.append(float(row[1]))
                 lead_s_energy.append(float(row[0]))
         lead_scattering = np.array([lead_s_energy,lead_s])
         lead_abs = [] #create a blank list for the x-sects
         lead_abs_energy = [] #create a blank list for the x-sects energies
         #this loop will only execute if the file opens
         with open('pb_radcap.csv') as csvfile:
             pbAbs = csv.reader(csvfile)
```

```
for row in pbAbs: #have for loop that loops over each line
    lead_abs.append(float(row[1]))
    lead_abs_energy.append(float(row[0]))
lead_absorption = np.array([lead_abs_energy,lead_abs])
```

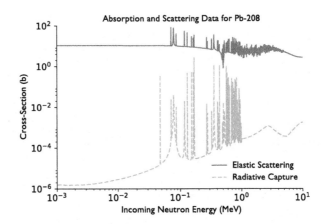

For the fission spectrum we will use the Watt fission spectrum. This spectrum is the relative likelihood of a fission neutron being born with energy E. The fission spectrum is typically denoted as $\chi(E)$ and given by

$$\chi(E) = 0.453e^{-1.036E} \sinh\left(\sqrt{2.29E}\right),$$

with E given in MeV.

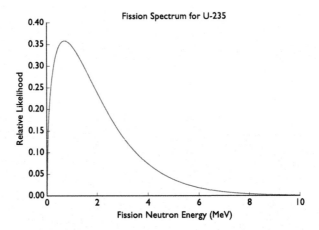

To solve this problem we will have to change our algorithm somewhat. Firstly, we will have to generate neutrons with energies sampled from the fission spectrum and then evaluate the cross-sections at the neutron's energy. Secondly, we will have to change how we scatter particles. When a particle scatters we will have to sample a new energy for the post-scattered neutron. We will tackle each of these next.

21.9 REJECTION SAMPLING

Previously, we discussed how to sample from a distribution by inverting the CDF and then using a random number between 0 and 1 to give us the inverse CDF value corresponding to our sample. With the fission spectrum we cannot do this easily. We do not have a CDF, we just have a distribution that gives a relative likelihood of a fission neutron being born with a certain energy. To sample from this we use rejection sampling. The idea of rejection sampling is to draw a box around the PDF of the distribution we want to sample from, and pick points in the box. If the point is below the curve, we accept the point, otherwise we reject it and sample again. The effect of this is that we will get more points where the PDF is large, and few points where the PDF is small.

We will demonstrate the idea of rejection sampling with our fission spectrum data. To do this we find the maximum and the minimum of the function over the energy range from the minimum energy in the lead scattering table to 10 MeV. This allows us to define the box, and we randomly pick points in the box. Then we check to see if a point is above or below the function.

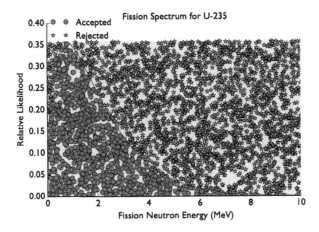

The histogram of the accepted samples should look like the fission spectrum we wanted to sample. The histogram will not be perfect because we have a finite number of samples.

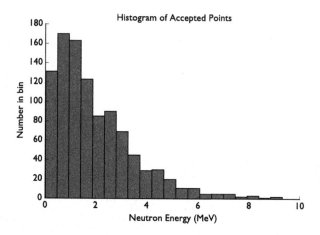

BOX 21.3 NUMERICAL PRINCIPLE

To generate samples from a distribution where you cannot invert the CDF, $F(x)$, you can use rejection sampling. To generate a sample, first pick a value $\hat{x}$ randomly between the minimum possible and maximum possible value of x. Then pick a value $\hat{y}$ randomly between 0 and the maximum value of the PDF. If $\hat{y} < f(\hat{x})$, where $f(x)$ is the PDF, then $\hat{x}$ is accepted as a sample. Otherwise, $\hat{x}$ is rejected, and we try again.

21.10 LOOKING UP ENERGIES

To look up the cross-section for lead at different energies we need a function that can return the value of the cross-section for a given energy. We can make this happen by finding the energy in the data set that is closest to the input energy. This can be done using the NumPy function argmin. This function returns the index of a NumPy array with the smallest value. Therefore, we pass to argmin the difference between a target energy and the vector of energies in the table. In effect for a given neutron energy this will give the point in the table closest to that energy. We define such an energy lookup function below.

```
In [16]: def energy_lookup(data_set, inp_energy):
             """look up energy in a data set and
             return the nearest energy in the table
             Input:
             data_set:   a vector of energies
             inp_energy: the energy to lookup

             Output:
             index: the index of the nearest neighbor in the table
             """

             #argmin returns the indices of the smallest members of an array
```

```
#here we'll look for the minimum difference
#between the input energy and the table
index = np.argmin(np.fabs(data_set-inp_energy))
return index
```

21.11 ELASTIC SCATTERING

When a particle scatters elastically, the energy of the scattered neutron, E', of a neutron with an initial energy E is governed by a probability given by

$$P(E \to E') = \begin{cases} \frac{1}{E(1-\alpha)} & \alpha E \le E' \le E \\ 0 & \text{otherwise} \end{cases},$$

where

$$\alpha = \frac{(A-1)^2}{(A+1)^2},$$

with A the mass of the nucleus. Therefore, we can sample the value of the scattered neutron's energy from a uniform distribution from αE to E. The neutron's angle would also be a function of the scattered energy, in order to preserve momentum. For simplicity we will say that the neutron's angle after the scatter is isotropic, though this is not completely correct. The proper scattering would give a particular value of μ for a given energy change.

21.12 LEAD SHIELDING OF REACTOR ALGORITHM AND CODE

The algorithm will be an enhanced version of the one before that concerned neutrons striking an absorbing and scattering slab. We will assume isotropic, elastic scattering and radiative capture as the only reactions in the lead, though this simplification could be modified using the additional cross-sections for inelastic scattering, $(n, 2n)$ reactions, etc. The algorithm now looks like:

1. Create a neutron with μ sampled from the uniform distribution $\mu \in [0, 1]$ and an energy sampled from the fission spectrum via rejection sampling. Set $x = 0$.
2. Sample randomly a distance to collision, l, from the exponential distribution.
3. Move the particle to $x = x + l\mu$.
4. Check to see if x is greater than the shield thickness, if so stop following the neutron. Check if $x < 0$, if so go to 1.
5. Sample a random number s in [0,1], if $s < \Sigma_s / \Sigma_t$, the particle scatters and sample $\mu \in [-1, 1]$ and an energy in based on the formula above and go to step 3. Otherwise, continue.
6. Go back to 1 until we have run "enough" neutrons.

The resulting function is given below.

```
In [17]: def slab_reactor(sig_s,sig_a,thickness,density,A,N,isotropic=False):
             """Compute the fraction of neutrons that leak through a slab
             Inputs:
             sig_s:      The scattering microscopic x-section array in form Energy,
                         X-sect
             sig_a:      The absorption microscopic x-section
             thickness: Width of the slab
             density:   density of material in atoms per cc
             A:          atomic weight of shield
             N:          Number of neutrons to simulate
             isotropic: Are the neutrons isotropic or a beam

             Returns:
             transmission:  energies of neutrons that leak through
             created:  energies of neutrons that were born
             """
             alpha = (A-1.0)**2/(A+1.0)**2
             Sig_s = sig_s.copy()
             Sig_a = sig_a.copy()
             Sig_s[1,:] = density/1e24*Sig_s[1,:]
             Sig_a[1,:] = density/1e24*Sig_a[1,:]
             #make rejection box
             min_eng = np.min([np.min(Sig_s[0,:]),np.min(Sig_a[0,:])])
             max_eng = np.max([np.max(Sig_s[0,:]),np.max(Sig_a[0,:])])
             max_prob = np.max(np.max(expfiss(Sig_a[0,:])))
             transmission = []
             created = []
             N = int(N)
             for i in range(N):
                 #sample direction
                 if (isotropic):
                     mu = random.random()
                 else:
                     mu = 1.0
                 #compute energy via rejection sampling
                 rejected = 1
                 while (rejected):
                     #pick x
                     x = random.uniform(min_eng,max_eng)
                     y = random.uniform(0,max_prob)
                     rel_prob = expfiss(x)
                     if (y <= rel_prob):
                         energy = x
                         rejected = 0
                 #initial position is 0
                 x = 0
                 created.append(energy)
                 alive = 1
                 while (alive):
                     #get distance to collision
                     scat_index = energy_lookup(Sig_s[0,:],energy)
                     abs_index = energy_lookup(Sig_a[0,:],energy)
                     cur_scat = Sig_s[1,scat_index]
                     cur_abs = Sig_a[1,abs_index]
```

```
Sig_t = cur_scat + cur_abs
l = -math.log(1-random.random())/Sig_t
#move particle
x += l*mu
#still in the slab
if (x>thickness):
    transmission.append(energy)
    alive = 0
elif (x<0):
    alive = 0
else:
    #scatter or absorb
    if (random.random() < cur_scat/Sig_t):
        #scatter, pick new mu and energy
        mu = random.uniform(-1,1)
        energy = random.uniform(alpha*energy,energy)
    else: #absorbed
        alive = 0
return transmission, created
```

This algorithm will be more time consuming to run because tracking each neutron now requires more work (e.g., we have to perform rejection sampling to get an initial energy), and the fact that the scattering ratio for lead appears to be high from the figure above. We will run the algorithm with only one million neutrons. Before running the code, we need to compute the number density for lead-208 because the algorithm takes the microscopic cross-section as an input and multiplies that by a number density to get the macroscopic cross-section.

```
In [18]: N = 1e7
         density = 11.34/208*6.022e23
         thickness = 150
         transmission,created = slab_reactor(lead_scattering,lead_absorption,
                                     thickness,density,208, N,
                                     isotropic=True)
```

The output of the function is an array of the transmitted particle energies and the initial energies. The distribution of the transmitted energies is

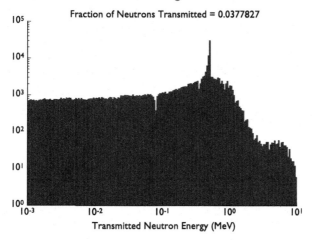

The result is that about 3.77% of the incident neutrons transmit through the shield (making this not a very good shield). There are some interesting phenomenon that we can observe in the transmitted neutrons. In particular, the peaks and valleys of the radiative capture cross-section are mimicked in the transmitted neutron energies.

The neutrons that entered the slab, as function of energy as sampled from the fission spectrum, are shown in the next histogram. With a logarithmic scale, the fission spectrum will look a bit different.

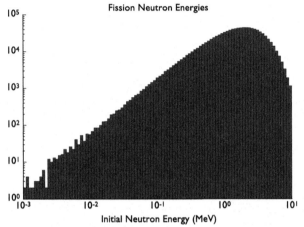

As we would expect, the transmitted energies are a blend of the incident energies and the cross-sections. We could modify this problem by adding other materials, making the shield thicker, or other modifications to improve the shield if we desired. The basics of the algorithm will not change.

CODA

We have demonstrated that we can solve complicated problems by "rolling dice" if we roll many, many dice and move particles around based on these random numbers. One feature of this approach is that it requires little in the way of mathematical sophistication, with the tradeoff that the convergence is slow (remember that the noise in the solution decays as the number of samples to the negative one-half power). Nevertheless, Monte Carlo methods are attractive and they are widely used in nuclear engineering and other fields. In the next two chapters we will expand our Monte Carlo capabilities. In the next chapter we will go over how to reduce the run-to-run variability of Monte Carlo calculations, and provide ways to estimate the scalar flux of neutrons in a system.

FURTHER READING

The Monte Carlo method is a rich subject in nuclear engineering. For a more detailed coverage of the topic we encourage the reader to read one of the monographs devoted to the topic. Two good examples are the recent book by Dunn and Shultis [25] or the work of Kalos and Whitlock [26].

PROBLEMS

Short Exercises

21.1. Consider a beam of neutrons striking a slab of thickness 5 cm and $\Sigma_t = 1.0$ cm^{-1}. Compute the transmission fraction and time how long the calculation takes using $N = 10^6$ neutrons and several different scattering ratios: $0, 0.1, 0.5, 0.9, 1.0$. Compare your computed transmission fractions as a function of scattering ratio to the expected trend as the scattering is increased.

21.2. Modify the shielding code to consider neutrons of a single energy impinging on the shield and to tally the energy of the absorbed neutrons. Assume the neutrons are all 2.5 MeV and are produced from the fusion of deuterium. Plot the distribution of transmitted and absorbed neutrons with a large enough number of sampled neutrons.

21.3. The Maxwell–Boltzman distribution, often called just a Maxwellian distribution, gives the distribution of speeds of particles in a gas by the formula

$$f(v) = \sqrt{\left(\frac{m}{2\pi kT}\right)^3} \, 4\pi v^2 e^{-\frac{mv^2}{2kT}}$$

where m is the mass of the particles, T is the temperature, and k is the Boltzmann constant. Consider a gas of deuterium at $kT = 1$ keV $= 1.60218 \times 10^{-16}$ J. Sample particle speeds from the Maxwellian using rejection sampling. From your sampled points, compute the mean speed and the square-root of the mean speed squared (i.e., compute the mean value of the speed squared and then take the square root, aka the root-mean square speed). The mean speed should be

$$\int_0^\infty dv \, v f(v) = \sqrt{\frac{8kT}{\pi m}},$$

and the root-mean square speed should be

$$\sqrt{\int_0^\infty dv \, v^2 f(v)} = \sqrt{\frac{3kT}{m}}.$$

Compute these quantities using sample numbers of $N = 10, 10^2, 10^3, 10^4, 10^5$ and discuss your results.

Programming Projects

1. Monte Carlo Convergence

In this exercise you will demonstrate the standard deviation of the estimates decays as $N^{-1/2}$. Solve the problem of a beam striking an absorbing slab, $\Sigma_t = \Sigma_a = 1.0 \text{ cm}^{-1}$, with thickness 3 cm. Solve this problem with $N = 10^2, 10^3, 10^4, 10^5, 10^6$. At each value of N estimate the solution 10 times, and take the standard deviation of the estimates. Plot the standard deviation as a function of N on a log-log scale. Compare your results to the expected trend.

2. Track-Length in Sphere

Consider a solid sphere of radius R that has neutrons born isotropically in angle and uniformly in space inside of the sphere. Assuming that no neutrons collide in the sphere, using Monte Carlo compute the average distance the neutron travels inside of the sphere before exiting the sphere. Compute this for several values of R and see if you can find a trend.

Monte Carlo Variance Reduction and Scalar Flux Estimation

"Man, this place is way too analog."

–"Trotter" in the television show The Upright Citizens Brigade

CHAPTER POINTS

- Abandoning analog tracking of simulated particles can have benefits for our calculations.

- Implicit capture allows neutrons to never be absorbed, rather we change a weight associated with the simulated particle.

- To estimate scalar fluxes we use one of two techniques based on either collisions or the track-length of particles in a mesh cell.

- We can play games with our sampling to randomly sample "better".

In the last chapter we talked about how to construct particles using a computer program that were analogs of real neutrons moving through a system. We saw that in many cases it took very many simulated neutrons to get reasonable answers. In this chapter we will see that by making our simulated particles behave in a non-analog way we can improve our answers.

Computational Nuclear Engineering and Radiological Science Using Python
DOI: 10.1016/B978-0-12-812253-2.00025-X

We will not improve the convergence rate of Monte Carlo methods. Recall that the statistical error as measured by the standard deviation of the estimate converges as $O(N^{-1/2})$, where N is the number of simulated particles. This means that the standard deviation of our estimate is, to leading order, $CN^{-1/2}$. What variance reduction does is reduce the magnitude of C, the constant in the convergence.

22.1 IMPLICIT CAPTURE AND PARTICLE WEIGHTS

In problems where radiative capture is important, it can be beneficial to remove this absorption process from the types of interactions considered via a process called implicit capture. To do this we first introduce a particle weight, w. The weight is defined so that each particle represents a collection of neutrons, rather than a single one. Consider a neutron source in a volume, Q, with units neutrons per cm^3 per second. We will use N particles each with a weight w_i to represent this source in steady state calculation. The weights must satisfy

$$\sum_{i=1}^{N} w_i = \int_V dV\, Q.$$

From this relation we see that the units of w_i are neutrons per second.

To use implicit capture, as a particle moves a distance s in a material, we reduce its weight by a factor of $\exp(-\Sigma_\gamma s)$. This is done because this factor represents the likelihood of the neutron traveling a distance s without having a radiative capture reaction. With implicit capture, our simple shielding calculation has its procedure changed to be

1. Create a counter, $t = 0$ to track the number of neutrons that get through.
2. Create neutron with μ sampled from the uniform distribution $\mu \in [0, 1]$. Set $x = 0$. Set the particle's weight to be $w = 1/N$.
3. Sample randomly a distance to scatter, l, from the exponential distribution.
4. Move the particle to $x = x + l\mu$.
5. Reduce the weight of the particle by a factor $\exp(-\Sigma_\gamma s)$.
6. Check to see if $x > 3$. If so $t = t + w$, and go to 2. Otherwise, if $x < 0$ go to step 2.
7. Go back to step 3.

A couple of notes on this new algorithm. Now each particle has a weight and that is what we sum up to get the fraction of neutrons that leak out per unit time. Also, when we sample a distance to collision, we only sample a distance to scatter.

One drawback of implicit capture is that it can result in the tracking of particles that have a very small weight. After traveling a large distance, the weight could be much less than the initial weight and the particle could contribute only a small amount to the result. This is wasted computational effort tracking these low weight particles; it would be better to stop tracking them. For this purpose we introduce a cutoff weight. After decreasing from the initial weight by more than the cutoff, we treat the particle using analog tracking so that it can die via absorption.

The code to implement implicit capture for the slab transmission problem is below.

```
In [1]: def slab_transmission(Sig_s,Sig_a,thickness,N,
                              isotropic=False,
                              implicit_capture = True,
                              cutoff = 1.0e-3):
            """Compute the fraction of neutrons that leak through a slab
            Inputs:
            Sig_s:     The scattering macroscopic x-section
            Sig_a:     The absorption macroscopic x-section
            thickness: Width of the slab
            N:         Number of neutrons to simulate
            isotropic: Are the neutrons isotropic or a beam
            implicit_capture: Do we run implicit capture
            cutoff:    At what level do we stop implicit capture
            Returns:
            transmission:  The fraction of neutrons that made it through
            """
            imp_input = implicit_capture
            Sig_t = Sig_a + Sig_s
            iSig_t = 1.0/Sig_t
            iSig_s = 1.0/(Sig_s + 1.0e-14)
            transmission = 0.0
            N = int(N)
            initial_weight = 1.0/N
            for i in range(N):
                if (isotropic):
                    mu = random.random()
                else:
                    mu = 1.0
                x = 0
                alive = 1
                weight = initial_weight
                while (alive):
                    if (weight < cutoff*initial_weight):
                        implicit_capture = False

                    if (implicit_capture):
                        #get distance to collision
                        l = -math.log(1-random.random())*iSig_s
                    else:
                        #get distance to collision
                        l = -math.log(1-random.random())*iSig_t
                    #make sure that l is not too large
                    if (mu > 0):
                        l = min([l,(3-x)/mu])
                    else:
                        l = min([l,-x/mu])
                    #move particle
                    x += l*mu
                    if (implicit_capture):
                        weight *= np.exp(-l*Sig_a)
                    #still in the slab?
                    if (math.fabs(x-thickness) < 1.0e-14):
                        transmission += weight
                        alive = 0
                    elif (x<= 1.0e-14):
                        alive = 0
                    else:
                        if (implicit_capture):
                            mu = random.uniform(-1,1)
                        else:
                            #scatter or absorb
                            if (random.random() < Sig_s*iSig_t):
                                #scatter, pick new mu
                                mu = random.uniform(-1,1)
                            else: #absorbed
```

```
                           alive = 0
                implicit_capture = imp_input
          return transmission
```

The example problem of a beam incident on a pure absorber should be exact using implicit capture and only one particle.

```
In [2]: N = 1
        Sigma_s = 0.0
        Sigma_a = 2.0
        thickness = 3
        transmission = slab_transmission(Sigma_s,Sigma_a, thickness,
                                   N, isotropic=False, implicit_capture=True)
        print("The fraction that made it through using implicit capture was",
              transmission, "with a percent error of",
              np.abs(transmission - np.exp(-6))/np.exp(-6)*100,"%")
        transmission = slab_transmission(Sigma_s,Sigma_a, thickness,
                                   N, isotropic=False, implicit_capture=False)
        print("The fraction that made it through using analog tracking was",
              transmission, "with a percent error of",
              np.abs(transmission - np.exp(-6))/np.exp(-6)*100,"%")
```

```
The fraction that made it through using
implicit capture was 0.00247875217667 with a percent error of 0.0 %
The fraction that made it through using
analog tracking was 0.0 with a percent error of 100.0 %
```

Indeed, implicit capture gave the correct answer For isotropic incident neutrons on the same slab, we expect implicit capture to be better than analog tracking, but not exact. In this test we will use 1000 simulated particles.

```
In [3]: true_sol = 0.000318257463690406727
        transmission = slab_transmission(Sigma_s,Sigma_a, thickness,
                                   N, isotropic=True, implicit_capture=True)
        print("The fraction that made it through using implicit capture was",
              transmission, "with a percent error of",
              np.abs(transmission - true_sol)/true_sol*100,"%")
        transmission = slab_transmission(Sigma_s,Sigma_a, thickness,
                                   N, isotropic=True, implicit_capture=False)
        print("The fraction that made it through using analog tracking was",
              transmission, "with a percent error of",
              np.abs(transmission - true_sol)/true_sol*100,"%")
```

```
The fraction that made it through using implicit capture was 0.000308930894796
    with a percent error of 2.93051065849 %
The fraction that made it through using analog tracking was 0.0
    with a percent error of 100.0 %
```

Implicit capture reduces our error from 100% to less than 5% with only 1000 particles. Previously, we needed about 2 million particles to get that accuracy.

The solution to the problem with scattering is below. In this figure we compare the results using the two approaches and different numbers of particles, up to 10^5.

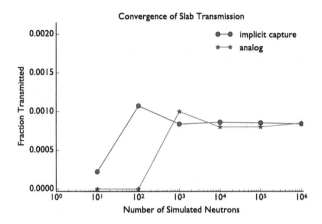

These results indicate that implicit capture converges to the correct solution with many fewer simulated particles. A reasonable question is whether extra effort in performing implicit capture (because particles are not absorbed above the cutoff), is worth the cost.

22.1.1 A Figure of Merit

A way of measuring the benefit of a variance reduction technique is through the quantity called the figure of merit (FOM)

$$\text{FOM} = \frac{1}{\sigma^2 T}.$$

It should be said that this is a figure of merit, not *the* figure of merit because others are possible. In the expression for FOM, σ^2 is the variance in an estimate and T is the time it takes to get the estimate. The benefit of the FOM is that it gives us a way to decide if the cost of a variance reduction technique is worth any increase in the computational time.

We can try this on our slab problem. We will run the problem 20 times with each number of particles, compute the time to solution, and the variance of the runs, and then plot the FOM. Remember that a larger number is better (lower variance and/or time). We expect that the implicit capture method should be better because it appears to give less error (see above) and the exponentials we evaluate should not be costly relative to the particle tracking.

```
In [4]: import time
        N_parts = [10,100,1000,2000,4000,8000,16000, 32000, 64000, 128000]
        Ntimes = 20
        solution_implicit = np.zeros((len(N_parts),Ntimes))
        solution_analog = np.zeros((len(N_parts),Ntimes))
        times_implicit = np.zeros(len(N_parts))
        times_analog = np.zeros(len(N_parts))
        var_implicit = np.zeros(len(N_parts))
        var_analog = np.zeros(len(N_parts))
        Sigma_s = 0.75
        Sigma_a = 2.0 - Sigma_s
        count = 0
        for N in N_parts:
            for replicate in range(Ntimes):
```

```
tmp = time.clock()
solution_implicit[count,replicate] = slab_transmission(Sigma_s,Sigma_a,
                                                        thickness, N,
                                                        isotropic=True,
                                                        implicit_capture=True,
                                                        cutoff=1e-2)
times_implicit[count] += (time.clock()-tmp)/Ntimes
tmp = time.clock()
solution_analog[count,replicate] = slab_transmission(Sigma_s,Sigma_a,
                                                      thickness, N,
                                                      isotropic=True,
                                                      implicit_capture=False)
times_analog[count] += (time.clock()-tmp)/Ntimes
var_implicit[count] = np.std(solution_implicit[count,:])**2
var_analog[count] = np.std(solution_analog[count,:])**2
count += 1
```

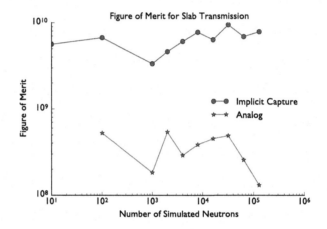

In this example, we see that the FOM for implicit capture is about an order of magnitude larger than analog tracking. This means that implicit capture can get the same variance as analog tracking in one-tenth the time.

22.2 ESTIMATING SCALAR FLUX

Now that we have introduced our first variance reduction technique, we will discuss scalar flux estimators before moving on to other techniques. The first that we will consider is the collision estimator.

22.2.1 Collision Estimators

Consider the reaction rate in a volume, defined by

$$R = \int_V dV \, \Sigma_t(\mathbf{r})\phi(\mathbf{r}).$$

If inside the region the cross-section is constant, we can compute the average scalar flux via the relation

$$\bar{\phi} = \frac{1}{V} \int_V dV\, \phi(\mathbf{r}) = \frac{R}{V \Sigma_t}.$$

Therefore, if we sum, or tally, the weight from each collision inside the volume and divide that count by the total cross-section, we get an estimate of the scalar flux.

Notice, however, that we cannot use this in voids. We can, however, use other processes to estimate the scalar flux. For example, we could compute the scattering rate and divide by Σ_s, for implicit capture this is what we will do.

The slab problem from above will be modified for this purpose. We will introduce a mesh onto the problem and count the reactions in each mesh cell. Also, instead of a source on the boundary we will add a volumetric source to the problem. The source will be uniform between a and b. Therefore, we need to sample a position of the neutron's birth as well as an angle in $\mu \in [-1, 1]$.

We hold off showing the code for the collision estimator and all of the other features we discuss until the end of the chapter in Section 22.4. This is done to avoid re-listing the code repeatedly with only minor changes.

If we run a calculation with the collision estimator on a slab of thickness 3 with $N = 1000$, $\Sigma_s = 2.0$, $\Sigma_a = 0.5$, and the source between 1 to 2, we get the following results for analog tracking and implicit capture with 100 mesh cells.

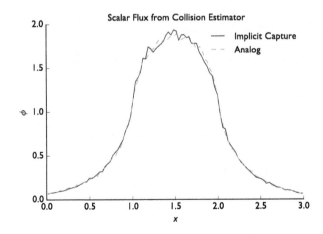

Note that using implicit capture, the collision estimator uses only scattering collisions.

If we use fewer mesh cells, 30 in this case, the answer looks better because we are averaging the scalar flux over a larger volume.

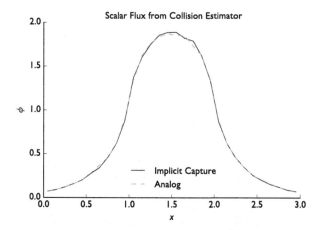

It is hard to tell which of the two methods (implicit capture or analog) is doing better. The figure of merit can help, but we need to select what quantity to measure the variance in. We could consider the variance in the scalar flux estimate in any of the cells as the quantity to estimate the variance in. In the following figure we look at the variance in the scalar flux at the left edge.

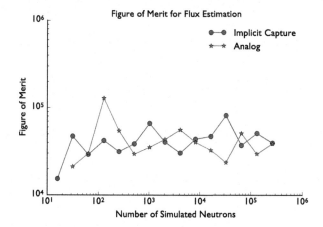

On this problem there are few collisions near the edge of the problem, so we do not see a large difference in the FOM. We expect implicit capture to outperform analog tracking because analog tracking will kill particles before they reach the edge of the problem. Implicit capture allows more particles to get to the edge, and therefore the estimation of the scalar flux will be better. However, we need to use something other than collisions to estimate the scalar flux here.

22.2.2 Track-Length Estimators

Another type of estimator for the scalar flux uses the definition of the scalar flux to estimate it. Recall that the scalar flux is the rate-density at which neutrons generate track length.

Therefore, for a given cell, every time a neutron moves inside it we sum the weight of the neutron times its path length in the cell. We then divide this by the volume of the region. We can write this in equation form as

$$\bar{\phi} = \frac{1}{V} \sum_{neutrons} weight \times path\,length.$$

For implicit capture, the weight is changing while the neutron moves in the region. Therefore, we integrate the weight over the track to decide the contribution. A neutron traveling a distance s inside a region will contribute

$$contribution = \int_0^s ds' w_0 e^{-\Sigma_a s'} = \frac{1}{\Sigma_a} w_0 (1 - e^{-\Sigma_a s}),$$

where w_0 is the initial weight before moving the track-length s.

To implement this estimator we will reformulate how we do our tracking. We will make our method work by checking the distance to collision against the distance to the edge of a cell. Whichever distance is shorter, that event occurs: either the neutron has a collision or it moves to the next cell and a new distance to collision is sampled. This means that the neutrons will step through the problem cell by cell. This will slow down the code, because now the particles can only take steps limited by the width of the mesh cells.

Comparing the figure of merit of the track-length estimator to analog Monte Carlo we see that, once again, on the edge of the problem the implicit capture result has a higher FOM.

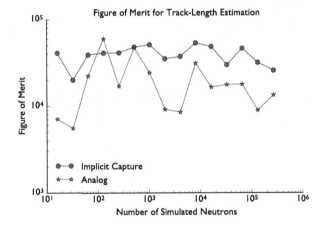

Moreover, the FOM for track-length estimation is higher than that for the collision estimator. In general, it can be beneficial to use both estimators for the scalar flux. By looking at discrepancies between the two methods we can tell where the statistical variance in the problem might be high.

22.2.3 Geometric Dependence of the FOM

As alluded to above, the Figure of Merit can change based on where we look in the problem. In particular, when neutrons have to cross an absorber to get to a particular cell, few of them will do that with analog tracking. Moreover, in a region of the problem where there are many neutrons, analog tracking may have low variance. We can demonstrate this by defining a problem that has a slab of thickness 3 cm $\Sigma_s = 1$ cm^{-1} everywhere and $\Sigma_a = 0.1$ cm^{-1} between $x = 1$ and $x = 2$, and $\Sigma_a = 2$ cm^{-1} otherwise. This problem has a strong absorber at the edges and a scatterer in the middle. We will source neutrons in the middle between $x = 1$ and $x = 2$. The following figure shows the estimate for the scalar flux in this problem using 10^4 simulated particles.

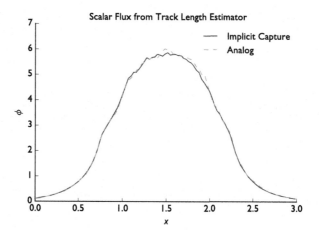

The figure of merit for this problem is quite different in the middle versus the edge.

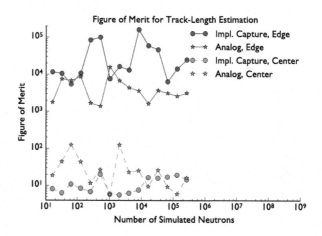

In the middle of the problem there are many particles because that is where the source is. Furthermore, the center region is scattering dominated so the difference between analog and implicit capture is small in terms of the estimate. At the edge, the neutrons have to cross at least two mean-free paths; this results in implicit capture having a higher FOM. This result tells us that depending on what one cares more about, the Monte Carlo methods chosen may have to change.

22.3 STRATIFIED SAMPLING

The idea behind stratified sampling is to control the randomness in the simulation. We want to use random numbers to simulate neutron interactions, but there is no guarantee that random numbers will not be close together. Stratified sampling is a way to spread out the numbers.

It is easiest to think about stratification in terms of a single random variable uniformly distributed between 0 and 1. There are several possible formulations, but the most straightforward to use divides the range between 0 and 1 into S bins of equal size. We then pick a bin at random by generating a random integer between 0 and $S - 1$. Then inside that bin we randomly pick a location. If we perform this sampling so that each bin has the same number of samples, we expect that random samples will do a better job of filling the space between 0 and 1 than simple random sampling.

For this example, the code to produce the stratified samples is straightforward.

```
In [5]:  def strat_sample(N,S):
             N = N + (N % S)
             assert(N%S == 0 )
             dS = 1.0/S
             bins = np.zeros(N,dtype=int)
             count = 0
             for i in range(N//S):
                 bins[count:count+S] = np.random.permutation(S)
                 count += S
             place_in_bin = np.random.uniform(-0.5*dS,0.5*dS,N) + (bins+0.5)*dS
             return place_in_bin
```

If we run an example with 100 samples and different numbers of strata, the histogram of the samples that we get is

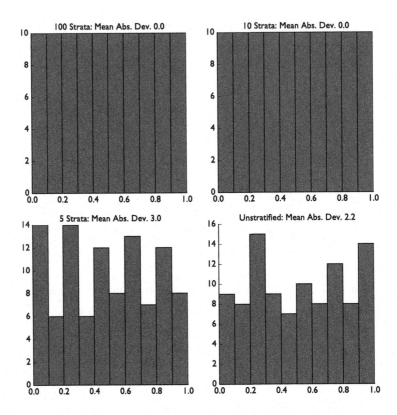

Each of these histograms has 10 bars in the range from 0 to 1. Using 100 or 10 strata has the samples uniformly distributed among these bars. The mean absolute deviation from the mean for each of the bars is reported in the figure; this is a measure of how far away from 10 each of the bar heights are. When we have fewer bins, or do not use stratification, the samples are not uniformly distributed and the average deviation from 10 is larger. This means that we sample some parts of space more than others. The result for a particle transport calculation means that there could be more variance if we use these unstratified samples in our Monte Carlo code. Note that inside of the bins in this figure, the samples are randomly chosen. This means we still have randomness, it is just controlled to live in a particular bin.

Using stratified sampling we can decrease the variance in our solution. This can be most effective when the number of bins is equal to the number of samples. One can show that this is the best case scenario for filling out the range because the farthest apart two samples can be is twice the bin width, and maximizing the number of bins, minimizes the bin width.

Stratified sampling applied to the problem with a scatterer in the middle and an absorber on the edges, results in the following FOM. The stratified results include the implicit capture while the analog do not. In this case we used stratified sampling to choose the location where the neutrons are born in the source region. This is will make the neutron birth locations more uniform.

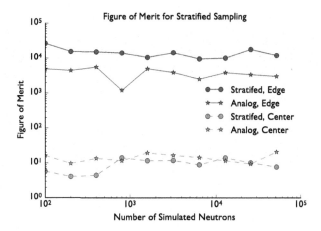

From these results, we can see that stratification provides an improvement in the FOM over previous sampling strategies, especially in the center. The stratification makes the neutrons be born more uniformly in the center of the problem. This makes the implicit capture results competitive with analog tracking. For the particles at the edge, the stratification also provided a benefit over the results from implicit capture with standard sampling.

We can generalize stratified sampling to multiple dimensions, though the number of strata increases as S^D where D is the number of dimensions, if the number of strata in each dimension is S. This can make it difficult to match the number of samples to the number of strata. The following code gives a stratification in two-dimensions. It will increase the number of samples to match the desired number of strata, if needed. It also allows the number of strata in each dimension to differ.

```
In [6]:  def strat_sample_2D(N,S1,S2):
             """Create N samples in S1*S2 strata.
             Inputs:
             N:              number of samples
             S1:             number of strata in dimension 1
             S2:             number of strata in dimension 2
             Returns:
             samples:        N by 2 numpy vector containing the samples
             """
             #number of bins is S1*S2
             bins = S1*S2
             #make sure we have enough points
             if (N<bins):
                 N = bins
             N -= (N % bins)
             Num_per_bin = N//bins
             assert(N % bins == 0)
             samples = np.zeros((N,2))
             count = 0;
             for bin_x in range(S1):
                 for bin_y in range(S2):
                     for i in range(Num_per_bin):
                         center = (bin_x/S1 + 0.5/S1,bin_y/S2 + 0.5/S2)
                         samples[count,0] = center[0] + random.uniform(-0.5,0.5)/S1
                         samples[count,1] = center[1] + random.uniform(-0.5,0.5)/S2
                         count += 1

             return samples
```

Sampling a 2-D space with 5, 10, and 50 strata in each dimension (for a total of $5^2 = 25$, $10^2 = 100$, and $25^2 = 2500$ strata), all with 2500 samples, are compared with unstratified sampling in the following figure.

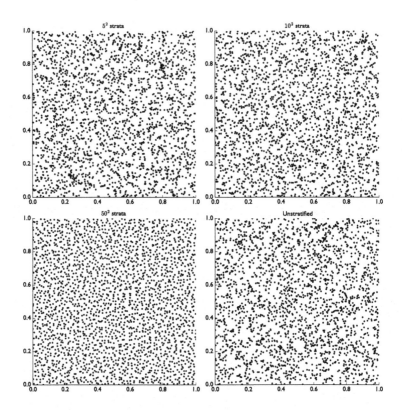

In this figures we see the benefit of stratification in 2-D. The 50^2 strata case fills the space nicely without the clumps that can be seen in the unstratified example. As such the 5^2 and the 10^2 strata cases do not appear to be much of an improvement over the unstratified case.

We can apply 2-D stratification by using it to pick the initial position and μ for the source particles in the slab. This should decrease the variance in our calculation when we increase the number of neutrons sampled to have a large number of strata. On the problem with a scatter in the middle and absorber on the edge, the FOM demonstrates that 2-D stratification does best when the number of simulated neutrons is high.

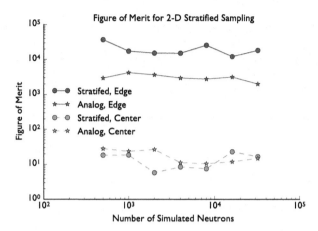

Figure of Merit for 2-D Stratified Sampling

In these results, the FOM for stratified sampling is consistently higher at the edge of the problem when using analog tracking compared with analog tracking. It does appear, however, that in the center there is less benefit due to the large number of neutrons that are born here.

22.4 COMPLETE MONTE CARLO CODE FOR SLABS

The following code listing has all of the features discussed in this chapter.

```
def slab_source(Nx,Sig_s,Sig_a,thickness,a,b,N,Q,
                implicit_capture = True,
                cutoff = 1.0e-3,
                stratified = [1,1]):
    """Compute the fraction of neutrons that leak through a slab
    Inputs:
    Nx:        The number of grid points
    Sig_s:     The scattering macroscopic x-section
    Sig_a:     The absorption macroscopic x-section
    thickness: Width of the slab
    a,b:       Endpoints of Source
    N:         Number of neutrons to simulate
    implicit_capture: Do we run implicit capture
    cutoff:    At what level do we stop implicit capture
    stratified: Use stratified sampling in space and angle
                Specify a list of length two with the number of
                strata in each dimension; default [1,1] for unstratified

    Returns:
    transmission:  The fraction of neutrons that made it through
    scalar_flux:   The scalar flux in each of the Nx cells
    scalar_flux_tl:  The scalar flux in each of the Nx cells
                     from track length estimator
    X:             The value of the cell centers in the mesh
    """
    imp_input = implicit_capture
    dx = thickness/Nx
    X = np.linspace(dx*0.5, thickness - 0.5*dx,Nx)
    scalar_flux = np.zeros(Nx)
    scalar_flux_tl = np.zeros(Nx)
    assert (Sig_s.size == Nx) and (Sig_a.size == Nx)
```

```
Sig_t = Sig_a + Sig_s
iSig_t = 1.0/Sig_t
iSig_s = 1.0/(Sig_s+1.0e-14)
iSig_a = 1.0/(Sig_a+1.0e-14)
leak_left = 0.0
leak_right = 0
N = int(N)
#make a vector of the initial positions and mus
samples = strat_sample_2D(N,stratified[0],stratified[1])
xs = samples[:,0]*(b-a) + a #adjust to bounds of source
mus = (samples[:,1]-0.5)*2 #shift to range -1 to 1
N = int(xs.size)
#the initial weight does not change
init_weight = Q*thickness/N
for i in range(N):
    mu = mus[i]
    x = xs[i]
    alive = 1
    weight = init_weight
    #which cell am I in
    cell = int(x/dx)
    implicit_capture = imp_input
    while (alive):
        if (weight < cutoff*init_weight):
            implicit_capture = False
        if (implicit_capture):
            l = -math.log(1-random.random())*iSig_s[cell]
        else:
            #get distance to collision
            l = -math.log(1-random.random())*iSig_t[cell]
        #compare distance to collision to distance to cell edge
        distance_to_edge = ((mu > 0.0)*( (cell+1)*dx - x) +
                            (mu<0.0)*( x - cell*dx) + 1.0e-8)/math.fabs(mu)
        if (distance_to_edge < l):
            l = distance_to_edge
            collide = 0
        else:
            collide = 1
        x += l*mu #move particle
        #score track length tally
        if (implicit_capture):
            scalar_flux_tl[cell] += weight*(1.0 -
                                    math.exp(-l*Sig_a[cell]))*iSig_a[cell]
        else:
            scalar_flux_tl[cell] += weight*l
        if (implicit_capture):
            weight *= math.exp(-l*Sig_a[cell])
        #still in the slab?
        if (math.fabs(x-thickness) < 1.0e-14) or (x > thickness):
            leak_right += weight
            alive = 0
        elif (x<= 1.0e-14):
            alive = 0
            leak_left += weight
        else:
            cell= int(x/dx) #compute cell particle collision is in
            if (implicit_capture):
                if (collide):
                    mu = random.uniform(-1,1)
                scalar_flux[cell] += weight*iSig_s[cell]/dx
            else: #scatter or absorb
                scalar_flux[cell] += weight*iSig_t[cell]/dx
                if (collide) and (random.random() < Sig_s[cell]*iSig_t[cell]):
                    #scatter, pick new mu
                    mu = random.uniform(-1,1)
                elif (collide): #absorbed
```

```
                      alive = 0
        return leak_left,leak_right, scalar_flux, scalar_flux_tl/dx, X, N
```

CODA

In this chapter we have expanded our Monte Carlo toolkit to estimate the scalar flux and to make better estimates via Monte Carlo. These techniques, along with those in the previous chapter, give us all the tools we need to solve source-driven neutron transport problems via Monte Carlo. There is still an important class of problems that we do not have to tools to solve, yet: k-eigenvalue problems. We will discuss Monte Carlo techniques for these problems in the next chapter.

PROBLEMS

Short Exercises

22.1. Repeat Short Exercise 21.3 using the 2-D stratified sampling to pick the proposal points with the number of strata being 5^2, 10^2, 50^2, 100^2, and 200^2. Only keep the accepted proposed points (this will mean that you get fewer samples than the number of strata). How do these results compare to standard rejection sampling?

Programming Projects

The problems below ask you to modify the iron-shielded reactor example from the previous chapter to include the methods discussed in this chapter.

1. Iron-Shielded Reactor: Implicit Capture

Modify the iron shielding code to include implicit capture. Compare the analog to the implicit capture results with $N = 10^2, 10^3, 10^4$, and 10^5. Compute the figure of merit for the average transmitted energy per neutron for these values of N.

2. Iron-Shielded Reactor: Scalar Flux

Inside the shield compute the scalar flux of neutrons in the range 100 keV to 1 MeV and the scalar flux of neutrons above 1 MeV. Compare the collision estimator and track-length estimator for several different mesh resolutions.

"Maybe that's what it is," said Somers. "That's a useful way of putting it. I can't help my aura colliding, can I?"

–D.H. Lawrence Kangaroo

CHAPTER POINTS

- We can use Monte Carlo to calculate the k-eigenvalues of a nuclear system.

- One approach is to record where simulated neutrons have a fission event, and use those fission sites as the birthplace for neutrons in the next generation. By

  properly defining weights, the ratio of neutrons in successive generations is the estimate of k_{eff}.

- Another approach forms a matrix based on a Monte Carlo simulation. The eigenvalues and eigenvectors of this matrix are the k-eigenvalues of the system.

23.1 FISSION CYCLES

As we have discussed, we often want to compute the multiplication factor, i.e., the k-eigenvalue, for a system under consideration. We can use Monte Carlo to do this. One definition of the multiplication factor is the following limit

$$k_{\text{eff}} = \lim_{n \to \infty} \frac{\text{\# of neutrons in generation } n}{\text{\# of neutrons in generation } (n-1)},$$

where a generation is a step of the chain reaction, for example the neutron that started the chain reaction is in generation 1 and the fission neutrons created from generation 1 neutrons are in generation 2, and so on. To complete this calculation we need to track neutrons in generations to compute the ratio of neutrons in successive generations. To do this we use fission cycles: that is we simulate generations of neutrons by tabulating where fissions take place.

The idea of a fission cycle is that we have all of the fission locations from the previous generation of neutrons. We then source neutrons from these locations with total weights that sum to the number of fission sites times ν, the average number of neutrons born per fission event. We then track these neutrons to get the fission sites for the birth of neutrons in the next generation. The ratio between the sum of the weights of neutrons born from fission between these generations is an estimate of the eigenvalue. If we run enough cycles, we can get an estimate of the eigenvalue.

One issue with this approach is the starting of the calculation. We typically will not know where the fission sites are for the first generation. Therefore, we need to guess the initial fission sites. If we guess these sites at random, and then run several cycles it is reasonable to believe that the initial distribution of fission sites will not matter. In other words the fission sites will relax to an equilibrium. The random initial sites are analogous to the initial guess we used for the eigenvector when we used power iteration to solve diffusion eigenvalue problems.

In practice, what we do is start with an initial distribution of neutrons randomly chosen in the system. We then take some number of fission cycles that we do not use to estimate k_{eff}. These initial cycles that we do not use are called inactive cycles. The active cycles are those cycles after the inactive cycles. From the estimate of k_{eff} from each cycle we compute the mean k_{eff} and the standard deviations.

In the following code the value of k_{eff} for each fission cycle is calculated for a homogeneous slab. It returns the estimate of k_{eff} for both in the inactive and active cycles.

```
In [1]: def homog_slab_k(N,Sig_t,Sig_s,Sig_f,nu, thickness,
                          inactive_cycles = 5, active_cycles = 20):
            Sig_a = Sig_t - Sig_s
            iSig_t = 1/Sig_t
            iSig_a = 1/Sig_a
            #initial fission sites are random
            fission_sites = np.random.uniform(0,thickness,N)
            positions = fission_sites.copy()
            weights = nu*np.ones(N)
            mus = np.random.uniform(-1,1,N)
            old_gen = np.sum(weights)
            k = np.zeros(inactive_cycles+active_cycles)
            for cycle in range(inactive_cycles+active_cycles):
                fission_sites = np.empty(1)
                fission_site_weights = np.empty(1)
                assert(weights.size == positions.size)
                for neut in range(weights.size):
                    #grab neutron from stack
                    position = positions[neut]
                    weight = weights[neut]
                    mu = mus[neut]
                    alive = 1
                    while (alive):
```

```
                    #compute distance to collision
                    l = -math.log(1-random.random())*iSig_t
                    #move neutron
                    position += l*mu
                    #are we still in the slab
                    if (position > thickness) or (position < 0):
                        alive = 0
                    else:
                        #decide if collision is abs or scat
                        coll_prob = random.random()
                        if (coll_prob < Sig_s*iSig_t):
                            #scatter
                            mu = random.uniform(-1,1)
                        else:
                            fiss_prob = random.random()
                            alive = 0
                            if (fiss_prob <= Sig_f*iSig_a):
                                #fission
                                fission_sites = np.append(fission_sites,position)
                                fission_site_weights
                                = np.append(fission_site_weights,weight)
            fission_sites = np.delete(fission_sites,0,axis=0)
            #delete the initial site
            fission_site_weights = np.delete(fission_site_weights,0,axis=0)
            #delete the initial site
            #sample neutrons for next generation from fission sites
            num_per_site = int(np.ceil(N/fission_sites.size))
            positions = np.empty(1)
            weights = np.empty(1)
            mus = np.random.uniform(-1,1,num_per_site*fission_sites.size)
            for site in range(fission_sites.size):
                site_pos = fission_sites[site]
                site_weight = fission_site_weights[site]
                positions = np.append(positions,
                site_pos*np.ones((num_per_site,1)))
                weights = np.append(weights,
                site_weight * nu/num_per_site*np.ones((num_per_site,1)))
            positions = np.delete(positions,0,axis=0) #delete the initial site
            weights = np.delete(weights,0,axis=0) #delete the initial site
            new_gen = np.sum(weights)
            k[cycle] = new_gen/old_gen
            old_gen = new_gen
    return k
```

To demonstrate how this works we will consider a homogeneous slab with single speed neutrons where $\Sigma_t = 1$ cm^{-1}, $\Sigma_s = 0.75$ cm^{-1}, $\Sigma_f = 0.10285$ cm^{-1}, and $\nu = 2.5$. The value of k_∞ for this system is 1.0285. This critical thickness for a slab of this material 11.331 cm [24].

Now we run a k-eigenvalue calculation for a slab of this material that is exactly 11.331 cm thick with 10^4 simulated neutrons per fission cycle and a slab thickness of 20. The expected value of k_{eff} is 1. What we expect to see is that the estimate for k_{eff} will change between iterations and that the early results could be far away from the results in later cycles because of the random initial fission sites.

The results for 10^4 simulated neutrons per cycle are given first. In the following figure we show the ratio between the number of neutrons in successive generations as an estimate of k_{eff}. The vertical line serves to indicate where the inactive cycles stop and the active cycles begin.

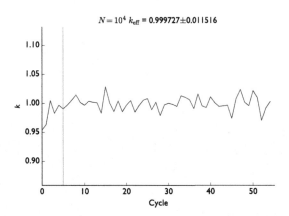

In the title of this figure we give the mean of k estimates for the active cycles, as well as the standard deviations. The estimate for k_{eff}, 0.999727, is within 27.3 pcm of the correct answer; note that error in the estimate is much smaller than one standard deviation of the k_{eff} estimates over the active cycles.

It is clear to see that early on in the calculation the estimate of k is away from the true value and eventually settles into a range about the mean value. Had we included the inactive cycles in the estimate of k_{eff}, we would have had a much lower of an estimate of the eigenvalue in this case.

Upon increasing the number of simulated neutrons per cycle by a factor of 10, we expect that the estimate will get better. The next figure compares the fission cycles using 10^4 and 10^5 neutrons per cycle.

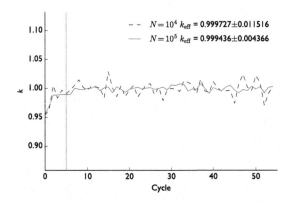

In these results we see that the estimate of k_{eff} went down, but the standard deviation of the estimates per fission cycle went down by a factor of $(0.011516/0.004366) \approx 2.64$. This decrease in the standard deviation is close to the expected decrease in the standard deviation by increasing the number of simulated particles by a factor of 10: $\sqrt{10} \approx 3.16$. The decrease in the variation in the estimate between cycles is obvious in the figure.

We can make the slab thinner and re-run the calculation. We would expect the solution to have a smaller value for k_{eff} because more neutrons will leak out of the system. To have a baseline for comparison, we can compare the Monte Carlo solution to another transport calculation based on the discrete ordinates (S_N) method [27] with high resolution. We expect that the Monte Carlo and S_N solution should agree to several digits. The S_N estimate of k_{eff} for this problem with a thickness of 1 cm is 0.312001. The result with 10^4 neutrons per cycle is shown next.

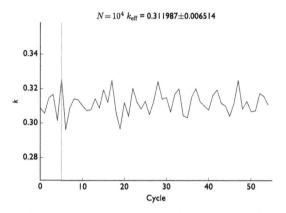

The result is $k_{eff} = 0.311987$. Given that this is such a leaky system (most of the fission neutrons leak out of the system), we may need more neutrons per fission cycle to suppress the noise in the estimates (as observed in the figure). Also notice that in this system the number of inactive cycles could be decreased because the eigenvalue seems to be near the mean from the first cycle.

With 10^5 neutrons per cycle, the variation in the estimates of k_{eff} decrease:

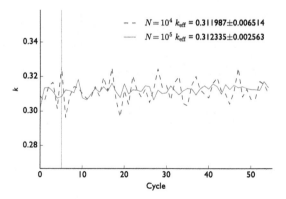

Once again, the decrease in the standard deviation is about a factor of $\sqrt{10}$; in this case the decrease is about 2.56.

23.2 FISSION MATRIX METHODS

Another way of estimating the eigenvalue of a system would directly estimate the eigenvalues of the transport operator. In particular, we will discretize the system in space and then use Monte Carlo to estimate the number of fission neutrons born in one region that cause fission in another. Upon tallying where neutrons are born and where they cause fission, we will have what is known as a fission matrix. With this matrix we can use standard linear algebra techniques to estimate the eigenvalues of that matrix.

The complete derivation of the fission matrix method is outside the scope of our study, primarily because it involves a firm grasp of transport theory. Here we will give a rough justification of the approach and rely on the numerical results to demonstrate the validity of the approach *faute de mieux*.

We start by considering the integral operator $\mathcal{H}$ defined as

$$\mathcal{H}s(\mathbf{r}, E) = \int dE' \int_{V'} dV' \, F(\mathbf{r}', E' \to \mathbf{r}, E) s(\mathbf{r}', E'),$$

where $F(r', E' \to r, E)$ is the expected number of fission neutrons created at position $\mathbf{r}$ and energy E from a fission neutron born at $\mathbf{r}'$ with energy E'. In this sense, $\mathcal{H}s(\mathbf{r}, E)$ is the rate density of neutrons born from fission at location $\mathbf{r}$ and energy E due to a generic density of neutrons $s(\mathbf{r}, E)$.

Now, if $s(\mathbf{r}, E)$ gives the density of fission neutrons born in a generation, then we can write the k-eigenvalue problem as

$$\mathcal{H}s(\mathbf{r}, E) = ks(\mathbf{r}, E).$$

Therefore, we could solve for k by computing

$$k = \frac{\langle \mathcal{H}s(\mathbf{r}, E) \rangle}{\langle s(\mathbf{r}, E) \rangle}.$$

The angle brackets, $\langle \cdot \rangle$ denote integration over space and energy. This is basically what we did with the fission cycle calculation: we started with a distribution of neutrons for given generation, computed the number of fission neutrons born in the next generation and looked at the ratio.

An alternative approach would be to discretize the $\mathcal{H}$ operator in space by defining a mesh of regions. We then can write the total number of fission neutrons in region i caused by neutrons that are born in region j as

$$\mathbf{H}_{ij} = \frac{\int dE \int dE' \int_{V_i} dV \int_{V_j} dV' \, F(\mathbf{r}', E' \to \mathbf{r}, E) \hat{s}(\mathbf{r}', E')}{\int dE' \int_{V_j} dV' \, \hat{s}(\mathbf{r}', E')}.$$

The quantities $\mathbf{H}_{ij}$ form a matrix. The elements of this matrix can be estimated via Monte Carlo. Typically, one does this by assuming a flat fission source in each region of the problem. This is equivalent to stating that the fission rate density is constant in the region. We can then solve the eigenvalue problem

$$\mathbf{Hs} = \hat{k}\mathbf{s}.$$

If the regions are small enough so that the error in making the source flat is negligible, then we can interpret $\hat{k}$ as an eigenvalue of the system. This means that the dominant eigenvalue $\hat{k}$ and its associated eigenvector are the fundamental mode for the system. We could also find the other modes in the system this way.

With the knowledge of the fundamental mode eigenvector, we then know the steady-state flux shape in the system. Furthermore, numerical experiments below demonstrate that the magnitudes of the imaginary parts of the eigenvalues are a measure of the uncertainty in the fundamental mode eigenvalue.

To compute the fission matrix we need to specify a mesh over the problem, and then emit neutrons in each region, and count the number of fission neutrons born in each other region. We will demonstrate this in our homogeneous slab problem.

```python
In [4]: def fission_matrix(N,Sig_t,Sig_s,Sig_f,nu, thickness,Nx):
            Sig_a = Sig_t - Sig_s
            H = np.zeros((Nx,Nx))
            dx = thickness/Nx
            lowX = np.linspace(0,thickness-dx,Nx)
            highX = np.linspace(dx,thickness,Nx)
            midX = np.linspace(dx*0.5,thickness-dx*0.5,Nx)
            for col in range(Nx):
                #create source neutrons
                positions = np.random.uniform(lowX[col],highX[col],N)
                mus = np.random.uniform(-1,1,N)
                weights = np.ones(N)*(1.0/N)
                #track neutrons
                for neut in range(positions.size):
                    #grab neutron from stack
                    position = positions[neut]
                    mu = mus[neut]
                    weight = weights[neut]
                    alive = 1
                    while (alive):
                        #compute distance to collision
                        l = -np.log(1-np.random.random(1))/Sig_t
                        #move neutron
                        position += l*mu
                        #are we still in the slab
                        if (position > thickness) or (position < 0):
                            alive = 0
                        else:
                            #decide if collision is abs or scat
                            coll_prob = np.random.rand(1)
                            if (coll_prob < Sig_s/Sig_t):
                                #scatter
                                mu = np.random.uniform(-1,1,1)
                            else:
                                fiss_prob = np.random.rand(1)
                                alive = 0
```

```
if (fiss_prob <= Sig_f/Sig_a):
    #find which bin we are in
    row = np.argmin(np.abs(position - midX))
    H[row,col] += weight*nu
return H, midX
```

With this function we will know how to build the fission matrix. Then we can use the NumPy function, eig to find all the eigenvalues of the system and take the largest of these as the value of k_{eff}. The other eigenvalues can be used to help analyze the behavior of the system during transients. The thin slab reactor analyzed above with fission cycles will have its fission matrix and eigenvalue computed in next. In the calculation we use 10^6 simulated neutrons. From these results, we plot the entries of the matrix in a color map and the eigenvalues in the complex plane.

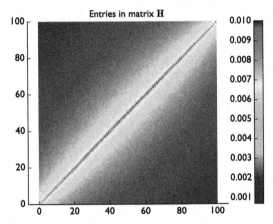

This plot of the matrix elements tells us about the physics of the reactor. The diagonal elements have the largest magnitude, and these represent fission neutrons that are born and cause fission in the same region. Also, the elements farther away from the diagonal are smaller because these represent neutrons that travel far from the birth region before causing fission. Also, there is clear statistical noise in the matrix elements.

Next, we will look at the eigenvalues of the fission matrix:

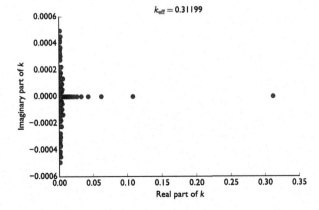

These are the eigenvalues in the complex planes. The fundamental mode eigenvalue is far to the right and is purely real. Its value is 0.31199. Closer to zero there is a cloud of complex eigenvalues. The actual k-eigenvalues of the system should all be real—the imaginary part of these eigenvalues is due to noise in the calculation of the fission matrix. Notice that the eigenvalue from the fission matrix agrees with the S_N transport calculation to five digits.

One benefit of the fission matrix approach is that we can also estimate the shape of the fundamental mode scalar flux from `np.eig`.

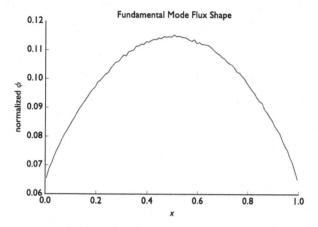

The shape of the fundamental mode scalar flux is the basic shape we would expect: peaked in the middle of the slab and falling off toward the edges of the slab. There is obvious noise present as well.

If we modify the problem to be thicker, i.e., make the slab have a thickness of 11.331, without increasing the number of regions, the noise decreases. This decrease is partially because the thickness of each region is larger, as well as the fact that fewer fission neutrons leak out. Recall that in this problem the exact answer is $k_{eff} = 1$.

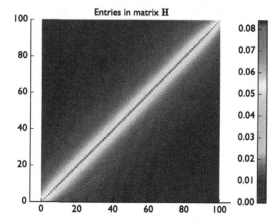

The matrix is more diagonally dominant, and neutrons appear to not move much between regions before causing fission. The k_{eff} estimate we obtain close to 1: within 4 pcm of the actual answer. If we look at the spectrum of the eigenvalues in the complex plane, we see that the imaginary part of the eigenvalues is still present.

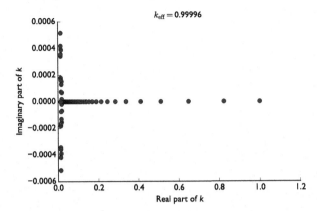

The fission matrix calculated for this system have fewer complex eigenvalues, Also, notice that the separation between the fundamental mode eigenvalue and the second largest eigenvalue is smaller than in the thinner system. This implies that power iteration for this problem should converge more slowly than for the thin system. Indeed, we see artifacts of this in the fission cycle calculation above: for the thicker system there was a clear need for several inactive cycles to settle on the fundamental mode. The thin system had no such slow approach to the fundamental mode.

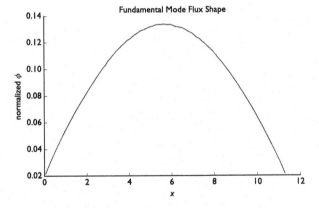

As we saw with the fission cycles, there is less noise in the scalar flux for this thicker slab with thicker regions.

There are several benefits to the fission matrix calculation. Firstly, it relies on Monte Carlo to move the neutrons and linear algebra software to estimate the eigenvalues, whereas the fission cycle calculation combines Monte Carlo and the eigenvalue estimation in the same calculation. In the fission matrix calculation we can rely on well-tested linear algebra libraries to get the eigenvalues after we use Monte Carlo to move neutrons around. As a result we get more information including all the eigenvalues and eigenvectors. The downside of this approach is that we must approximate the fission source as flat in each region.

CODA

Monte Carlo codes are an important tool in nuclear engineering and in any application where radiation physics are important. In this chapter we applied the Monte Carlo principles from the previous two chapters to solve eigenvalue problems. We neglected any energy dependence and only solved homogeneous problems in the examples. It is straightforward, if not simple, to make these extensions to the codes above. The codes and techniques we have discussed here are only a sample of the features available in production Monte Carlo codes such as MCNP, Geant 4, and keno. Nevertheless, the material we have covered will give the reader a strong foundation to either explore Monte Carlo methods in more detail or to run production codes confidently.

PROBLEMS

Programming Projects

1. Pure Plutonium Reactor

In Chapter 18 we defined a pulsed reactor made of pure plutonium with the following cross-sections from the report *Reactor Physics Constants*, ANL-5800:

Quantity	Value
σ_f [b]	1.85
σ_a [b]	2.11
σ_{tr} [b]	6.8
ν	2.98

For the density of plutonium use 19.74 g/cm^3; you may assume that $\sigma_t \approx \sigma_{tr}$.

- Compute k_{eff} for a slab of thickness 7 cm made from pure plutonium-239 using fission cycles and the fission matrix method.
- Compute k_{eff} for a solid sphere of plutonium-239 with a radius of 7 cm using either method discussed in this chapter. You will have to modify the codes above to handle transport in a sphere.
- Finally, compute k_{eff} for a spherical shell of inside radius of 2 cm and an outer radius of 6 cm. You may consider the hollow part of the shell a void. In this case you will have to modify your code to handle the fact that a collision cannot take place in the hollow part of the sphere.

Bibliography

[1] Scopatz A, Huff KD. Effective computation in physics. 1st ed. O'Reilly Media. ISBN 978-1-4919-0153-3, 2015.

[2] Guttag JV. Introduction to computation and programming using Python. MIT Press. ISBN 978-0-2625-2500-8, 2013.

[3] Bornemann F, Laurie D, Wagon S, Waldvogel J. The SIAM 100-digit challenge: a study in high-accuracy numerical computing. SIAM e-books. Society for Industrial and Applied Mathematics. ISBN 9780898717969, 2004.

[4] Kalos MH, Whitlock PA. Monte Carlo methods. John Wiley & Sons; 2009.

[5] LaTeX: a document preparation system. 2nd ed. Pearson Education. ISBN 9788177584141, 1994.

[6] Kopka H, Daly P. Guide to LaTeX. Tools and techniques for computer typesetting. Pearson Education. ISBN 9780321617743, 2003.

[7] Lewis E. Fundamentals of nuclear reactor physics. Nuclear energy ebook collection series. Elsevier Science. ISBN 9780080560434, 2008.

[8] Butcher P. Debug it!: Find, repair, and prevent bugs in your code. Pragmatic Bookshelf series. Pragmatic Bookshelf. ISBN 9781934356289, 2009.

[9] Trefethen L, Bau D. Numerical linear algebra. Society for Industrial and Applied Mathematics. ISBN 9780898713619, 1997.

[10] Stacey W. Nuclear reactor physics. John Wiley & Sons. ISBN 9783527406791, 2007.

[11] Li X, Demmel J, Gilbert J, Grigori L, Shao M, Yamazaki I. SuperLU users' guide. Tech. Rep. LBNL-44289, Lawrence Berkeley National Laboratory; 1999. Available from: http://crd.lbl.gov/~xiaoye/SuperLU/ [last update: August 2011].

[12] Saad Y. Iterative methods for sparse linear systems. Computer science series. PWS Publishing Company. ISBN 9780534947767, 1996.

[13] Hastie T, Tibshirani R, Friedman J. The elements of statistical learning: data mining, inference, and prediction. 2nd ed. Springer series in statistics. New York: Springer. ISBN 9780387848587, 2009.

[14] Gelman A, Hill J. Data analysis using regression and multilevel/hierarchical models. Analytical methods for social research. Cambridge University Press. ISBN 9781139460934, 2006.

[15] Taleb N. Fooled by randomness: the hidden role of chance in life and in the markets. Incerto. Random House Publishing Group. ISBN 9781588367679, 2008.

[16] Taleb N. The black swan. A Random House international edition. Random House. ISBN 9780812979183, 2009.

[17] Taleb N. Antifragile: things that gain from disorder. Incerto series. Random House Publishing Group. ISBN 9780812979688, 2014.

[18] Atkinson K. An introduction to numerical analysis. Wiley. ISBN 9780471029854, 1978.

[19] Martins JRRA, Sturdza P, Alonso JJ. The complex-step derivative approximation. ACM Transactions on Mathematical Software 2003;29(3):245–62.

[20] Shu C. Differential quadrature and its application in engineering. London: Springer; 2000.

[21] Ganapol BD. A highly accurate algorithm for the solution of the point kinetics equations. Annals of Nuclear Energy 2013;62:564–71.

[22] Brunner TA, Mehlhorn T, McClarren R, Kurecka C. Advances in radiation modeling in ALEGRA: a final report for LDRD-67120, efficient implicit multigroup radiation calculations. Tech. Rep. SAND2005-6988, Sandia National Laboratories; 2005.

[23] Case KM, Zweifel PF. Linear transport theory. Reading, Massachusetts: Addison-Wesley; 1967.

[24] Bell GI, Glasstone S. Nuclear reactor theory. Malabar, Florida: Robert E. Kreiger Publishing; 1970.

[25] Dunn W, Shultis J. Exploring Monte Carlo methods. Elsevier Science. ISBN 9780080930619, 2011.

[26] Kalos M, Whitlock P. Monte Carlo methods. Wiley-Blackwell. ISBN 9783527407606, 2008.

[27] Lewis E, Miller W. Computational methods of neutron transport. John Wiley and Sons; 1984.

Index

Printed in the United States
By Bookmasters